Electricity Marginal Cost Pricing

Electricity Marginal Cost Pricing

Applications in Eliciting Demand Responses

Monica Greer, Ph.D

AMSTERDAM • BOSTON • HEIDELBERG • LONDON
NEW YORK • OXFORD • PARIS • SAN DIEGO
SAN FRANCISCO • SINGAPORE • SYDNEY • TOKYO

Butterworth-Heinemann is an imprint of Elsevier

Butterworth-Heinemann is an imprint of Elsevier
225 Wyman Street, Waltham, MA 02451, USA
The Boulevard, Langford Lane, Kidlington, Oxford, OX5 1GB, UK

Notices
Knowledge and best practice in this field are constantly changing. As new research and experience broaden
our understanding, changes in research methods, professional practices, or medical treatment may become
necessary.

Practitioners and researchers must always rely on their own experience and knowledge in evaluating and using
any information, methods, compounds, or experiments described herein. In using such information or methods
they should be mindful of their own safety and the safety of others, including parties for whom they have a
professional responsibility.

To the fullest extent of the law, neither the Publisher nor the authors, contributors, or editors assume any
liability for any injury and/or damage to persons or property as a matter of products liability, negligence or
otherwise, or from any use or operation of any methods, products, instructions, or ideas contained in the
material herein.

Library of Congress Cataloging-in-Publication Data
Greer, Monica.
    Electricity marginal cost pricing: applications in eliciting demand responses / Monica Greer.
        p. cm.
    ISBN 978-0-12-385134-5
1. Electric utilities–Rates–United States. 2. Electric utilities–United States–Costs–Econometric models.
3. Electric power consumption–United States–Econometric models. I. Title.

    HD9685.U5G756 2012
    333.79'32310973–dc23

                                  2011042180

British Library Cataloguing-in-Publication Data
A catalogue record for this book is available from the British Library.

For information on all Butterworth–Heinemann publications
visit our website at *www.elsevierdirect.com*

Typeset by: diacriTech, Chennai, India

Printed in the United States
12 13 14 15 16 17   7 6 5 4 3 2 1

Contents

Preface   xv
Acknowledgements   xix

1. Introduction   1

Introduction   1
Competitive Paradigm   2
Marginal Cost Pricing Doctrine   2
     Theory   3
A Brief Overview of the United States Electric Market   4
     Structure of the United States Electricity Industry   4
     Players and Their Incentives   4
Objective Functions: The Players   5
     Investor-Owned Utilities: Profit Maximization   5
     Publicly-Owned Firms   5
     Cooperatively-Owned Firms   6
     Other   7
Reducing Carbon Emissions   7
     Regulation/Rate Making   7
     The United States Electric Power Industry—Regulation   7
Regulation of Investor-Owned Electric Utilities in the United States   8
     Aside: Issues with Rate-of-Return Regulation—
         The Averch–Johnson Effect   9
Internalizing the Cost of Reducing Carbon Emissions   10
     Current Policy   11
Optimal Rate/Tariff Design and Tax Credits to Promote Efficient
     Use of Energy and a Reduction in Carbon Emissions   12
Tariff Design and Rate-Making Issues   13
     Marginal Cost Pricing for Electric Utilities   13
Conclusion   14

2. The Theory of Natural Monopoly and Literature Review   15

The Natural Monopoly Conundrum   15
Defining Natural Monopoly   16
     For a Single-Output Market   17
     Average Cost   17
Economies of Scale   19
     Aside: Necessary Condition versus Sufficient Condition   19

Efficient Industry Structure   21
Degree-of-Scale Economies   21
Economies of Scale Applied to the Electric Utility Industry   21
Literature Review—Economies of Scale in Generation:
   Single-Output Models   22
Network Economies   24
For a Multiple-Output Natural Monopoly   26
Multiproduct Natural Monopoly   27
Ray Average Costs   27
Degree-of-Scale Economies   28
Product-Specific Economies of Scale   28
Economies of Scope   30
Subadditivity of the Cost Function   31
Electricity as a Multiple-Output Industry and/or Economies of
   Scope and Subadditivity   31
Economies of Vertical Integration and Separability   33
Vertical Integration of Electric Utilities   34
Defining Vertical Integration   35
Separability   35
Relevant Literature Review—Vertical Integration and Separability   36
Conclusion   37
Exercises   37

3. U.S. Electric Markets, Structure, and Regulations   39

Introduction   39
The U.S. Electric Industry Structure   39
Deregulation   41
Market Participants   42
Vertically Integrated Model   42
Investor-Owned Utilities   42
Independent Power Producers   44
Municipal Utilities and Other Publicly-Owned Utilities   44
Federal Power Agencies   45
Tennessee Valley Authority   47
Power Marketing Administrations   49
Rural Electric Cooperatives   50
Nonutility Power Producers   52
Power Pools   52
Industry Restructuring and the Competitive Electric Market   52
Merchant Generators   53
Transmission Companies   53
Independent System Operators and Regional Transmission
   Organizations   54
Electric Power Marketers   56
Utility Distribution Company   56
Energy Service Companies   57

Regulation of the Electric Utility Industry   58
    History of Regulation in the U.S. Electric Utility Industry   58
    The Rise of Regulation   58
    Public Utility Holding Company Act (PUHCA) of 1935   60
    The Era after PUHCA   61
    Focus on Reliability: The North America Reliability Council   62
    The 1970s—A Time of Change   64
    Federal Energy Regulatory Commission   65
    National Energy Conservation Policy Act of 1978   66
    Public Utility Regulatory Policy Act of 1978   66
    Industry Restructuring in the 1980s and 1990s   69
    Natural Gas Utilization Act of 1987   71
    The 1990s—The Pace of Industry Restructuring Accelerates   71
    Energy Policy Act (EPACT) of 1992   72
    FERC Orders 888 and 889 (1996)   75
    FERC Order 2000 and Grid Regionalization   78
    Energy Policy Act (EPACT) of 2005   82
    Energy Independence and Security Act (EISA) of 2007   83
    FERC Order 719   84
    Consolidated Appropriations Act of 2008   84
    FERC Order 980   84
    American Recovery and Reinvestment Act (ARRA) of 2009   85
    American Energy and Security Act (ACES) of 2009   85
    American Clean Energy Leadership Act of 2009   86
    FERC Order 1000   87
    Background   87
    Cost Allocation Reforms   88
    State Regulations   88
Looking Forward: Renewable Resources and
    Generating Technologies   89
    Renewable Portfolio Standard   89
    Green Power Purchasing and Aggregation Policies   91
    Interconnection Standards   91
    Utility Green Power Consumer Option   92
    Net Metering   92
    Public Benefit Funds and System Benefit Charges   92
    Rebate Programs   93
    Renewable Energy Access Laws   93
    Renewable Energy Production Incentives   93
    Tax Incentives   93
    Feed-in Tariffs   94
Future of the Electric Industry   95

4. The Economics (and Econometrics) of Cost Modeling   101

General Cost Model   101
    Cobb–Douglas Cost Function   102
    Translogarithmic Cost Function   103

The Econometrics of Cost Modeling: An Overview   104
   Ordinary Least-Squares Estimation   104
   Regression Analysis and Cost Modeling   104
   Examples: Examining Data—An Illustration of Salient Points   105
   Estimation Results: Basic Cost Model   108
   Consequences of Heteroscedasticity   109
   Impure Heteroscedasticity   109
   Estimation Results—Quadratic Cost Model   111
A Brief History of Cost Models and Applications to the Electric Industry 111
   Cobb–Douglas Functional Form   111
   Returns to Scale   112
   Nerlove's Cobb–Douglas Function   112
   Economies of Scale   114
   Minimum Efficient Scale   114
   Nerlove's Results   114
   A Priori Expections   115
   Elasticities   116
   Constant Elasticity of Substitution Functional Form   117
   Generalized Leontief Cost Function   117
   ASIDE: Leontief Production Technology   118
   Leontief Cost Function   119
   Hicks–Allen Partial Elasticities of Substitution   120
   Translogarithmic Cost Function   120
   Digression: Use of Zellner's Method (Seemingly Unrelated
     Regressions Method)   123
   Quadratic Cost Models   123
   Digression: Reasons That the Quadratic Cost Functional Form
     Is Well Suited for Modeling Industry Structure   124
   Multiple-Output Quadratic Cost Function   124
   Degree-of-Scale Economies   125
   Ray Average Cost   125
   Product-Specific Returns to Scale   125
   Economies of Scope   127
   Cubic Cost Models   128
Conclusion   129
Appendix   129
Exercises   130

5. Cost Models   133

Introduction   133
**Determination of an Appropriate Objective Function: A Brief
   Overview of the Literature**   133
Rural Electric Cooperatives   135
   Reasons That Cooperatively-Owned Utilities Are Different   135
Differences between Coops and IOUs   136
   Urban versus Rural   136
   Institutional   137

Regulatory 138
Philosophical 138
Literature Review: Cost Studies on Rural Electric Cooperatives 138
Data 140
Review of the Literature: Cost Function Estimation in
  the Electric Utility Industry 140
Economies of Scale 140
Nerlove's Cobb–Douglas Cost Model 141
Further Considerations 143
End of Section Exercises: Basic Cost Model versus Nerlove Cost Model 143
Flexible Functional Forms 144
Translogarithmic Cost Function 144
Cost-Share Equations 145
A Priori Expectations 146
Discussion of Estimation Results: Single-Output Translog
  Cost Equation 147
Substitution Elasticities among Inputs: Hicks–Allen Partial Elasticities
  of Substitution 148
Price Elasticities 149
Homotheticity 150
End of Section Exercises: Translogarithmic Cost Function 150
Aside: Translogarithmic Cost Model Details: Calculating
  Average and Marginal Cost 152
Proof 152
Multiproduct Cost Functions 153
Literature Review 153
Economies of Scale and Integration 153
Distributed Electricity as a Multiproduct Industry 154
Multiproduct Cost Models 155
Multiproduct Translogarithmic Cost Model 156
Estimation Results 157
Multiproduct Cost Concepts (Revisited) 157
Ray Average Costs 157
Product-Specific Economies of Scale 160
Economies of Scope 160
Quadratic Cost Functions 161
A Properly Specified Quadratic Cost Function 162
Aside: Nonlinear Least-Squares Estimation 162
Reasons That the Quadratic Form Is the "Best" Suited for Modeling
  Industry Structure 164
Estimation Results 165
Discussion of Table 5.8 Results 166
Discussion of Table 5.9 Results 167
End of Section Exercises 167
Cubic Cost Models 168
A Priori Expectations 169
Back to the Example at Hand 170
Discussion 170

Multiple-Output Models   171
  Discussion   172
More Complex Multiple-Output Models   173
  Discussion   173
Other Issues   174
Measures of Efficiency for Multiple-Output Models   175
Ray Cost Output Elasticity   175
Degree-of-Scale Economies   175
  Corollary   176
Product-Specific Economies of Scale   176
Economies of Scope   177
Cost Complementarity   178
  Discussion of Table 5.14 Results   179
Conclusion   179
End of Section Exercises: Multiple-Output Cost Models   180
Appendix: Generalized Method of Moments (GMM)   180
  Introduction   180
  Consistency   181
  Asymptotic Normality   183
  Conditions   184
  Efficiency   184
  Implementation   185
  J Test   186
Appendix: Proofs   187

6. Case Study   191

Introduction   191
Theory of Efficient Pricing   193
Study Design   194
Reasons That Cooperatively-Owned Utilities Are Different   194
Literature Review   196
Estimating Cost Models   197
Data   198
Cost Models   199
  A Properly Specified Cost Model   200
  Discussion of Cost Models   201
Estimation Results   202
  Discussion of Estimation Results   203
Efficiency Measures   204
Discussion of Figure 6.4—Average Incremental Cost and
    Marginal Cost   205
General Implications of Estimation Results   205
Conclusion   205
Appendix A: Panel Data   206
  A Caveat   206
Appendix B: Heteroscedasticity-Consistent Covariance
    Matrix Estimation   207

Contents (xi)

7. Case Study: Cost Models to Illustrate KLEM Data   209

Cobb–Douglas Cost Model   209
Elasticities of Substitution for Cobb–Douglas   210
Translogarithmic Cost Function   211
Substitution Elasticities for the Translog Form: Hicks–Allen Partial
     Elasticities of Substitution   215
Price Elasticities   216
     Discussion of Results   216
Generalized Leontief Cost Function   216
Empirical Estimation   217
Iterated Zellner-Efficient Estimator   219
Generalized Leontief Cost Function—Elasticities   220
     Discussion of Results   221
Quadratic Cost Model   224
Nonlinear Quadratic Cost Model   224
     Estimation   224
     Discussion of Results   224
Elasticities   225
     Price Elasticities   225
     Substitution Elasticities   226
Morishima Elasticities of Substitution   226
Conclusion   227
Appendix: Quasiconcavity in Input Prices   227
Exercises   228

8. Efficient Pricing of Electricity   231

Introduction   231
Theory of Efficient Prices   231
Debate on the Optimal Pricing of Electricity:
     A Brief History   232
     II. Demand Response Proposal More Than a Century Old   233
     III. A Renewed Interest in Demand Response, but "Whither the
         Economic Rationale for Efficient Pricing?"   234
Rate Design   236
More About Rate Design: In Theory   236
Overview of Rate Design Process   237
     Total Revenue Requirements   237
     Functionalization and Classification of Costs   237
     Identification of Rate Classes   237
     Design of End-User Rates   239
Efficient Public Utility Pricing   239
Ramsey Prices: A Second-Best Option   241
Ramsey Pricing—The Second-Best Option   243
Another Option: Average Cost Pricing   244
Two-Part Tariffs   246
Two-Part Tariff with Different Customer Classes   247
     Optimal Two-Part Tariff: The Solution   249
     Aside: The Issue of Cross-Subsidization in Utility Rate Making   250

Multipart Tariffs   251
Nonuniform Pricing: Block Rates   251
Time-of-Use Rates   254
A Brief History of Time-of-Use Pricing   254
Real-Time Pricing   256
Understanding Electric Utility Customers   257
    Approach   258
Assessment of Rate Structure Options   258
    Approach   259
    Impact   259
    How to Apply Results   259
Conclusion   260
Exercises   261

9. Price and Substitution Elasticities of Demand:
   How Are They Used and What Do They Measure?   263

Introduction   263
Price Elasticity of Demand   263
A Brief Review of the Literature: Energy Demand, Elasticities of
    Demand in Energy Markets, and Functional Forms   265
    The Demand for Energy: Econometric Models   265
    Functional Forms   267
    Long-Run Elasticities versus Short-Run Elasticities   267
    Expenditure Share Models   270
Price and Substitution Elasticities   272
    Cross-Price Elasticities   273
    Remarks   273
Dynamic Models   274
Structural Models   276
Expenditure System Models   276
Econometric Issues: Identification and Systems Bias   277
    Aside: The Identification Problem   279
Simultaneous Equations   281
Consistent Parameter Estimation   282
More Advanced Estimation Methods   283
Additional Econometric Issues (Berndt, 1991)   285
A Brief Survey of the Literature: Price Elasticity of Demand   287
    Aside: The Necessity to Separate Short-Run from Long-Run
        Effects and Methodologies   289
Substitution Elasticities   289
    The California Debacle: A Decade Later   291
Federal Legislation   292
    EPACT and EISA   292
    American Recovery and Reinvestment Act   293
    Policies and Programs   294
Technology—the "Smart Grid"—How It Works   295

Elasticities 296
  Models for Estimating Price and/or Substitution Elasticities:
    The Constant Elasticity of Substitution Functional Form 296
  Discussion of Estimation Results 299
  More Complex Models: Price and Substitution Elasticities Using
    Constant Elasticity of Substitution Model 300
  Overview of Results—Faruqui and Sergici Study 300
  Conclusion 301
  For Interested Readers 301
  Recommended Reading 301
  Exercises 302

10. Time-of-Use Case Study 303

  Introduction 303
  The Electricity Crisis: Summer 2000 304
    Factors Precipitating the Crisis 304
    Chronology of the California Electricity Crisis:
      Lessons Learned (or What Not to Do) 304
    Mistake #1 306
    Mistake #2 307
    Mistake #3 307
    Mistake #4 307
  Chapter Overview 309
  Literature Review 310
    Aside: Difficulties in Price Elasticity Estimations 312
  Effect of Time-of-Use Pricing on Peak Utility Load 313
  Recent Experience in California 313
  The California Debacle: A Decade Later 314
  Using Real-Time Pricing to Estimate Price Elasticity of Demand 317
  Further Implications of the Residential Responsive
    Pricing Pilot 320
  Peak-Load Pricing 321
    Discussion of Results 321
    What Is Going on Here? 323
    Economic Theory 323
  Substitution Elasticities 325
    A Generalized Leontief Model to Calculate Substitution Elasticities 325
    Empirical Estimation 326
  Elasticities of Substitution 327
  Conclusion 328
  Recommended Reading 328
  Exercises 329

References 331
Index 339

Acknowledgements

Like my first book *Electricity Cost Modeling Calculations*, this book has been quite an endeavor and would not have been possible without the loving support of my family: my husband David and new daughter Samantha; my parents, David and Helen, who instilled in me the value of an education and strong work ethic; and my Graduate school professors, who believed in me and encouraged me when times were tough. They are: Partha Deb, David Freshwater, Roy Gardner, Robert Kirk, and Frank Scott.

Chapter 1

Introduction

INTRODUCTION

The issue of global climate change and its consequences has become one of growing concern in recent years. In fact, in December 2009 the Environmental Protection Agency announced that emissions of greenhouse gases (GHG) are an endangerment to public health.[1] Despite the fact that there is no specific regulation, one is expected in the near future and will affect utilities profoundly, especially those with coal-fired generation. According to *Public Utilities Fortnightly*, utilities will have to build 250 GW of new generation capacity to offset retirements and meet new demands through 2035, and this new supply will be composed primarily of natural gas and renewable capacity. Subsequently, there has been an increased focus on energy efficiency and on the development of alternative sources of energy, particularly renewable resources, but also nuclear and clean coal technologies, such as carbon capture and storage (CCS).[2] "Going green" has become the buzzword of the early 21st century.

As a result, much of the work that is being performed at utilities has been focused on the potential impacts of conservation and energy efficiency on load forecasts, resource planning that includes retirements of generating older units with replacement by renewable resources, and, in the case of investor-owned utilities, shareholder value. What seems to be missing are well-designed rate-setting mechanisms that will provide the proper incentives to consumers to make the appropriate choices in energy efficiency; in other words, the majority of the methodologies by which electric rates are set in the United States (and some other countries that regulate rates paid by end users) provide neither the proper incentive to consumers nor the reward for "doing what is right." The bottom line is that rates are not based on economic efficiency, which occurs when fixed costs are recovered via fixed charges (i.e., customer- or demand-related costs) and variable costs (i.e., marginal costs) via an energy charge. Instead, other motivations tend to guide the rate-making process, politics among them.

---

[1]This finding was driven by the U.S. Supreme Court decision that greenhouse gas emissions should be covered under the Clear Air Act and defined as air pollutants.

[2]Renewable technologies have the added benefit of not being subject to the price volatility of fossil fuels, but may have drawbacks that include intermittent availability and high initial capital costs.

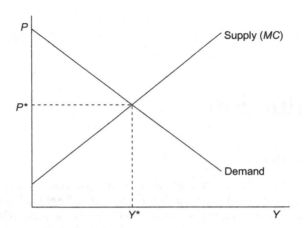

FIGURE 1.1   A competitive market in equilibrium.

COMPETITIVE PARADIGM

In competitive markets the intersection of supply and demand sets the market-clearing price. You may recall from introductory economics texts that the supply curve represents the marginal cost of production in a competitive paradigm. As such, it is the case that P^*, the market-clearing price, is equal to marginal cost, which is efficient both allocatively and productively. This is displayed in Figure 1.1.

As such, marginal cost pricing yields a welfare-maximizing outcome in which both the consumer and the producer receive the maximum benefit possible. Also known as a pareto-optimal outcome, this is the classic model introduced in the economic textbooks. This is discussed in much more detail in chapters on pricing and regulation.

However, and as we all know, the generation, transmission, and distribution of electricity require large and highly sunk investments in capital equipment. As such, the market structure certainly cannot be described as competitive (except possibly the generation stage), which is described in much detail in Chapter 2. Nonetheless, and as we will see, marginal cost pricing could be used in pricing the variable component (i.e., the amount of electricity consumed by end users), which yields a significant increase in efficiency compared to the average Cost Pricing mechanism often used by utilities. This is discussed later and in more detail throughout this book. For now, an excerpt from "Marginal Cost Pricing for Utilities: A Digest of the California Experience" makes this point well.

MARGINAL COST PRICING DOCTRINE

The "marginal cost pricing doctrine" is shorthand for the proposition that utility rates should be predicated upon marginal costs for the purpose of attaining economic efficiency by means of accurate price signals.

The doctrine stems from Professor Alfred E. Kahn's hugely influential two-volume book *The Economics of Regulation* (1970, 1971). Kahn espoused marginal cost pricing as a means of bringing "economic efficiency" to regulated utilities. This pricing would result in "price signals" to consumers of sufficient accuracy so that they could evaluate the appropriate economic level and timing of their use of utility services. Thus, the buying decisions of consumers would be the means by which the end purpose of economic efficiency would be reached.

Theory

Quoting Professor Kahn, normative/welfare microeconomics concludes that "under pure competition, price will be set at marginal cost" (the price will equal the marginal cost of production), and this results in "the use of society's limited resources in such a way as to maximize consumer satisfactions" (economic efficiency) (Kahn, 1970, pp. 16–17).

The basis for the theory is clear cut: because productive resources are limited, making the most effective use of these limited resources is a logical goal. In a competitive economy, consumers direct the use of resources by their buying choices. When they buy any given product, or buy more of that product, they are directing the economy to produce less of other products. The production of other products must be sacrificed in favor of the chosen product.

From this point, marginal cost theory takes a giant step. In essence, it states that if consumers are to choose rationally whether to buy more or less of any product, the price they pay should equate to the cost of supplying more or less of that product. This cost is the marginal cost of the product. If consumers are charged this cost, optimum quantities will be purchased, maximizing consumer satisfaction. If they are charged more, less-than-optimum quantities will be purchased: the sacrifice of other foregone products will have been overstated. If they are charged less, the production of the product will be greater than optimum: the sacrifice of other foregone products will have been understated. A price based on marginal costs is presumed to convey "price signals," which will lead to the efficient allocation of resources. This is the theory, drawn from the microeconomic model of pricing under perfect competition, upon which the doctrine rests (Conkling, 1999).

To be fair, the reticence to adopt marginal cost pricing is due in large part to the thus-far inability to accurately estimate/calculate the marginal cost of distributing electricity to various types of end users. This is also the aspect of the puzzle that has been ignored thus far and the primary motivation of this book: How do we accurately estimate the true cost of providing electric service so that rates can be set in an efficient manner, which will provide the proper incentives to both producers and consumers to make the appropriate investments in energy efficiency, demand-side management, and conservation in general. This is addressed in a case study in which an appropriately specified cubic cost model is used to estimate the marginal cost of serving end-use customers so that price could be set efficiently.

Before delving further into the contents of this book, the author would like to provide a brief overview of the U.S. electric power industry, including the types of players (i.e., suppliers) and a general overview of the regulatory environment and its relationship to greenhouse gasses.

A BRIEF OVERVIEW OF THE UNITED STATES ELECTRIC MARKET

Structure of the United States Electricity Industry

The majority of the electricity distributed in the United States is by *investor-owned utilities*, which tend to be *vertically integrated*, which means that the same entity generates, transmits, and distributes electricity to end users in its service territory. In the case of such investor-owned firms, traditional rate making is that a return to investors is earned on every kilowatt-hour sold, thus providing the incentive to sell as much as possible. Figure 1.2 displays the structure of the electric industry in the United States.

Players and Their Incentives

To assess the impact of various policies and rate-making schemes intended to affect climate change, it is necessary to distinguish each type of electric supplier and to examine the incentives that each type faces. Unlike investor-owned utilities of which the objective is profit maximization, publicly and cooperatively-owned utilities face their own set of circumstances and have their own objectives. Nonetheless, the ability to accurately estimate the true cost of providing service to various types of customers is tantamount to

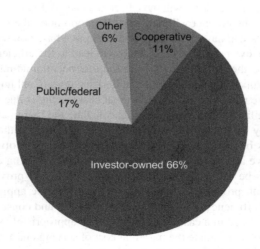

FIGURE 1.2   Structure of the electric industry in the United States. Breakdown of 2006 sales by type of supplier.

designing effective legislation, despite the different objective functions faced by each, which are described later.

First and foremost, all utilities in the United States have an obligation to serve that which is part of their franchise agreement, which means that they have been given an exclusive right to supply utility service to the customers that reside within that service territory. Whether a supplier is subject to certain types of regulation depends on the type of supplier, the state in which they are operating, and whether they are vertically integrated or not. Each has its own objective function, which is discussed in the next section.

OBJECTIVE FUNCTIONS: THE PLAYERS

Investor-Owned Utilities: Profit Maximization

All investor-owned utilities in the United States are subject to some type of regulation, typically price and performance (e.g., an obligation to serve native load and reliability in providing service). The objective function of the *regulated* investor-owned utility is to maximize profit (π), which is equal to total revenue (TR) less total cost (TC), subject to a breakeven constraint under a regulated price, P_r, while procuring (or generating) enough electricity to satisfy market demand, Y_m; that is,

$$\text{maximize } \pi = TR(P_r, Y_m) - TC(Y_m) \tag{1.1}$$

subject to

$$TR \geq TC$$

and

$$Y \geq Y_m,$$

where Y is total output.

Under the type of regulation to which the utility is subject, which is discussed in more detail later (and in subsequent chapters), the price allowed by the regulator (P_r) includes an appropriate rate of return to investors. The intent here is to compensate shareholders for risks involved in holding the stock issued by the utility.

Publicly-Owned Firms

Under the umbrella of publicly-owned utilities are nonprofit organizations, which have been established to serve their communities at cost. While some of them generate their own electricity, many others serve to transmit and distribute power purchased from other wholesale generators, which are mostly federally-owned entities, such as the Tennessee Valley Administration and the Bonneville Power Administration. (Some other power administrations include the Southeastern Power Administration and the Southwestern Power Administration.) This being said, some publicly-owned entities do purchase from investor-owned or cooperatively-owned entities. To best serve the public interest, the objective function is cost minimization subject to a breakeven constraint (i.e., that total revenues cover total costs). This is given by Equation (1.2):

$$\text{minimize } C = f(Y, P) \qquad (1.2)$$

subject to

$$\text{total revenue} \geq \text{total cost},$$

where Y is output and P is price.

Organizational types include municipals, public power districts, and state authorities. Publicly-owned utilities are exempt from certain taxes and can obtain new financing at lower rates than investor-owned utilities typically can. In addition, they are given priority in the purchase of the less-expensive power produced by federally-owned generators. These are discussed in much more detail in Chapter 3.

Cooperatively-Owned Firms

Rural electric cooperatives are owned by members of the cooperative and were established to provide electricity to their members, which reside in rural areas deemed too costly to serve by investor-owned entities. (This is discussed in more detail in the case studies presented in Chapter 6.) Like publicly-owned utilities, cooperatives enjoy benefits that investor-owned utilities do not: they are able to borrow directly from various federal agencies created especially to serve them, predominantly the Rural Utilities Service, which allows them to obtain financing that carries a lower interest rate than the market does. In addition, they enjoy certain tax exemptions and are given preference in the purchasing of lower-cost federally-produced power.

Presumably, the cooperatives' incentives are welfare maximization (W), which is equal to consumer surplus (CS) plus producer surplus (PS), which is due to the coincidence of sellers and buyers. The objective function is displayed in Equation (1.3) (they are also subject to satisfying market demand, Y_m):

$$\text{maximize } W = PS + CS \qquad (1.3)$$

subject to

$$Y \geq Y_m,$$

where PS is the area below P^* above the supply curve and CS is the area above P^* below the demand curve.

Cooperatively-owned utilities are interesting in that they are not subject to price regulation in all states. In fact, fewer than 20 states regulate rates charged by rural electric cooperatives, which are organized as either generation and transmission (G&T) or distribution only (also known as member coops). While not truly vertically integrated, member coops are typically bound contractually to a G&T coop to supply their power needs. While this is not always the case, it is far more common than not. This particular organizational structure was the impetus for the paper entitled "A Test of Vertical Economies for Non-Vertically Integrated Firms: The Case of Rural Electric Cooperatives" published in *Energy Economics* in 2008. This paper was also presented as a case study in *Electricity Cost Modeling Calculations* (Greer, 2011).

Other

Other types of suppliers include power marketers, independent power producers, public power agencies, power pools, and energy service providers. These are discussed in more detail in Chapter 3.

REDUCING CARBON EMISSIONS

Regulation/Rate Making

There is little doubt that any meaningful limit or reduction of carbon dioxide emissions will have a significant impact on the electric supply industry. For example, in the United States, the electric power sector accounted for about 40% of the total carbon dioxide emissions in 2006, which increased by 2.3% in 2007. Emissions declined in 2008 and 2009 due to the recession but increased by over 5% in 2010 (U.S. Energy Information Administration). Also, these emissions increased by over 14% from 1996 to 2006 as the demand of electricity has increased. Over 97% of carbon dioxide emissions come from burning coal and natural gas for generating electricity. This is not surprising, as together these fuels account for 69% of the fuel used to generate electricity in 2006[3] (Energy Information Administration, 2007).

The United States Electric Power Industry—Regulation

In addition to the scale of the emissions and importance of fossil fuel in generating electricity, a complicating factor is that electric generation in the United States is regulated by a complex mix of federal and state laws and regulations. These laws and regulations have an influence on the generation resource choices that suppliers make. At the federal level, generation is subject to oversight by the Federal Energy Regulatory Commission (FERC). Since the mid-1990s, the FERC has increasingly relied on market mechanisms to determine prices and generation resources for the wholesale regions they regulate. Also, at the state level, 20 states have modified or "restructured" their regulation of their electric utilities and now permit some or all utility customers the opportunity to choose their own supplier. However, 30 states remain regulated in the "traditional" or cost-based/rate-of-return manner that has been used for over a century. Also,

---

[3]Based on data from the U.S. Department of Energy, Energy Information Administration, "Emissions of Greenhouse Gases Report" and "Electric Power Annual," 2007.

while the mix of federal and state regulations may be unique to the United States, many features of markets and regulations apply to other countries as well.

REGULATION OF INVESTOR-OWNED ELECTRIC UTILITIES IN THE UNITED STATES

Average-cost pricing is the typical regulatory mechanism employed by states that are price regulated in the United States, which is displayed in Figure 1.3. Not only does this permit the utility to recover its prudent costs, but also compensates shareholders for the risk that they bear by holding the stock of the utility. Typically initiated by the utility when rates fall below average cost, a rate case is the formal procedure for determining the price of electricity sold to various types of end user (i.e., residential, commercial, industrial, or other). More specifically, this process involves the establishment of the utility's *revenue requirement*, which is the amount of dollars that must be collected from ratepayers to recover the utility's expenses (and required rate of return, in the case of investor-owned utilities) for the period during which such rates would be in effect. Once the revenue requirement is determined, it is then multiplied by the allowed rate of return (i.e., return on equity, or ROE) set by the public regulatory commission. From here, allocations ("base rates") are made among the various rate or revenue classes served by the utility based on cost of service, price elasticity of demand, and politics. This is described in much more detail in Chapter 8.

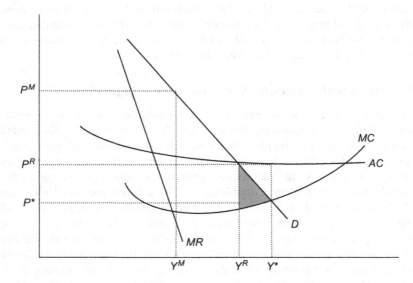

FIGURE 1.3   Average-cost pricing. Rate-of-return regulation creates a dead-weight loss, as price (P^R) is set above marginal cost, which yields a price of P^*. (Dead-weight loss is approximated by the shaded triangle.) However, this loss is de minimus when compared to the lost consumer surplus from monopoly pricing without regulation, which is given by P^M.

Aside: Issues with Rate-of-Return Regulation—
The Averch–Johnson Effect

Despite its prevalence in the United States, rate-of-return regulation creates an inefficient use of resources because it provides an incentive for the utility to overinvest in capital (and hence increase its rate base, which is one component of its revenue requirement), thus earning higher returns for shareholders of investor-owned firms.

To show this, I appeal to Averch and Johnson (1962), which is reproduced in Rothwell and Gomez (2003) as Exercise 4.4.

Averch and Johnson assume that the utility maximizes profit subject to the rate-of-return constraint:

$$\max \pi = P*Q(L,K) - w*L - r*K \tag{1.4}$$

subject to

$$s = [P*Q(L,K) - w*L]/K, \tag{1.5}$$

where $Q(L, K)$ is output, a function of capital (K) and labor (L); r is utility's cost of capital; and s is allowed rate of return.

It is further assumed that $s > r$; that is, its rate of return is higher than its cost of capital. Using the Lagrangian multiplier technique, Equations (1.4) and (1.5) become

$$\max \pi = P*Q(L,K) - w*L - r*K + \lambda*[s*K + w*L - P*Q(L,K)], \tag{1.6}$$

where λ is the Lagrangian multiplier.

The profit-maximizing levels of capital (K^*) and labor (L^*) are obtained by setting the derivative of profit with respect to each input and λ equal to zero and then solving for that level of input; that is,

$$\partial\pi/\partial K = P*Q_K - r + \lambda*(s - P*Q_K) = 0 \tag{1.7}$$

$$\partial\pi/\partial L = P*Q_L - w + \lambda*(w - P*Q_L) = 0 \tag{1.8}$$

$$\partial\pi/\partial\lambda = s*K + w*L - P*Q = 0, \tag{1.9}$$

where Q_K is the marginal product of capital and Q_L is the marginal product of labor. If the constraint is binding, then $s*K + w*L - P*Q = 0$.

Because $\partial\pi/\partial\lambda = 0$, the rate-of-return constraint is satisfied at maximum profit. Averch and Johnson show that the marginal rate of technical substitution (MRTS) of labor for capital is given by

$$\text{MRTS} = \{r - [(s-r)*\lambda/(1-\lambda)]\}/w = (r-\alpha)/w. \tag{1.10}$$

If the allowed rate of return is equal to the cost of capital (i.e., $s - r = 0$), then MRTS $= r/w$; that is, the profit-maximizing level of capital and labor to be employed occurs where the ratio of input prices equals the marginal rate of technical substitution.

However, if $s > r$, then the higher return on capital motivates the firm to increase investment in capital beyond K^* (the efficient level of capital) so that

$$K > K^*,$$

a nonoptimal outcome.

In addition, and also from an investor-owned utility's perspective, the nature of rate-of-return regulation is such that the more electricity sold, the more money (i.e., profit) earned. According to a recent National Action Plan for Energy Efficiency (U.S. Environmental Protection Agency, 2007) report,

"Between rate cases utilities have a financial incentive to increase retail sales of electricity (relative to forecast or historical levels, which set "base" rates) and to maximize the "throughput" across their wires since there is often a significant incremental profit margin on such sales."

Furthermore, this report indicates that three impediments exist to the pursuit of energy efficiency under traditional regulation:

1. Negative impact on cash flow and earnings if expenditures on energy efficiency and demand-side management are not recovered in a timely fashion.
2. Reduction in sales and revenues could lead to underrecovery of fixed costs.
3. Unlike supply-side investments (i.e., new generation), investments in energy efficiency do not earn a return.

Addressing the concerns raised by suppliers without causing harm to ratepayers will require compromises among regulators, utilities, and consumers. This is a global issue that transcends regulatory structure—whether one is an electric customer in a deregulated European market or in the United States. The bottom line is that the end result will be the same: decarbonizing electricity will entail extra costs that will be reflected in rates. But what is important here is *how* those rates are structured—unlike traditional rate making, which does not necessarily differentiate rates by the actual cost to serve (meaning the per-kilowatt-hour charge versus the demand charge) but rather is based on the elasticity of demand, politics, etc. In other words, rates are not set efficiently, which means that energy charges should be based on the marginal cost of providing electricity and demand charges based on fixed costs. This is discussed in more detail in Chapter 8.

INTERNALIZING THE COST OF REDUCING CARBON EMISSIONS

To affect a reduction in the amount of carbon emitted into the atmosphere, at least two events must occur. First, from an economic efficiency perspective, those who are causing the problem should pay for it. In the case of electricity, we can identify two sources:

1. Producers of electricity—because carbon emissions and greenhouse gas (GHG) are the result of using fossil fuels to generate electricity.

2. Consumers of electricity—because without them there would be no need to produce electricity, which generates GHG.

Subsequently, prices must rise, which could result in two things:

1. Conservation and investment in energy efficiency (on the part of consumers).
2. Seeking, investing in, and obtaining regulatory approval for energy-efficiency programs, including demand-side management (DSM), renewable sources of power supply, clean coal, and nuclear, and in generating technologies, such as carbon capture and storage.

As stated previously, this entails a compromise among the utility, the state regulatory commission, and the ratepayers.

Current Policy

Currently, policy measures being used or discussed to internalize the costs of carbon emissions include a carbon tax and cap and trade and mandates including renewable portfolio standards and incentives for energy-efficiency investments. With a tax, a specific carbon "price" is imposed directly on the producer, typically on a per-ton basis, while a cap-and-trade system sets a price indirectly by establishing an emissions limit with trading rights to emit so that the forces of supply and demand determine the price (at least theoretically). Both will result in an increase in costs, which will be borne by producers and passed onto consumers as allowed by state regulatory commissions. Clearly rates will rise. How much depends on the relative price elasticities (producer vs. consumer) and on how much the state regulatory commission will allow the utility to pass on to ratepayers and in what fashion.[4]

The bottom line is that both producers and consumers will be affected; as rates rise, consumers will likely reduce their consumption and begin to invest in energy efficiency (appliances, home weatherization, and other improvements), especially given the tax credits that have been made available by both federal and state governments. The subsequent reduction in utility sales will impact shareholder returns, thus necessitating a rate case filing and an additional increase in rates.[5] Before long, a vicious cycle emerges, which could be obviated by simply charging the marginal cost of the power that is consumed at the time that it is consumed. This will be expounded upon in subsequent chapters.

In a cap-and-trade system, the government sets a cap on total emissions of greenhouse gases from various industrial and utility sources, including power

---

[4]For example, will it be incorporated into the base rate or be a "below-the-line" item, such as environmental cost recovery or demand-side management mechanisms. In states such as Kentucky, where construction work in process is allowed, such items are recovered "below the line" until a rate proceeding in which they become part of the base rate.

[5]Also known as decoupling, it is a method by which utility revenues are not tied to throughput so that a reduction in sales of electricity may not impact the bottom line.

plants burning fossil fuels to generate electricity. It then issues allowances to polluters allowing them to emit carbon dioxide and other greenhouse gases; total emissions are meant to stay under the cap. Over a period of time, the government reduces the cap and the number of allowances gradually until it reaches its target. If companies' emissions exceed their allowances, they must buy more.

Economists like the system because companies can choose to either lower their emissions, such as by investing in new technology, or buy more allowances from the government or from companies that don't need them, whichever makes the best economic sense. It is meant to create a carbon market, putting a value on emissions. Also, unlike a tax, no dead-weight loss is created. This appears to be the method of choice given the creation of the Regional Greenhouse Gas Initiative, whereby allowances can be bought or sold. Currently, 10 states in the United States participate (along with several provinces in Canada) in this initiative, which is a cap-and-trade system for CO_2 emissions from power plants in participating states, including Connecticut, Delaware, Maine, Maryland, Massachusetts, New Hampshire, New Jersey, New York, Rhode Island, and Vermont.

OPTIMAL RATE/TARIFF DESIGN AND TAX CREDITS TO PROMOTE EFFICIENT USE OF ENERGY AND A REDUCTION IN CARBON EMISSIONS

The bottom line is that any policy implemented to encourage energy efficiency must be crafted carefully to protect both shareholders and consumers and not one at the other's expense. However, rates should be structured to encourage consumers to use energy wisely.

Economic theory dictates that as rates rise, consumers' incentives to increase investments in energy efficiency also rise. As a result, not only would the quantity demanded of electricity decline, but also does the demand for it (ceteris paribus). The end result is that not only do emissions fall (a good thing), but also the utility's need for additional generating capacity is obviated (or at least delayed), a move that typically results in the utility filing a rate case that subsequently ends in an upward adjustment in rates. However, investments in energy efficiency on the part of the consumer will allow the utility to recover only (and earn a return on) its current (and possibly shrinking) rate base, which may diminish the utilities' ability to attract and retain appropriate levels of capital and to provide safe, reliable service at a reasonable cost. As a result, it is likely that a series of rate increases will ensue. Also, this is the crux of the matter for regulators to consider in this critical matter. It is the regulators after all that hold the key to the success of this endeavor by structuring rates and tariffs in such a way to motivate consumers to use energy wisely and producers to make prudent investment choices that earn appropriate returns, either supply side or demand side.

Up to this point, much of the focus has been on appropriate policies that will encourage producers to invest in energy efficiency and DSM. However, as noted, given the type of regulation that utilities face, there is clearly an inclination to *not* make such investments under the current regulatory scheme as they may impede the utilities' fiduciary duty to its shareholders. One solution offered has been that of decoupling of revenues from throughput, which likely means that rates simply adjust (upward) so that the utility is held harmless. This raises several questions, however. For example, how can behavior be altered to reduce usage during peak hours when power is the most expensive? Is it possible to devise a pricing mechanism to accomplish the objectives set forth in this chapter?

From the consumer's perspective, one could argue that the pricing signal is adequate to motivate investments in energy efficiency and conservation. However, it has been argued that, in relatively low-cost areas, electric rates are not sufficiently high to promote any real change in behavior. This is because the price is in the inelastic portion of the demand curve so that a small price change has little impact on the quantity demanded. However, incorporating the effects of carbon legislation for controlling emissions of greenhouse gases, developing alternative technologies and fuel sources (renewable, even nuclear), and investing in more efficient appliances and equipment could be enough to precipitate some of the changes required to make a difference in the amount of carbon dioxide emitted into the atmosphere. But there are some fundamental issues that will not be addressed by uniform increases in rates (i.e., rates that change by a certain percentage without regard to the amount of electricity distributed).

TARIFF DESIGN AND RATE-MAKING ISSUES

It was stated earlier that affecting any real changes and minimizing losses to both consumers and producers will require compromise among consumers, producers, and public regulatory commissions. This means that tariffs must be designed to motivate investments in energy efficiency on the parts of both producers and consumers. Economic theory suggests that setting price equal to marginal cost maximizes total welfare, which is equal to consumer plus producer surplus. However, given the nature of the industry (a natural monopoly), this first-best outcome is often deemed infeasible, as all costs may not be recovered (see Figure 1.3), which gives rise to the use of a second-best pricing scheme. Also known as Ramsey prices, this set of uniform prices serves to maximize total surplus (and minimize dead-weight loss) subject to a breakeven constraint. (This is described more in Chapter 8.) To provide this in the electric utility industry, a modification must be made: instead of a schedule of uniform prices, a nonuniform pricing scheme in which prices *increase* with usage could be used to approximate marginal cost and promote more efficient behavior and energy usage.

Marginal Cost Pricing for Electric Utilities

With this all said, pricing at marginal cost is not new. An excerpt from "Marginal Cost Pricing for Utilities: A Digest of the California Experience" (Conkling, 1999) informs us that this methodology was in place until 1996 when California became the first state to embark on the restructuring of its electric market (this is also known as the California debacle, which is discussed in some detail in a later chapter). More specifically, Conkling advises that

California was not the first state to consider marginal costs. That distinction goes to the Wisconsin PUC. Its August 8, 1974 decision re Madison Gas and Electric Company is recognized as the first to hold that marginal costs, together with peak load pricing, were appropriate rate design considerations. The New York PUC was only five months behind California's March 16, 1976 adoption of marginal costs. The New York PUC's Opinion 76-15, of August 10, 1976, issued when Professor Kahn was chairman, held that marginal costs, as distinguished from average costs, are the most relevant costs for rote-setting (sic) and should be utilized to the greatest extent practicable. Other early states were Florida, North Carolina, and Connecticut.

CONCLUSION

This book is divided into 10 chapters, including this introductory chapter. Chapter 2 is theoretical, encompassing the theories of natural monopoly and the related topics of scale and scope economies, network economies, and vertical integration. Also included is a review of the literature pertaining to the electric industry and these topics. Chapter 3 offers a review of the types of regulation that have transpired and the problems therewith, along with a discussion of market structure. Currently proposed legislation is included, along with suggestions as to the direction that rate design should go in terms of the efficient pricing of electricity.

Chapters 4 through 7 are similar in that they introduce cost models and provide examples and exercises so that the user may have a "hands-on" experience with estimating the various models that have been used to estimate economies of scale, scope, and vertical integration in the electric industry.

Chapters 8 through 10 involve the consumer side of the equation—pricing and elasticity—and include a case study on real-time pricing of electricity. More specifically, Chapter 8 is all about pricing electricity efficiently, which is a key element in promoting the efficient use of energy and in the subsequent reduction of greenhouse gas emissions into the atmosphere. Chapter 9 delves into price and substitution elasticities and provides a series of examples and exercises designed to illustrate the computation of these elasticities. Chapter 10 is a case study on real-time pricing and a discussion of how to design a meaningful pilot program.

Chapter 2

The Theory of Natural Monopoly and Literature Review

THE NATURAL MONOPOLY CONUNDRUM

Historically, conventional wisdom has held that certain markets were "naturally monopolistic," which means that, due to the presence of high fixed costs of which the average declines with increases in output, efficiency is best obtained when there is only one supplier. According to Kahn (1970, p. 15),

... the public utility industries are preeminently characterized in important respects by decreasing unit costs with increasing levels of output. That is indeed one important reason why they are organized as regulated monopolies: a "natural monopoly" is an industry in which the economies of scale are such that one company supplies the entire demand. It is a reason, also, why competition is not supposed to work well in these industries.

Included herein are the markets for electricity, natural gas, telephone, and water services. It has often been argued that what is driving this phenomenon is the irreversibility of the initial investment required to produce a particular good or service in a naturally monopolistic industry. More specifically, the underlying production technology of this product is such that a level of output exists for which average cost is minimized; at levels of output below this level average costs decline, and at levels above they rise. This is known as economies of scale and is investigated further in context to its relationship with the theory of natural monopoly.

Economists have spent many years attempting to assess that level of output at which the minimum efficient scale occurs. In some industries, such as the generation of electricity, a consensus has been reached that, at least in 1970, most firms were producing in and around this level, given a particular production technology (Christensen and Greene, 1976). In other words, economies of scale in the generation of electricity had been exhausted.

Until recently, no one questioned that the production of electricity was, in fact, a natural monopoly, as, like telephony, what is required here is a network—a complex, interactive, interdependent connection of wires (by which individuals gain access to the local distribution company, which is connected to the transmission grid at various nodes). This network represents an irreversible investment, which is characterized both by economies of scale and by those of network planning, and as such, yields a natural monopoly. Because this network leads to externalities, vertical integration has traditionally yielded the most efficient organization of the industry, especially for larger firms. But it is due to the vertical nature of electricity production that questions have arisen concerning whether any aspect of the production process may not be a natural monopoly. If this is the case, questions then become: would the market be better served by allowing competition into that component and would the gains from competition exceed the lost economies that would result? This is the critical element that needs to be explored.

Things are not always so clear, however. While there has been little work done in the areas of testing whether the transmission and distribution processes are natural monopolies, they are usually assumed to be so, as both are characterized by what is known as network economies. Network economies arise due to the interconnectedness of the national transmission grid so that significant saving in inputs and direct routing yield both economies of scale and economies of scope. These are defined later in this chapter, along with a review of the relevant literature.

DEFINING NATURAL MONOPOLY

Older industrial organization theory cited that the presence of scale economies determined whether an industry was a natural monopoly. It is important to note that much of the theory of natural monopoly is concerned with the precise meaning of "increasing returns" or, equivalently, decreasing average costs. Scale economies exist when a proportionate increase in output leads to a less-than-proportionate increase in cost. Mathematically, a cost function (one output) is said to exhibit global (local) economies of scale if

$$C(\lambda q) < \lambda C(q) \tag{2.1}$$

for $\lambda > 1$, $q \geq 0$.

According to Marshall (1927), increasing returns can be either internal or external to the firm and, similarly, internal or external to the industry. A natural monopoly tends to arise due to high fixed costs, which tend to be asset specific and, as such, are largely sunk. As a result, average cost tends to decline as output is expanded over a large range, thus rendering a single provider socially optimal. In addition, economies of scale can be either technical (relating to the production process) or pecuniary, that is, related to the prices paid for inputs.

One of the difficulties in testing for natural monopoly is the practical application of testing for subadditivity of a firm's cost function, which is critical, as local (global) subadditivity is a necessary and sufficient condition for local (global) natural monopoly (Evans, 1983). In addition, it is necessary to distinguish between single-output and multiple-output natural monopolies, which is done next.

For a Single-Output Market

An industry is said to be a natural monopoly if one firm can produce the desired market demand at a lower cost than two (or more) firms can. More specifically, it is defined in terms of a single-firm's efficiency relative to the efficiency of other firms in the industry (as opposed to a firm being the controller of an essential resource or having a patent on a particular product). In other words, economies of scale may exist in the production of a particular product. Some characteristics of a natural monopoly, which are attributable to economies of scale, include:

1. Decreasing long-run average cost.
2. High fixed costs.
3. Subadditivity of its cost function.

Although interrelated, the most important of these is subadditivity of the firm's cost function, which means that it is less expensive for one firm to produce the total output demanded than it would be for several firms to produce proportions of it. This can be expressed as

$$C(Y) < \sum C(y^i), \qquad (2.2)$$

where $\sum y^i = Y$.

If this holds, then the cost function is strictly subadditive at output level Y (Sharkey, 1982). For a single-output firm, subadditivity is both necessary and sufficient for a natural monopoly, as subadditivity implies that it is more efficient for a single firm to produce all of the output in the market. It is important to note that subadditivity is a local concept; that is, just because the cost is subadditive at one level of output does not necessarily mean that it is subadditive at all output levels, or globally subadditive. This implies that the total cost of production must be evaluated at all levels of output up to that level that satisfied market demand.

Average Cost

Certainly, declining average cost throughout the relevant range of outputs is an indicator that the cost function is subadditive and that it is more efficient for one firm to supply the entire industry output, that is, a natural monopoly. What this requires, however, is that marginal cost also declines throughout a subset of this range of outputs. And necessary for this is a twice-differentiable cost function, which yields the appropriately shaped average and marginal cost curves. Such a cost function is displayed in Figure 2.1.

A cubic cost function yields the appropriately shaped average and marginal cost curves. For $Y < Y^*$, cost increases at a decreasing rate. In this range, both marginal and average costs are declining. However, once diminishing returns set in, costs begin to increase at an increasing rate; it is in this range that

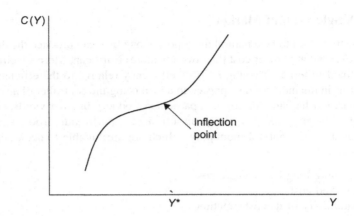

FIGURE 2.1 Total cost curve generated by a cubic cost function.

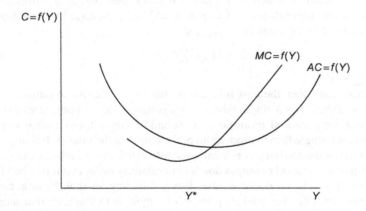

FIGURE 2.2 Average and marginal cost curves generated by cubic total cost function.

marginal costs begin to rise and total cost increases at an increasing rate, which causes average cost to begin rising and yields the U-shaped average cost curve displayed in Figure 2.2.

A cubic cost function generates this particular shape and is of the general form:

$$C(Y) = a + bY + cY^2 + dY^3 \qquad (2.3)$$

so that average cost is given by

$$AC(Y) = a/Y + b + cY + dY^2, \qquad (2.4)$$

where $AC(Y) = C(Y)/Y$ and marginal cost is given by

$$\partial C(Y)/\partial Y = b + 2cY + 3dY^2. \qquad (2.5)$$

estimate economies of scale in generation are Nerlove (1963), who used 1955 data on 145 utilities and found that cost function was characterized by increasing returns to scale but that returns to scale tended to decline with the size of the firm. His study is discussed in more detail in Chapter 4.

Concerning the generating stage, it is necessary to distinguish between studies that use the plant as the sample unit and those that use the firm itself. In their seminal paper, Christensen and Greene (1976) used both Nerlove's 1955 data and also 1970 data on those same firms, and found that by 1970, most firms were generating electricity at a point on the average cost curve in which economies of scale had been exhausted. Using a different cost specification than Nerlove, they found that the minimum efficient scale was attained at 3800 megawatts (MW) and that some firms were producing even beyond this level of output (i.e., in the diseconomies of scale region of the long-run average cost curve). This implied that the generation of electricity was not a natural monopoly, and led to the realization that competition in the generation component was not only feasible, but may also be more efficient. It was this realization that precipitated the eventual deregulation of the generation component of the industry, which is discussed in Chapter 3. In addition, the translogarithmic cost function employed in this particular study is the subject of further analysis in Chapters 4, 6, and 7.

Joskow and Schmalensee (1983) present a summary of different studies carried out in the United States based on econometric estimations and engineering methods. The minimum efficient scale (MES) for conventional electricity generation is around 800 MW and around 2000 MW for nuclear energy. In a later study, Huettner and Landon (1997), using yet another cost specification and 1971 data on 74 electric utilities, confirmed the Christensen and Greene results, although they found that scale economies were exhausted at an even lower level of output. As they pointed out, relationships observed at plant level, particularly scale economies, are often modified by interrelationships at higher levels of decision making, such as the firm level.

Greene (1983) studied economies of scale using panel data on investor-owned utilities from 1955 to 1975 and found that scale economies actually declined over that period of time. Technical change, he argued, was a significant factor in the decreasing average costs that firms were experiencing throughout the majority of the study period. Thermal efficiencies were being exhausted while, at the same time, the demand for electricity was rising thanks to declining power prices. Using 1970 data on 123 privately-owned firms, Atkinson and Halvorsen (1984) employed yet a different cost model and found that most of the firms in the sample were operating in the downward sloping portion of the long-run average cost curve.

For the most part, these studies consider cost functions in which the output is kilowatt hours of electricity generated (Atkinson and Halvorsen, 1984; Christensen and Greene, 1976; Huettner and Landon, 1977). However, others, namely Kamerschen and Thompson (1993) and Thompson and Wolf (1993), studied possible cost differences between conventional electricity-generating technology (fossil fuel generation) and nuclear electricity generation.

This was not the case, however, in the years (and decades) that followed, which were extremely turbulent ones for the industry. Rapidly rising fuel prices, double-digit inflation, and rising capital prices led to a decline in the demand for electricity, which caused financial distress for a number of utilities, which were saddled with excess capacity (Thompson, 1995).

Later studies include Maloney (2001), who estimated the MES at 321 MW and 260 MW for coal- and gas-fired plants, respectively, but found that the average cost curve is flat at this level. Kleit and Terrell (2001) and Hiebert (2002) found increasing scale economies for most observations. Hiebert found that the degree-of-scale economies was 20% in coal-fired plants and 12% for natural gas–fired plants for average sample values (780 MW and 284 MW, respectively). This work also found that major economies can be attained by producing with more than one plant for each kind of generation. This latter aspect highlights the importance of distinguishing between plant and company in generation.

While these studies focus on the generation component, a few studies focus on either transmission or distribution alone, two of the three components, and all three components. Of those that focus on some combination of the components, most do so to study the economies associated with vertical integration, which are discussed later in this chapter. In virtually all of these studies, the consensus is that distributed electricity is not a homogeneous good. This is discussed further in Chapter 4, but for now suffice it to say that different end users have different elasticities of demand and some users are more costly to serve than others.

Economies of Scale and Density in Transmission and Distribution

Some studies estimated the economies of scale for transmission and distribution elements, such as Huettner and Landon (1977), who found the minimum efficient scale occurred at around 2600 MW capacity. Kaserman and Mayo (1991) also found specific economies of scale for these phases and situated the minimum efficient scale at around 5 GWh. Greer (2003) found that none of the rural electric cooperatives distributed anywhere near the minimum efficient scale in 1996.

The network elements and the costs involved in these activities can be studied in greater depth by studying economies of density. This concept explains the evolution of average costs when production is increased and when some of the characteristics that define the product are maintained constant, for example, the size of the service area or the number of consumers.

Network Economies

For electricity, a quintessential element is the transmission network grid by which electricity, once generated, is transmitted to local distribution companies and then to final end users. Because of economies of scale, the per-mile cost of transmitting electricity along a longer, interconnected grid is much less than doing so along a series of shorter grids (assuming, of course, that line losses are minimal). Furthermore, because some electricity is sold in bulk while the rest is sold to various classes of end users, both economies of scale and of scope arise due to the fact that these multiple outputs utilize this interconnected transmission grid jointly. Furthermore, the very nature of this grid yields additional savings due to the network economies, or economies of density, which play a critical role in such an industry. According to Salvanes and Tjotta (1994, p. 23), who examined the distribution function in Norway, "the characteristics of the network affect all costs and should be included by a measure of the number of nodes supplied." They asserted that

In industries where output is delivered via a network to spatially distributed points with distinct demand characteristics and thus a continuum of outputs exists, a traditional approach with a single output to represent firm size to facilitate econometric estimation may have serious implication for measuring productivity differences.

Hence, no longer is it sufficient to measure only returns to scale; returns to density must also be considered if one is to obtain precise and relevant measures of industry structure and form appropriate public policy. Employing the definition of Caves et al. (1984), returns to density (RTD; for the translogarithmic cost specification) are given by

$$\text{RTD} = 1/(\partial\ln C/\partial\ln Y), \tag{2.17}$$

where $\partial\ln C/\partial\ln Y$ is the cost elasticity with respect to output.

Returns to density are increasing, constant, or decreasing for RTD greater than, equal to, or less than unity. Thus, returns to density measure the economies of increasing the number of kilowatt hours produced where the size of the network is fixed.

While only a few studies have attempted to measure economies in transmission and distribution functions, few dispute that they exist. Schmalensee (1978, p. 270) asserted that

Total distribution cost depends on the cost of transmitting services and on the spatial pattern of demand. Everywhere-decreasing average cost of transmission is found to be sufficient, but not necessary, for natural monopoly.

Nonetheless, Schmalensee developed a model to show that economies in transmission at all service flows are sufficient, but not necessary, for distribution to be a natural monopoly. Furthermore, pricing at marginal cost fails to cover total cost, and even in the presence of economies of scale in transmission, average distribution cost may rise with total demand.

Of those (few) studies that attempt to quantify such economies, Huettner and Landon (1977) employed nonconventional (in that some variables are in natural logarithms, whereas others enter as quadratics) cost functions for both transmission and distribution. They found that, for transmission, both long-run and short-run average variable cost curves (oddly, they do not include the fixed costs of transmitting electricity) were inverted U-shaped with the maximum occurring at a capacity of 4000 trillion megawatts (long-run curve) and a utilization rate of 94% (short-run curve). Neither the capacity nor the utilization variables' coefficients were significant, however. For distribution, another non-conventional cost function was utilized with the finding that coefficients of the capacity variables were statistically significant and of the appropriate sign, generating the appropriate U-shaped, long-run average variable cost curve with the minimum point occurring at a firm size of 2600 MW. However, they go on to indicate that "this U-shaped curve is somewhat L-shaped over the range of observed firm sizes." Upon examining the short-run average variable cost curve for distribution they again found an inverted U-shaped curve with its maximum occurring at a 54% utilization rate. Finally, they included a measure for the density of the distribution network and found higher unit costs for more densely populated areas (this is not what they had expected to find). They concluded that higher congestion costs associated with higher density overwhelm any economies that may have been present. What is interesting is that they included fixed costs in the generation component but not in either the transmission or the distribution function. As stated previously, economies of scale, scope, and density are primarily the result of the highly sunk capital investments required in both transmission and distribution.

Another study that sought to measure scale economies in the distribution of electricity is that of Giles and Wyatt (1989), who examined the presence of economies of density in New Zealand. They found that the number of firms operating in the industry at the time was greater than that which was consistent with average cost minimization. They found that the cost-minimizing level of output was 2315 gigawatt hours, which could have been produced efficiently by about 20 firms, 40 fewer than there actually were at the time of this study.

For a Multiple-Output Natural Monopoly

It is well established that distributed electricity is not a homogeneous good; that is, the electricity distributed to different types of end users can be differentiated by voltage level. For example, many industrial customers can accept electricity at much higher voltage levels than either commercial or residential customers, which is one reason that rates are set in the fashion that they are set; that is, different rate/revenue classes pay different base rates (i.e., energy charges) and it is often the case that residential customers do not pay demand charges. The structure of rates is discussed in more detail in Chapter 8. Given this, numerous studies recognize that distributed electricity should be modeled as a multiproduct industry, which motivates the concepts described in the following section.

Multiproduct Natural Monopoly

While single-output scale economies imply single-output natural monopoly, this is not necessarily the case for multiple-output (or multiproduct) firms. The subadditivity conditions for a multiple-output natural monopoly are far more complex than those of a single-output monopolist. In this case, economies of scale are not equivalent to decreasing average cost, as the firm may not operate along a linear expansion path. For a multiproduct firm, cost analysis requires the examination of not one but several concepts.

Ray Average Costs

Ray average costs (RACs) describe the behavior of the cost function as output is expanded proportionally along a ray emanating from the origin. Baumol et al. (1982) offer the following definition: in the two-product case, one considers the behavior of costs along a cross section of the total cost surface. Defining a composite good, this measure allows a calculation of the average cost of this particular bundle and is given by

$$RAC = C(tY^0)/t, \tag{2.18}$$

where Y^0 is the unit bundle for a particular mix of outputs and t is the number of units in the bundle such that $Y = tY^0$ (Baumol et al., 1982, p. 49). This is displayed in Figure 2.5.

Consider the behavior of costs along a cross section of the total cost surface obtained by dropping a perpendicular plane along a ray that emanates from the

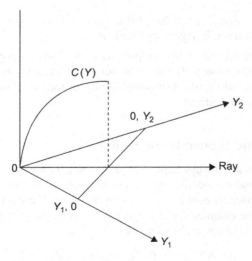

FIGURE 2.5 Ray average. *(Source: Reproduced with permission from Baumol et al. (1982; Figure 3A1).)*

origin. The RAC at any point on $C(Y)$ is equal to the slope of the cost function at that point. Note: In the case of Figure 2.5, as drawn, the slope of the cost function $C(Y)$ at $C(Y_1, Y_2) = 0$.

Degree-of-Scale Economies

As the analog to the single-output concept of economies of scale, the degree-of-scale economies, S_N, is equal to the ratio of average cost to marginal cost. In the multiple-output case, we have

$$S_N(Y) = C(Y)/Y_i C_i(Y), \qquad (2.19)$$

where $C_i(Y)$ is the marginal cost with respect to Y_i for $i = 1, \ldots, n$.

Baumol et al. (1982, p. 51) have shown that

$$S_N = 1/(1 + e), \qquad (2.20)$$

where e is the elasticity of RAC (tY) with respect to t at point Y (t is a scalar).

Corollary Returns to scale at output point y are increasing, decreasing, or locally constant ($S_N > 1$, $S_N < 1$, $S_N = 1$, respectively) as the elasticity of RAC at y is negative, positive, or zero, respectively. Moreover, increasing or decreasing returns at y imply that RAC is decreasing or increasing at Y, respectively.

As such, S_N (the degree-of-scale economies) may be interpreted as a measure of the percentage rate of decline or increase in RAC with respect to output (Baumol et al., 1982).

Cost Concepts Applicable to Multiproduct Cases for Nonproportionate Changes in Output

As said, RAC are relevant when outputs move in fixed proportions, which is quite often not the case in the distribution of electricity. For this, several concepts are required to establish subadditivity of the cost function, which is discussed here in more detail.

Product-Specific Economies of Scale

Because output is not always expanded proportionally for a multiproduct firm, the concept of product-specific economies of scale must be examined. That is, to assess the impact on cost of a change in one output, holding other outputs constant, one must examine the average incremental cost (AIC) of the product of which output is being varied. This is defined as

$$\text{AIC}(y_i) = [C(Y_N) - C(Y_{N-i})]/Y_i, \qquad (2.21)$$

where $C(Y_{N-i})$ is the cost of producing all N of the multiproduct firm outputs except product i.

This specification allows the identification of returns to scale that are specific to a particular output. Hence, product-specific returns to scale are given by

$$S_i(y) = \text{AIC}(y_i)/(\partial C/\partial y_i), \tag{2.22}$$

where $\partial C/\partial y_i$ is the marginal cost with respect to product i. Returns to scale of product i at y are said to be increasing, decreasing, or constant as $S_i(y)$ is greater than, less than, or equal to unity, respectively.

If product 2 (Y_2) as shown in Figure 2.6 has no output-specific fixed costs, then the total cost surface rises continuously above ST (curve AE). However, if some special fixed cost exists that must be incurred to begin production of Y_2 as an addition to the firm's line of other products, then the cross section of the cost surface will contain a vertical fixed-cost segment AB, which results in a jump discontinuity of $C(Y)$ above the Y_1 axis. Thus, the height CE in Figure 2.6 measures the total incremental cost of Y_2 at output vector T, which is the addition to the firm's total cost resulting from the decision to add Y_2 to the firm's product mix. The average incremental cost of Y_2, $\text{AIC}_2 (Y_1*, Y_2*)$, is clearly given by the slope of the line from A to E. What is also clear from Figure 2.6 is that the average incremental costs of product 2 are declining with Y_2, at least between 0 and Y_2*. This suggests, by analogy to the single-output case, the novel and useful concept of product-specific scale economies.

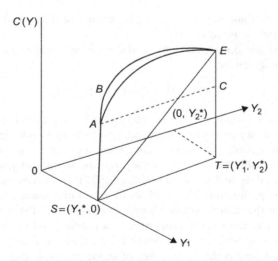

FIGURE 2.6 Product-specific returns to scale. *(Source: Reproduced with permission from Baumol et al. (1982; Figure 4A2).)*

Economies of Scope

The multiproduct cost concepts that have been discussed prior to this relate to the behavior of cost along a cross section of the cost output space. In addition to economies that result from the size or scale of a firm's operations, cost savings exist that can result from the production of several outputs at the same time; that is, in many cases, and certainly in the case of electricity, fixed costs are utilized jointly in production of the firm's outputs. These common costs, as they are also known, give rise to the concept of economies of scope (or economies of horizontal integration) and provide a basis for determining whether an industry is a multiproduct natural monopoly.

Mayo (1984, p. 209) argued that

In addition to measures of scale, efficient industry structure is determined by the behavior of costs as the scope of the firm is altered. The cost savings or dissavings that result from multiproduct versus specialized firm operations are given by the notion of economies and diseconomies of scope.

Therefore, economies of scope (also known as economies of joint production) are said to exist if a given quantity of each of two or more goods can be produced by one firm at a lower cost than if each good were produced separately by two different firms or even two different production processes. That is, for a two-product case, weak economies of scope are given by

$$C(Y_1, Y_2) \leq [C(Y_1, 0) + C(0, Y_2)] \qquad (2.23)$$

for all $Y_1, Y_2 > 0$. If not, then there are diseconomies of scope, and separate production of outputs is more efficient.

As in the single-output case, we will define the degree of economies of scope, which is given by

$$S_c = [C(Y_1, 0) + C(0, Y_2) - C(Y_1, Y_2)]/C(Y_1, Y_2). \qquad (2.24)$$

The importance of economies of scope cannot be overstated: economies of scope are a necessary condition for natural monopoly in a multiple-output firm.

Both economies of scale and of scope tend to occur due to specialization. As stated previously, the latter can arise from the sharing or joint utilization of inputs. According to Panzar and Willig (1977), if a given input is imperfectly divisible, production of a small set of goods may leave excess capacity in utilization of that input. Another way that economies of scope can arise is that the input may have some properties of a public good so that when it is purchased for one production process it can then be freely available to another. A third way is that economies of scope can arise due to the economies of networking (recall the discussion of network economies and returns to density).

Subadditivity of the Cost Function

However, even if a cost function exhibits both economies of scale and economies of scope, it is not necessarily subadditive. A sufficient condition must now be established for natural monopoly in a multiproduct industry. Cost complementarity, which requires that marginal or incremental costs of any output decline when that output or any other outputs increase, provides such a condition. Mathematically, cost complementarity for a twice differential multiproduct cost function exists if

$$\partial^2 C(Y)/\partial Y_i \partial Y_j < 0 \qquad (2.25)$$

for $i \neq j$ and for all $Y_i, Y_j > 0$.

If the aforementioned is satisfied, then the cost function exhibits cost complementarity, which is a sufficient condition for subadditivity in a multiproduct cost function. An industry is said to be a natural monopoly if, over the entire relevant range of outputs, the firm's cost function is subadditive.

Electricity as a Multiple-Output Industry and/or Economies of Scope and Subadditivity

While treating generated electricity as a homogeneous good may seem appropriate, it is certainly not appropriate treatment for distributed electricity. A quote from Joskow and Schmalensee (1983, pp. 54–55) summarizes this principle nicely:

... treating diverse power systems as single-product firms is likely to produce error. The cost of an optimally designed power system depends in complex ways on the distribution of demand over time and space. No two power systems produce the same mix of products and product mix differences affect the magnitude and form of optimal investments in transmission and in distribution.

Among the first to actually model electricity as a multiple-output function was Neuberg (1977), who examined the market for distribution employing four interdependent outputs. The predominant output in his analysis was the number of customers served; the other outputs were the number of megawatt hours sold, the size of distribution territory and the miles of overhead distribution line. Karlson (1986) tested for and found that the multiproduct characterization of electricity was appropriate, having treated residential and commercial electricity as distinct outputs. He found that the marginal cost of any one output depended on the levels of all other outputs and on all other inputs. Furthermore, he rejected the hypothesis of separability between inputs and outputs, which implies that the marginal rate of substitution between any two inputs is not independent of the quantities of outputs, nor is the marginal rate of transformation between any two outputs independent of the quantities of inputs.

Some studies have attempted to identify the existence of scope economies in this industry. Mayo (1984) employed a multiproduct quadratic cost function to estimate the cost of producing both electricity and gas for 200 public utilities. Using 1979 data, he confirmed the presence of economies of scope for smaller firms. However, as output is expanded, the absence of competitive pressure leads to cost inefficiencies and eventual diseconomies of scope. His finding led to the realization that the regulated utilities in his sample were characterized by interproduct discomplementarities, as his empirical results confirmed that

$$\partial^2 C(Y)/\partial Y_1 \partial Y_2 > 0. \qquad (2.26)$$

This particular result, which was anticipated by Kahn, can be attributed (at least in part) to the type of regulation imposed upon these firms; that is, average cost pricing in which firms are certainly not incentivized to minimize costs. Furthermore, the Averch–Johnson effect,[1] which is also a result of rate-of-return regulation, is still at play, as firms have an incentive to overinvest in capital as a mechanism to increase rates and hence profits.

Like Mayo's 1984 study, Sing (1987) employed a different cost specification (a generalized translog cost function) to estimate whether a sample of U.S. electric and gas utilities were natural monopolies. He found that the average combination utility exhibited diseconomies of scope but that other output combinations were associated with economies of scope.

Roberts (1986) and Thompson (1997) differentiated output according to voltage level. Their results suggest that there are economies of density; that is, for a given network size and a fixed number of customers, average costs fall when the quantity of power supplied increases. Roberts defined a measure of economies of output density as

$$R_D = 1/(\partial \ln C/\partial \ln Y_L + \partial \ln C/\partial \ln Y_H), \qquad (2.27)$$

where Y_L and Y_H denote low-voltage and high-voltage output, respectively.

He rejected the hypothesis of separability of the generation and transmission functions from distribution. This was due predominantly to the lack of separability between the inputs required to perform all three functions, which is the reason that the majority of the utilities in the United States are integrated vertically. (This confirmed Karlson's finding.) The concept of economies of vertical integration is explored later in the chapter and throughout this book as it is integral to the appropriate cost modeling and public policy making for

---

[1]Traditional ratemaking provides an incentive to overinvest in capital (i.e., the rate base). For an investor-owned utility, this is a large component of the revenue requirement upon which the utility is allowed to earn a return to its investors. Known as the Averch–Johnson effect, this is the tendency of companies to engage in excessive amounts of capital accumulation to expand the volume of their profits. If a firm's profits-to-capital ratio is regulated at a certain percentage, then there is a strong incentive for companies to overinvest to increase profits overall. This goes against any optimal efficiency point for capital that the company may have calculated, as higher profit is almost always desired over and above efficiency.

electric utilities. In addition, it provides the subject of a case study presented in Chapter 8.

As stated previously, most of the studies of this nature have focused on investor-owned utilities. However, and as stated previously in Chapter 1, other types of entities are worthy of such analysis. Yachew (2000) estimated the costs of distributing electricity using data on municipal electric utilities in Ontario, Canada, for the period 1993–1995. Data reveal substantial evidence of increasing returns to scale with minimum efficient scale being achieved by firms with about 20,000 customers. Larger firms exhibit constant or decreasing returns. Utilities that deliver additional services (e.g., water/sewage) have significantly lower costs, indicating the presence of economies of scope.

Greer (2003) estimated economies of scale and scope for U.S. distribution cooperatives. Distributed electricity (i.e., output) was differentiated by voltage level, with 1000 kVA being the distinction between "small" users and "large" users. She found that the cost function exhibits product-specific economies, as well as economies of scope, and that substantial cost savings could be realized via mergers between distribution coops. [Note: This study and the cost model used in the analysis were the subject of two case studies presented in *Electricity Cost Modeling Calculations* (2011).]

Fraquelli and coworkers (2004) studied Italian public utilities that provided in combination gas, water, and electricity. They confirmed the presence of global scope and scale economies only for multiutilities with output levels lower than ones characterizing the "median" firm. This indicates that relatively small specialized firms would benefit from cost reductions by evolving into multiutilities providing similar network services such as gas, water, and electricity. However, for larger-scale utilities, the hypothesis of null cost advantages is not rejected. Thus, it is possible that the recent diversification waves of leading companies are explained by factors other than cost synergies so that the welfare gains reasonably expected from such examples of horizontal integration, if any, are likely to be very low.

Economies of Vertical Integration and Separability

The issue with which we are dealing is the appropriate modeling of costs to formulate public policy that maximizes the total welfare of the players (both consumers and producers) involved. Thus far, we have been concerned with the separate stages (or processes) required to supply electricity to end users and whether each stage (or process) may be a natural monopoly. What has been established is that the generation component, due mostly to technological change, is no longer a natural monopoly and that there could be societal gains by allowing competition into that component of the process, which is what deregulation of the industry was all about. Unfortunately, what was essentially ignored was the network (or wires) aspect of the business; that is, unlike telephony (voice, data, and fax), water, and natural gas, electricity cannot be stored economically and, once generated, flows according to Kirchoff's law (i.e., the path of least resistance).[2]

Given this, what now needs to be established is the relationship among these three functional components. After all, part of the notion of deregulation was that generation could be separated (lack of scale economies so competition deemed feasible) from the transmission and distribution functions (irrefutably natural monopolies). More specifically, what is necessary is to establish the existence (or lack thereof) of economies of vertical integration, which are another critical and distinguishing aspect of this industry.

Vertical Integration of Electric Utilities

Lee (1995, p. 45) states that "Landon (1983) argued that 'the electricity industry has special characteristics such as close coordination of each process, transaction costs, and idiosyncratic capital requirements, which all favour vertical integration.'"

Vertical integration makes sense when a product is produced sequentially, such that the output from the first stage of production is employed as an input in successive stages, which is the case of electricity. When a firm is integrated vertically, it owns the entire production process, controlling both the upstream (input) supply and the downstream (output) production processes. Needless to say, the electric utility industry in the United States was organized in this fashion for a number of years by investor-owned firms, who were willing to supply power to the larger, more densely populated areas of the country. Vertical integration makes sense because it provides an alternative to market transactions, which tend to be costly given the nature of the industry, which requires specialized assets and sunk costs. It would have been extremely difficult to foresee the input price increases experienced since the mid-1970s; as such, were the industry not vertically integrated but rather contractually related, the financial difficulties experienced by utilities in the late 1970–1990s would have been far greater, as it is unlikely that these price increases were foreseen and could be written into the contracts, which were typically of longer duration.

Vertical integration is especially appealing in industries characterized by bottlenecks, which tend to occur when exclusive ownership of a resource exists that is necessary to the production of the good but of which the cost is prohibitive so that it is not economically feasible for separate firms to invest. This type of investment yields a market that approximates a natural monopoly in the sense that its cost is sunk and that its duplication would be wasteful. In the case of electricity, the bottleneck is that which yields access to the transmission mechanism that delivers electricity from generation to the local distribution system.

---

[2]This is a critical point that needs to be kept in mind. In the author's opinion, it is the reason that deregulation of the industry was such an utter failure.

There are additional benefits attributable to vertical integration as well, namely:

1. Elimination of the "wedge" that results when the upstream firm sells its product to the downstream firm at a price above economic cost.
2. Mitigation of certain problems that arise due to the separation of ownership of the firm from whoever actually controls it, which is also known as a principal-agent problem.

For the presence of economies of vertical integration in the supply of electricity, it must be the case that successive stages of production (generation, transmission, and distribution) are less costly for a single firm to perform than it would be for these functions to be performed by separate producers. Both of these issues are relevant in the production of electricity, the underlying production technology of which not only lends itself to economies of scale but also to economies of vertical integration.

Defining Vertical Integration

Mathematically, economies of vertical integration exist if the following is satisfied:

$$C(G,D) < C(G,0) + C(0,D), \tag{2.28}$$

where $G > 0$ represents the first stage of production (upstream production) and $D > 0$ represents the latter stage (downstream production) so that $C(G, D)$ is the cost of production for a vertically integrated firm. If this is less than the sum of the cost of separate production by separate entities, given by $C(G, 0) + C(0, D)$, then it is said that economies of vertical integration exist or, expressed in percentage terms,

$$S_v = [C(G,0) + C(0,D) - C(G,D)]/C(G,D). \tag{2.29}$$

For

$S_v > 0$ there are economies of vertical integration
$S_v < 0$ there are no economies of vertical integration

Separability

Because the marginal cost of any one output depends on the levels of all other outputs and all other inputs, the issue of separability must be considered upon the formation of appropriate policy. Karlson (1986, p. 78) states that

Separability between inputs and outputs requires that the marginal rate of substitution between any two inputs is independent of the quantities of outputs, and the marginal rate of transformation between any two outputs is independent of the quantities of inputs. ... The rejection of the hypothesis of separability between inputs and outputs implies that the relative marginal costs of electricity sold to different consumer classes depend on the product and input mixes; furthermore, it is impossible to construct some homogeneous aggregate output called "electricity" to be sold to consumers.

Karlson rejected the hypothesis of such separability, as do Henderson (1985), Roberts (1986), and Lee (1995). These are discussed in more detail here.

Relevant Literature Review—Vertical Integration and Separability

Several studies have tested for the presence of vertical economies in the supply of electricity. Virtually all of them test for and reject the separability of the functional components. In fact, it has been demonstrated empirically that economies of vertical integration exist in the production of electricity. Such studies include Henderson (1985), who finds that downstream costs are dependent on input usage at the generation stage, hence the cost function (which is translogarithmic) fails the test for separability between generation and distribution. Roberts (1986) concurs, as do Hayashi et al. (1997) and Thompson (1985).

More recent studies include those by Kaserman and Mayo (1991) and Gilsdorf (1994, 1995). As an extension to their testing for vertical economies, both Kaserman and Mayo (1991) and Gilsdorf (1994, 1995) employ a multiproduct cost function to determine whether vertical integration and economies of scale together constitute a natural monopoly. In fact, Kaserman and Mayo also test for multistage economies between generation and transmission/distribution. They too reject the separability of inputs and outputs (what is generated is an input to what is transmitted/distributed) in the cost function. It is important to note that separability is not the same thing as economies of vertical integration, whereby it is output–output interactions that matter. Lee (1995) estimated a production function and performed more direct tests for vertical integration and economies of scale. All reject separability of all three functional components of electricity production.

Kwoka (1996) employs the Kaserman and Mayo approach to test for multistage economies between generation and distribution. He too rejects separability (especially for "larger" systems) and argues that these vertical economies are precisely the reason that most investor-owned utilities are integrated vertically, whereas most "smaller" systems (i.e., publicly-owned utilities and rural electric cooperatives) are not. In addition, he finds that vertical integration achieves significant cost efficiencies, in some cases sufficient to offset diseconomies of scale in generation and distribution separately.

More recent studies include Nemoto and Goto (2004), who tested the technological externality effects of generation assets on the costs of transmission and distribution stages in their study of vertically integrated Japanese utilities. Their results show that downstream costs depend on the generation capital, suggesting significant economies of vertical integration. Fraquelli et al.

(2005) analysis of Italian municipal electric utilities finds significant vertical economies for average size and large utilities while failing to find any significant effects for smaller-than-average-size utilities. Efficiencies associated with vertical integration are largest for fully integrated utilities, confirming results found in most other studies. Greer (2008) estimated the lost economies of vertical integration due to the rural electric cooperatives' choice of market structure. As indicated in Chapter 1, cooperatives are organized as either generation and transmission or member coops (distribution only). Greer found that cost savings of close to 40% could be realized had they adopted a truly vertically integrated structure.

CONCLUSION

This chapter provided an overview of electric utility industry structure and some relevant cost concepts, as well as a brief survey of the literature pertaining to this industry. In subsequent chapters, these concepts are expounded upon and examined in much more detail. But first, a few exercises will assist in understanding some of the concepts described in the early part of this chapter.

EXERCISES

1. Let the demand and cost equations be given by

$$P = 20 - 0.5Y$$

and

$$C = 18 + 4Y.$$

a. What is the monopolist's profit-maximizing level of output?
b. What price will prevail in the marketplace for this product?
c. What is the monopolist's profit?

2. Let the demand and cost curves be given by

$$P = 20 - 0.5Y$$

and

$$C = 0.04Y^3 - 1.94Y^2 + 32.96Y.$$

What are the monopolist's profit-maximizing levels of output?

3. Note the use of the word "are" in Exercise 2. In fact, there are two solutions, but only one is optimal.
a. Which level of output maximizes profit? Why?
b. What price will prevail in the marketplace for this product?
c. What profit will the monopolist earn?

4. Mayo's 1984 article was mentioned in this chapter. In it, he tested for product-specific economies of scale and scope in the distribution of electricity and natural gas for a sample of utilities. The single-output version of his cost model is given by

$$C = (\alpha_0 + \alpha_y Y + 1/2\, \alpha_{yy} Y^2)\prod \beta_i p_i e^e, \qquad (2.30)$$

where Y is output, p_i is input prices, and α_y, α_{yy}, and β_i are parameters to be estimated.

   a. Derive the average cost for Mayo's model.
   b. Derive the marginal cost for Mayo's model.
   c. Calculate the degree-of-scale economies. Under what conditions would the degree-of-scale economies be increasing? Decreasing? Constant?

Chapter 3

U.S. Electric Markets, Structure, and Regulations

INTRODUCTION

Since the early 1990s, the U.S. electric utility industry has gone through sweeping changes associated with deregulation, regulations, reregulation, and the rate-making process. Due to the largely sunk capital investment and the well-established presence of economies of scale, economies of scope, and vertical integration, conventional wisdom has held that competition is infeasible (at least in the transmission and distribution segments). This same wisdom holds that price regulation is necessary to ensure that consumers pay a fair price and that producers and shareholders are compensated appropriately for the risk associated with holding the stock of the utility [in the case of investor-owned utilities (IOUs), which supply approximately two-thirds of the power to end users in the United States]. This chapter discusses the electric market structure, how utilities recover their costs, the different recovery mechanisms, federal and state regulations, and how the regulatory process can impact rates.

THE U.S. ELECTRIC INDUSTRY STRUCTURE

In recent years, the industry has been evolving from vertically integrated monopolies that provide generation, transmission, and distribution service at cost-based rates (regulated model) to an industry where the operation of generation, transmission, and distribution assets has been increasingly unbundled and even divested [in the case of generating assets (deregulated model)]. In certain markets the wholesale and retail price of electricity is determined competitively under a regulatory framework that promotes competition. Although transmission and distribution markets are still monopolistic and follow traditional cost-based rates, the Federal Energy Regulatory Commission (FERC) and a number of states have implemented rate-making approaches that give regulated utilities financial incentives to expand their transmission and distribution systems cost effectively and reliably. Fourteen states—Maine, New Hampshire, Massachusetts, Rhode Island, Connecticut, New York, New Jersey, Pennsylvania,

Delaware, Maryland, Ohio, Michigan, Illinois, Texas—and the District of Columbia, have retail markets where customers can access alternative power suppliers or continue to purchase power from their historic supplier. In these markets, competitive bidding is used to determine a portion of the entire retail price for electricity. However, a number of states have either suspended deregulation or amended laws and regulations governing competition due to the lack of competition and resulting increases in price. States that have suspended retail competition include Virginia, Arkansas, New Mexico, Arizona, Nevada, California, Oregon, and Montana. In 2008, Delaware, Illinois, and Ohio enacted legislation allowing utilities to once again build their own generation capacity based on a formalized competitive procurement process. Illinois went one step further and created a new governmental entity that can build new capacity and procure power.

In the most basic sense, the electric industry is divided into generation, transmission, and distribution functions. Prior to deregulation, the term *electric utility* traditionally denoted an investor-owned company or government agency that produced, transmitted, distributed, and sold electricity to the end user. Also prior to deregulation, vertically integrated utilities that served all four market segments were granted a monopoly franchise by the state or local government, which gave them the right to produce and sell electricity in that service territory. In return for this monopoly position, the utility accepted the obligation to serve all customers in that territory regardless of profitability or ease of access and was subjected to regulatory oversight regarding its operations and pricing. Underregulation profits were constrained. However, utilities used regulated tariffs to pass costs onto customers, and incentives to minimize costs or take on unrecoverable risk were largely absent. (Recall earlier discussions on the Averch–Johnson effect.)

In addition to traditional vertically integrated utilities, generation and transmission cooperatives (G&Ts) produced electricity in bulk and transmitted and sold it in bulk (wholesale) to other utilities. Often G&Ts provide bulk power within a region to local distribution companies (LDCs) and municipal utilities. LDCs only owned and operated the local distribution network and sold power to the end user. They also provided retail sales and service to local customers who often viewed these entities and the "utility company." LDCs were investor-owned, were operated by the local municipality, or were part of a rural cooperative.

The Public Utility Regulatory Policy Act (PURPA) of 1978 created two other types of organizations: the independent power producer (IPP) and the nonutility generator (NUG). Both IPPs and NUGs are privately-owned firms that own, operate, and sell power into the market. IPPs typically sell power to local LDCs, whereas NUGs are usually industrial companies that use much of the power they generate internally, but sell excess power back into the market. PURPA required traditional utilities to buy power from these entities if the power was priced less than the utility's own cost of generation.

DEREGULATION

This vertically integrated model has changed over the years through the introduction of deregulation. Today, regulated, unregulated, and partially regulated markets exist with a host of different entities serving various market elements. Utilities can be viewed by the segment of the electric market they serve (generation, transmission, distribution, or sales) or by their ownership type (investor owned, government agency, or cooperative) in either a regulated or an unregulated market. As part of deregulation and industry restructuring, the vertically integrated company concept has effectively been unbundled in some markets. In theory, deregulation promotes the interaction of many sellers and buyers to create economically efficient market pricing that is equal to the cost of producing the last unit sold (i.e., marginal cost). It should be noted that transmission and distribution are still considered natural monopolies that require regulation to ensure fair access for all market participants and to take advantage of the inherent economies of scale, scope, and vertical integration.

The intent of deregulation and the resulting industry restructuring was to protect the short- and long-term interests of consumers by creating an efficient market through the introduction of competition. Reasons often cited for deregulation included advances in combined-cycle gas turbines that produced economies of scale at low capital costs (lower-generation market entry price), global competition, ability of private sector companies to respond more quickly to economic and technology changes, and improved information and communication systems that could better manage the markets. Although many of these reasons have a certain degree of merit, they have not been enough to support universal deregulation. For example, although advanced combined-cycle gas-turbine plants are considerably less expensive to build than comparably sized coal-fired plants, the fact that fuel costs (natural gas) contribute to a greater extent to the cost of electricity production is ignored. Yes, the lower entry price created by this technology allowed new unregulated competition into the market, but this also introduced additional risk into the market associated with volatile fuel prices. As a result, when natural gas prices increased significantly in 2005 and again in 2008 (as displayed in Figure 3.1), some of these assets became stranded and the production cost of electricity increased in local markets dominated by natural gas turbine generation. In the case of information and communication systems, these technologies are facilitating day-ahead and online electricity markets between market participants and different transaction types. Real-time metering, billing, load management, and quality control are also being offered under deregulation and are an integral part of the smart grid process.

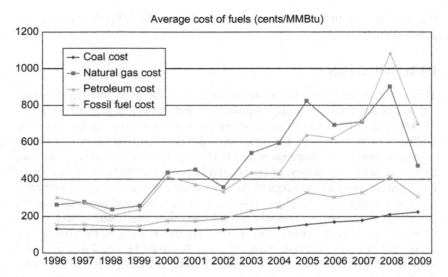

FIGURE 3.1   Energy price trends. *(Source: U.S. Energy Information Administration (EIA) (http://www.eia.gov/cneaf/electricity/epa/epaxlfile3_5.xls).)*

MARKET PARTICIPANTS

As stated previously, entities within the electric utility industry can be classified by ownership, which is an important distinction, as different types of ownership classes can be regulated somewhat differently. In addition to ownership classes, entities can be viewed in either a regulated or an unregulated context. In 2007 there were 3273 investor-owned, cooperatives, publicly-owned, and federal electric utilities, as well as retail and wholesale power marketers, as shown in Figure 3.2.

VERTICALLY INTEGRATED MODEL

The vertically integrated market model is represented by a monopoly that controls the different functions of generation, transmission, and/or distribution needed to serve the end user. There are a number of monopoly ownership types in the United States, including investor-owned, federal, cooperative (coop), municipal (muni), public power agencies, power pools, energy service providers, and IPPs. Each is discussed in some detail.

Investor-Owned Utilities

Investor-owned utilities are owned by stockholders that typically seek to maximize profits within the framework of regulations governing these types of utilities. These entities tend to be large organizations that try to take advantage of

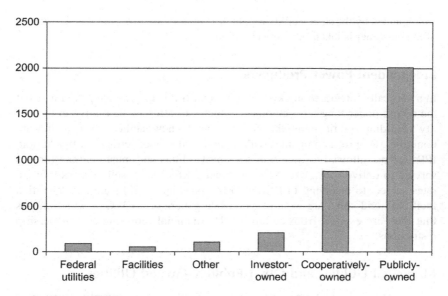

FIGURE 3.2 Ownership types.

economies of scale, as exemplified by recent mergers such as Duke and Progress Energy, Allegheny and First Energy, Northeast and NSTAR, and the possible purchase of Constellation by Exelon. Investor-owned utilities can exist as individual corporations or holding companies as part of a parent company that owns one or more operating utility. Most IOUs sell power at retail rates to various classes of customers and at wholesale rates to other utilities, including federal, state, and local government utilities, public utility districts, rural electric cooperatives, and even other investor-owned utilities. They are typically characterized as having high-density service territories. As monopolies, IOUs are regulated and required to provide (distribution) service to all customers in their franchised territory, charge reasonable and comparable prices to similar classifications of consumers, and provide consumers access to services under similar conditions. Most IOUs that operate in regulated retail states operate on a vertically integrated basis, providing generation, transmission, and delivery service at a bundled price to retail customers.

The 211 IOUs in the United States represent roughly 6% of the total number of electric utilities and approximately 38% of installed capacity, as shown later. They generate approximately 42% of the power produced and generate 66% of the sales and 67% of the revenue. Investor-owned utilities serve about 100 million consumers, representing about 71% of the total U.S. market and operate in all states except Nebraska. They are also referred to as privately-owned utilities and typically earn a return for their investors that is either distributed to stockholders as dividends or reinvested in the company. Because of their for-profit nature, they are regulated by the state's utility regulatory commission to ensure that the interest of the customer is taken into consideration.

Independent Power Producers

In a vertically integrated market the role of an IPP is to generate power and sell the output under long-term contracts, which can offer an alternative to the utility's building and financing the construction of a new facility. As such, they are nonutility, for-profit companies with no assigned service territories. In addition, IPPs are not allowed to own transmission facilities and must contract for this service to deliver power to their customers, which they sell at market-based rates subject to receiving FERC authorization. Finally, IPPs are not "qualified facilities" and some are exempt wholesale generators (EWGs), which means that they are exempt from certain FERC financial reporting and ownership restrictions.

Municipal Utilities and Other Publicly-Owned Utilities

Municipal utilities and other publicly-owned utilities are nonprofit government entities that serve at either the local or the state level. There are 2009 publicly-owned electric utilities in the United States, representing approximately 61% of the players in the power industry, 9% of the generating capability, 8% generation, 15% of retail sales, and 13% of the industry's revenue in 2007. This group of utilities consists of municipal, public utility and power districts, state authorities, irrigation districts, and joint municipal action agencies. Publicly-owned utilities have certain advantages, such as access to tax-free financing (municipal treasuries), ability to issue low-cost tax exempt debt to finance construction, and generally are not subject to state and federal taxes. Often they are financed by general obligation bonds and from revenue bonds secured by the sale of electricity. This can result in lower retail rates than IOUs.

Municipal utilities are owned and operated by local communities and often operate within the local municipal public works department. They can also be characterized as having a concentrated service territory similar to many IOUs. Municipal utilities are concentrated largely in the midwest and southeast and are located in every state except Hawaii. According to 2007 Energy Information Administration (EIA) records, there are nearly 1950 municipally-owned utilities in the United States. Municipal utilities can own and operate their own generation and distribution system, such as those in Austin, Texas; Jacksonville, Florida; and Colorado Springs, Colorado. However, more than half of the municipal utilities only own and operate the local distribution system and purchase their power wholesale, either from federal agencies or from IOUs or other entities. This class of utility is often not regulated by state or federal agencies, and municipalities may operate the utility as a tool to promote local economic expansion or lower local tax burdens.

Municipal utilities can range in size from one customer to over a million customers (e.g., the city of Los Angeles).

In some regions of the country, municipal utilities are run by a number of cities or a county and are called public utility districts (PUDs). PUDs and projects are more prevalent in Nebraska, Washington, Oregon, Arizona, and California, where voters elect commissioners or directors to govern the district independent of any local government. Smaller municipal utilities frequently ban together to create public utility districts that share ownership in generation and transmission assets.

Other publicly-owned utilities include municipal authorities, state authorities, and irrigation districts. State authorities, such as the New York State Power Authority and Santee Cooper (South Carolina Public Service), are utilities that function under a state charter and can generate or purchase electricity from other utilities and sell power into the wholesale market or to groups of other utilities within their states, as well as distribute power to local customers. The New York State Power Authority supplies wholesale power to municipal and cooperative utilities but also provides power to certain industrial customers. Santee Cooper provides both retail and wholesale electric service. Irrigation districts, such as the Salt River Agricultural and Improvement District, are controlled by a board of directors apportioned according to the size of landholdings. The Salt River Project provides both retail electric and water services. Irrigation districts are located primarily in the western United States and were initially formed for agricultural purposes by local farmers to manage water resources. Some states have created entities called joint municipal action agencies for the purpose of constructing power plants and purchasing wholesale power for resale to municipal distribution utilities. Some of these entities include the Massachusetts Municipal Wholesale Electric Company, the Indiana Municipal Power Agency, and the Municipal Electric Authority of Georgia.

Federal Power Agencies

Federal power agencies were initially established by the federal government to market the power from federal hydropower projects. There are nine federal power agencies controlled by various government agencies that operate in all areas except the northeast, the upper midwest, and Hawaii. These include:

- U.S. Army Corps of Engineers
- Bureau of Indian Affairs and Bureau of Reclamation in the Department of the Interior
- International Boundary and Water Commission in the Department of State
- Power Marketing Administrations (PMAs) in the Department of Energy [Bonneville Power Administration (BPA), Southwestern Power Administration (SWPA), Southeastern Power Administration (SEPA), Western Area Power Administration (WAPA)]
- Tennessee Valley Authority (TVA)

The TVA, the U.S. Army Corps of Engineers, and the U.S. Bureau of Reclamation are the only federal agencies that own and operate generating facilities. These federal power agencies were largely established based on the premise that publicly supported electricity was essential to provide electricity to large parts of rural America. The primary purpose of the TVA, BPA, and other PMAs was to market the surplus output of hydroelectric facilities that was generated as part of river-way navigation, flood control, and irrigation requirements.

U.S. federal involvement in the electric market which began with the Reclamation Act of 1902 and Town Sites and Power Development Act of 1906. With the Reclamation Act of 1902, the federal government became involved with the reclamation of arid lands largely through the development of irrigation projects that then generated electricity as a by-product. At the turn of the last century, hydroelectric power was the dominant source of electricity. At the time, the sale of surplus power, preferentially to local communities as defined in the 1906 act, was viewed as a way of repaying the costs associated with reclamation.

In the 1930s, the role of the federal government in marketing electricity from federally-owned facilities grew rapidly largely due to the Great Depression and the need to create jobs and stimulate the economy. At the time, the power produced by these entities was sold primarily to municipals and cooperatives. During the Great Depression, some of the world's largest hydroelectric power plants were constructed, including the Hoover Dam in 1936, the Bonneville Dam in 1938, and the Grand Coulee Dam in 1941. In the years leading up to World War II, nearly half of all new generating capacity was built by the federal government, which mitigated the effects of the depression to some degree through electrification and jobs. Due to the federal dam projects during this era, federal utilities produce today more hydroelectric power than other types of utilities, which makes the power they produce relatively inexpensive. With the exception of parts of the midwest and northeast, federal power is sold throughout the nation. States in the Pacific Northwest and the Tennessee River Valley receive the largest share of federal power.

Federal electric utilities primarily generate power from federally-owned facilities and transmit and sell their power to statutorily defined preferential customers, including municipal utilities, cooperatives, Indian tribes, state utilities, irrigation districts, state governments, and federal agencies. As required by law, they operate as not-for-profits and are required to recover the cost of operation and repay the U.S. treasury for funds borrowed to construct generation and transmission facilities. After meeting these statutory customer commitments, federal power agencies can and do sell surplus electricity to IOUs or directly to large, power-intensive industries (i.e., aluminum industry) in wholesale markets. Federal agencies do not sell directly to residential or commercial customers. These agencies also own transmission lines from their power generation facilities to other utility-owned grids.

The Tennessee Valley Authority, which is the largest federal power producer, operates its own power plants and sells both wholesale and retail power into the Tennessee Valley region markets. Power generated by the Bureau of Reclamation and the U.S. Army Corps of Engineers (except for the North Central Division in areas such as Saint Mary's Falls at Sault Ste. Marie, Michigan) is marketed by the various federal power marketing administrations. These administrations also purchase energy from other electric utilities for resale into wholesale markets. Federal power authorities represent less than 1% of all electric utilities in the United States, yet provide approximately 7% of all generating capability and about 4% of the generated electricity. Federal utilities are not subject to rate regulation but must submit their rates to the FERC to demonstrate that they are at a level sufficient to repay debt owed to the federal government.

Tennessee Valley Authority

The TVA was established during the Great Depression on May 18, 1933 under the Tennessee Valley Act as an "experiment" in social planning as part of Franklin Delano Roosevelt's (FDR) "First New Deal." (The "First New Deal" occurred during the first 100 days of FDR's administration and resulted in the passage of dozens of congressional acts and executive orders.) During that period of the Great Depression, the nations' economic peril was so severe (nearly 25% unemployment, bank closings, mortgage defaults, and 50% drop in farm crop prices) that Congress felt it did not have the time to seriously debate FDR's various acts before voting.

Supporting the creation of the TVA was the belief by many at the time that privately held power companies were charging too much as directed by their owners, the utility's holding company. These private utility holding companies controlled 94% of generation by 1921 and were largely unregulated (Public Utility Reports, Inc., 1988). The eight largest utility holding companies controlled 73% of the investor-owned electric industry by 1932 (Hyman, 1994), which led to the Public Utility Holding Company Act (PUHCA) of 1935 enacted during the "Second New Deal" era of 1934/1935. As a result of the creation of the TVA, many private companies in the Tennessee Valley were purchased by the federal government.

The TVA was largely formed to promote economic development in the Tennessee Valley, improve navigation, aid in flood control, and provide fertilizer manufacturing. In the early years the TVA was financed through federal appropriations. The 1959, the TVA act authorized the TVA to "self-finance," giving the TVA more freedom in making investment decisions. The TVA act limited how much power the TVA could sell outside of its jurisdiction, which was defined as the geographic area of the distributors served by the TVA in 1957. Direct appropriations for the TVA power program ended in 1959, and appropriations for TVA's stewardship, economic development, and multipurpose activities ended in 1999. Since 1999, the TVA has funded all of its operations almost entirely from the sale of electricity and power system financings. The Energy Policy Act of 1992 provided the TVA with an exemption from the Federal Power Act[1] and FERC authority to order utilities to provide transmission service. This exemption, referred to as the "anti-cherry picking" advantage, limits competition to the TVA by limiting access by others to its transmission lines and customers within TVA's defined service territory. TVA's rates are not subject to state or FERC regulation but are set by the TVA's board of directors.

Like other federal entities, the TVA sells power to municipalities and cooperatives that resell the power to their customers at a retail rate. The TVA also sells power to federal agencies, to customers with large or unusual loads, and to exchange power customers (systems that border TVA's service area). It is the largest federal power agency and supplies power to most of Tennessee, northern Alabama, northeastern Mississippi, and southwestern Kentucky, as well as sections of northern Georgia, western North Carolina, and southwestern Virginia. In 2008, TVA's revenues were $10.4 billion, virtually all from their power programs, including wholesale power contracts with 159 municipalities and cooperatives, which represented nearly 83.4% of total operating revenues in 2008 (SEC 2008 10-K filing). All of these contracts require customers to purchase all of their electric power and energy requirements from the TVA under a 5-, 10-, or 15-year notice of termination agreement, which provides for stability in TVA's revenue from electricity generation.

Today, the TVA operates 3 nuclear plants, 11 fossil fuel–fired plants, 29 hydroelectric plants, 6 combustion-turbine plants, and 1 pumped-storage plant. The TVA's Green Power Switch renewable program includes 16 solar sites, 1 wind-energy site, 1 methane gas facility, and 1 biomass/coal cofiring program. Fossil fuel plants produce about 60% of TVA's power, nuclear another 30%, and hydropower dams about 10%. Green power contributes less than 1% to the generation mix.

---

[1] From Wikipedia, the free encyclopedia: "The Federal Power Act is a law appearing in chapter 12 of Title 16 of the United States Code, 'Federal regulation and development of power.' Enacted as the Federal Water Power Act in 1920, its original purpose was coordinating hydroelectric projects in the United States. Representative John J. Esch (R-Wisconsin) was the sponsor. The act created the Federal Power Commission (FPC) (now the Federal Energy Regulatory Commission) as the licensing authority for these plants. The FPC regulated the interstate activities of the electric power and natural gas industries and coordinated national hydroelectric power activities. The commission's mandate called for it to maintain reasonable, nondiscriminatory, and just rates to the consumer. It was ensured that 37.5% of the income derived from hydroelectric power leases given out under the Water Power Act of 1920 went to the state in which the dam was built in. In 1935 the law was renamed the Federal Power Act, and the FPC's regulatory jurisdiction was expanded to include all interstate electricity transmission. Subsequent amendments to the law include the following statutes:

- Public Utility Regulatory Policies Act (PURPA)(Public Law 95-617)
- Energy Security Act (P.L. 96-294)
- Electric Consumers Protection Act of 1986 (PL 99-495)
- Energy Policy Act of 1992 (PL 102-486)"

Power Marketing Administrations

The Bonneville Project Act of 1937 created the Bonneville Power Administration for the purpose of generating hydropower from the Columbia River system and to promote regional economic development. The BPA is the largest PMA and second largest federal utility in terms of assets after the TVA. The Western Area Power Administration is the second largest PMA and was created in 1977 by the Department of Energy Organization Act of 1977. The WAPA markets hydropower in the western United States, including power from the Hoover Dam (built in 1935). Both the Southwestern Power Administration and the Southeastern Power Administration were created by the Pick–Sloan Flood Control Act of 1944. According to Section 5 of this act:

Electric power and energy generated at reservoir projects under the control of the War Department and in the opinion of the Secretary of War not required in the operation of such projects shall be delivered to the Secretary of the Interior, who shall transmit and dispose of such power and energy in such manner as to encourage the most widespread use thereof at the lowest possible rates to consumers consistent with sound business principles, the rate schedules to become effective upon confirmation and approval of the Federal Power Commission. Rate schedules shall be drawn having regard to the recovery (upon the basis of the application of such rate schedules to the capacity of the electric facilities of the projects) of the cost of producing and transmitting such electric energy, including the amortization of the capital investment allocated to power over a reasonable period of years. Preference in the sale of such power and energy shall be given to public bodies and cooperatives.

The federal government provides federal utilities, such as the Power Marketing Administrations (and cooperatives), that participate in the Rural Utility Services (RUS) electric program access to capital at reduced interest rates. The PMAs sell about 5% of the nation's electricity into the wholesale electric markets, including the sale of power to municipalities and cooperatives, state agencies, IOUs, public utility districts, federal agencies, and industrial customers. The service territory of the Bonneville Power Administration covers Washington, Oregon, and small pieces of western Montana and western Wyoming. The SWPA serves part of Kansas and Texas, Missouri, and Oklahoma, Arkansas, and Louisiana. The SEPA serves Illinois, West Virginia, Kentucky, Tennessee, Mississippi, Alabama, Georgia, the Florida panhandle, North and South Carolina, and Virginia. The WAPA covers California, Nevada, Utah, Arizona, New Mexico, most of Montana and Wyoming, parts of Texas, North and South Dakota, Nebraska, western and southern Kansas, and the western edges of Minnesota and Iowa.

Based on the 1944 Flood Control Act, PMA electricity is sold "at the lowest possible rates consistent with sound business principles," which is generally less than the price of power would be under competitive market conditions. Essentially, PMAs pass lower prices on to statutorily defined preference customers in lieu of profits to particular groups of preferred customers representing a level of price supports. An exception to this is the TVA, which was estimated by the Energy Information Administration to have had higher wholesale prices than neighboring utilities in 2006.

Rural Electric Cooperatives

In the 1920s and 1930s as the electric grid evolved in the United States, it became readily apparent that investor-owned utilities had little interest in building distribution systems in sparsely populated rural areas. To promote agriculture and quality of life in these rural areas, the federal government created the Rural Electrification Administration in 1936, which provided for the creation of rural electric cooperatives. Rural electric cooperatives are customer-owned electric utilities that provide electricity to end users in their service territories. They are largely in rural areas, and cooperatives are organized under state law and subject to the following:

- Cost-based operations (i.e., no profit incentive).
- Members (owner–customers) are entitled to receive a return of, but not a return on, capital contributed to the organization.
- Governance is based on a one-member-one vote. (The board of directors is elected by the membership.)

In addition to customer-owned cooperatives, some coops may be owned by a number of other coops. Three types of cooperatives exist: (1) distribution only, (2) distribution with power supply, and (3) generation and transmission. Some distribution cooperatives resemble municipal utilities in that they often do not generate electricity, but purchase it from other utilities and federal generation agencies. In the event that there is not enough federal-provided electricity, groups of coops have come together regionally to create generation and transmission cooperatives that own facilities on behalf of the distribution coops. These generating and transmission cooperatives are usually referred to as *power supply cooperatives*.

Generation and transmission coops were largely brought about by regulations in the 1970s that made it more difficult for distribution coops to purchase electricity from entities other than the federal government. Total nameplate capacity for G&Ts is 38 GW and they supply approximately 5% of the nation's power needs (U.S. Department of Agriculture, Rural Utilities Service, 2008). There are 841 distribution cooperatives and 65 generation and transmission cooperatives operating in 47 states. (The NRECA web site is http://www .nreca.org/members/Co-opFacts/Pages/default.aspx.)

Many cooperatives operate as qualified tax-exempt organizations under Section 501(c)(12) of the Internal Revenue Code. Under this tax-exempt status, cooperatives must receive at least 85% of their revenue from business conducted with members. Cooperatives that meet Rural Utilities Service eligibility requirements have access to low-cost federal government loans and loan guarantees. Cooperatives account for roughly 10% of electricity sales to ultimate consumers, 12% of the customers, 10% of the revenue, and nearly 43% of the miles of distribution lines in the United States.

An important facet of rural electric cooperatives is their ability to obtain financing through the RUS loan program, which was established under the Federal Crop Insurance Reform and Department of Agriculture Reorganization Act of 1994 to provide financial and technical assistance and facilitate electrification of rural America. The RUS is the successor to the Rural Electrification Administration, which was created under the Rural Electrification Act of 1936. This act provided for direct loans and loan guarantees to electric utilities serving customers in rural areas and also provided for low-interest, long-term loans from the federal government, which were made at a 2% interest rate until 1973 and increased to 5% between 1973 and 1993 with up to a 35-year term to maturity.

Today, RUS program loans and guarantees are used to finance the construction and improvement of electric generation, transmission, and distribution facilities in rural areas. In addition to cooperatives, entities eligible to apply for loan and loan guarantees include states, territories, public power districts, and agencies such as municipalities that provide retail electric service to rural areas or supply the power needs of distribution borrowers in rural areas. To qualify for loans and loan guarantees, borrowers must show that the loans will be repaid in accordance with their terms and provide adequate security pursuant to the RUS mortgage and loan contract. In 2008 the RUS loan program was redirected in recognition of the deregulation of the wholesale markets so that the focus of the RUS became the provision of financial assistance for transmission and distribution facilities and that G&Ts were to consider commercial capital markets for funding new generation (although upgrades to existing generation would still be provided).[2]

Although the RUS loan program supports the coops' need for capital, it is interesting to note that many coops face competitive challenges that, due to lack of economy of scale, they find difficult to overcome. As an example of these competitive pressures, in late 2009 American Electric Power's subsidiary AEP Southwestern Electric Power Co. (SWEPCO) offered to purchase the Valley Electric Membership Corporation (VEMCO—a coop), and the VEMCO board unanimously approved a voluntary dissolution of the cooperative, which serves 30,000 member customers in eight Louisiana parishes with 7000 miles of distribution and 90 miles of transmission lines. As a result, VEMCO's members are expected to save nearly 20% on their retail rates from this change of ownership; in addition, SWEPCO will provide payment for the VEMCO patronage capital (return of equity to the membership) at closing subject to the

---

[2]Office of Management and Budget, *Budget of the United States Government, Fiscal Year 2008—Appendix*, Department of Agriculture, p. 146 (see http://www.whitehouse.gov/omb/budget/fy2008/appendix.html).

Louisiana Public Service Commission, the Arkansas Public Service Commission, the Rural Utilities Service, and the National Rural Utilities Cooperative Finance Corporation approval of this transaction.

Nonutility Power Producers

Nonutility power producers are also referred to as qualifying facilities. Qualifying facilities were established under the Public Utility Regulatory Policies Act of 1978 and include combined heat and power plants and small power producers. Combined heat and power plants cogenerate (produce process/district heat and electricity) for primarily business purposes, but also produce electricity for sale into the market. Other nonutility power producers include entities that use renewable resources to generate electricity. These entities have been a small but growing segment of the overall power market.

Power Pools

Prior to the creation of independent system operators (ISOs) and regional transmission organizations (RTOs), power pools were created by groups of utilities to merge the scheduling and dispatch function for their generation assets. Multiple utilities within a region could then share their assets, producing higher reliability and lower costs for the pool. This allowed higher-cost utilities access to lower-priced power and lower-cost utilities to receive additional revenue. By pooling scheduling and resources, utilities relied on the pool to minimize costs as opposed to power trading. Prior to deregulation, power pools were used extensively in the northeast with the largest being PMJ (New England and New York), prior to being replaced by ISOs.

INDUSTRY RESTRUCTURING AND THE COMPETITIVE ELECTRIC MARKET

As mentioned earlier, the electric market in the United States has evolved into a hybrid regulated, unregulated, and partially regulated landscape. This restructuring has allowed competition in the generation and retail sales sides of the business in unregulated and partially regulated markets, providing opportunities for new players. As noted earlier, this restructuring first took place in the generation market, but other activities, such as system operations and retail sales to end-use customers, have moved away from the monopoly (regulated utility) in the restructured power industry. This has created distribution and transmission organizations that provide only a service and do not trade or sell power but distribute power on behalf of other market players. In the evolving competitive markets, these new players include merchant generators, transmission companies (Transcos), ISOs, RTOs, transmission owners, distribution companies, electric power marketers, and energy service companies (ESCOs).

Merchant Generators

In a deregulated market, the role of the IPP to generate and sell power has evolved to that of a merchant generator. In a restructured market, merchant generators are independent for-profit organizations that own and operate generation assets outside of the regulated utility. As opposed to IPPs, who sell their power output under long-term contracts, merchant generators are more likely to sell to a variety of market participants and are generally more at risk to market prices. Markets for merchant generators include utilities, marketers, ISOs, or directly to end-use customers. Many merchant generators were formed through the acquisition of existing generation assets of traditional vertically integrated regulated utilities during a state's restructuring process. Other merchant generators were formed as unregulated subsidiaries of utility holding companies. Examples of merchant generators include Mirant (MIR), NRG Energy (NRG), Reliant (RRI), and Dynegy (DYN).

Transmission Companies

Transmission companies are investor-owned and operate as for-profit companies. FERC Order 2000 explicitly allowed for Transcos to own and operate their transmission facilities (unlike ISOs that only operate the system), allowing for a profit structure subject to some regulatory constraints. As competition is generally limited, they are regulated by FERC. Again, as with merchant generators, Transcos typically acquire their transmission lines from formerly vertically integrated utilities or are required to build their own new transmission lines. An example of a Transco is ITC Transmission, the nation's first and currently largest fully independent transmission business, which was established in 2003. ITC Transmission consists of three operating companies—ITC Transmission, Michigan Electric Transmission Company, LLC (METC), and ITC Midwest—serving nearly 80,000 square miles in five states with approximately 15,000 miles of overhead and underground transmission lines. Through a subsidiary, ITC Great Plains, new transmission lines are being built in Kansas and Oklahoma to support the development of the growing wind industry in the region.

In late 2009, the American Electric Power Company (AEP) announced formation of a new Transco to cover at least 11 states as part of a three-part national transmission strategy. According to AEP,

Pursuing these activities in a Transco, with formula rates adjusted annually by the Federal Energy Regulatory Commission (FERC), benefits customers by enhancing AEP's access to capital. This enables the company to undertake substantial new investment while relieving our operating company balance sheets of the burden of meeting those capital demands, thereby allowing them to put capital to work on distribution and generation needs.

Independent System Operators and Regional Transmission Organizations

The FERC proposed the creation of ISOs in 1996 in response to the Energy Policy Act of 1992. FERC's Order 888 in 1996 provided for the creation of ISOs to consolidate and manage the operation of transmission facilities to provide non-discriminatory open transmission service for all generators and transmission customers. FERC Order 2000 supported the role of RTOs to oversee electric transmission and operate wholesale markets across a broad territory (multi-states). More recently, FERC Order 1000 closed some of the gaps in FERC Order 890, which is discussed in more detail later. Both ISOs and RTOs are independent entities, not affiliated with other market players, and the functions of each include day-to-day grid operations, long-term regional planning, billing and settlements, and other wholesale electric market services. ISOs tend to be smaller in geographic size and some are not subject to FERC jurisdiction (e.g., Canada and central Texas).

There are 10 ISO/RTOs in North America serving two-thirds of electricity consumers in the United States and more than 50% in Canada and include the following (see also Figure 3.3):

- Alberta Electric System Operator (AESO, an ISO)
- California Independent System Operator (California ISO)
- Electric Reliability Council of Texas (ERCOT, an ISO)
- Ontario's Independent Electricity System Operator (IESO, an ISO)
- ISO New England (ISO-NE, an RTO)
- Midwest Independent Transmission System Operator (Midwest ISO, an RTO)
- New York Independent System Operator (NYISO)
- PJM Interconnection (PJM, an RTO)
- Southwest Power Pool (SPP, an RTO)
- New Brunswick System Operator (NBSO, an ISO)

Independent system operators/regional transmission organizations coordinate generation and transmission across a wide geographic area, matching generation instantaneously to the market demand for electricity. Maintaining an optimal transmission grid requires management of the flow of power across the power grid, an understanding of the capabilities of the system, and the management of payments among producers, marketers, transmission owners, buyers, and others. In ISO and RTO regions, the owner of transmission assets is referred to as the transmission owner (TO), which can be a Transco or utility distribution companies (UDCs). The TOs own, maintain, and can expand the transmission system when appropriate. ISOs/RTOs control operation of the system and provide the TOs compensation for ownership and operation of the transmission lines.

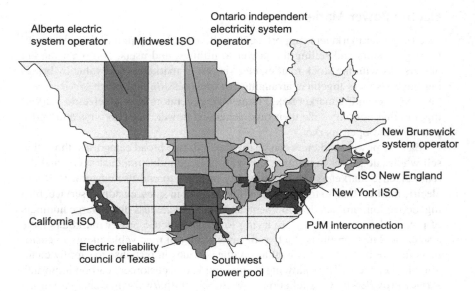

FIGURE 3.3 North American regions serviced by ISOs/RTOs. *(Source: ISO/RTO Council (IRC) 2009 state-of-the-market report.)*

Independent system operators/regional transmission organizations are more than transmission operators, as they provide more extensive grid reliability and transaction support services than offered previously in the market. In addition to nondiscriminatory transmission access, ISOs/RTOs facilitate competition among wholesale suppliers, provide regional planning, energy and/or capacity market operation, outage coordination, transactions settlement, billing and collections, risk management, credit risk management, and other ancillary services. Across large regions, they schedule the use of transmission lines, manage the interconnection of new generation, and provide market monitoring services to ensure fair market operations for all participants.

Independent system operators/regional transmission organizations also play a large role in grid reliability to avoid the types of blackouts experienced in the eastern United States in the past (e.g., the northeastern blackouts of 1965 and 2003). In addition, they forecast load and schedule the order that generation is dispatched to assure that sufficient power is available in the event that demand rises or a system failure occurs. Regional planning is another important function in that ISOs/RTOs take a broad view of the market to plan intra- and interregional infrastructure expansion for reliability and economic improvement. ISOs/RTOs play a pivotal role in planning for transmission lines associated with new generation, especially from renewable sources such as wind, geothermal, and solar.

Electric Power Marketers

Electric power marketers support a competitive market by purchasing electricity from generators and selling the power to utilities, end users, and other market players. As with the stock market, electric power marketers add value by bringing buyers/sellers together, arranging for transmission and other services, and (at times) accepting market risk. Frequently, generators have an electric marketing arm that not only sells their own generated power, but also buys and trades power in the open market.

Electric power marketers can be divided into two broad categories: those that sell wholesale power to utilities, the grid, or large industrial customers, and the retail electric marketers that focus solely on sales to end-use customers. Retail electric markets tend to focus much of their efforts in sales, customer service, billing, collection, product development, and brand awareness due to large numbers of potential customers they are trying to reach. Electric power marketers have also created new products, such as the availability of renewable energy to customers in diverse markets, as well as unique services in response to rapidly changing market needs. The following is an example of a customer "carbon footprint" service provided by Constellation New Energy (http://www.newenergy.com):

The service provides a structured platform that helps prepare customers for participation in GHG emissions reporting programs, including the U.S. EPA Climate Leaders® program, the Carbon Disclosure Project, Global Reporting Initiative (GRI), and other emerging regulatory initiatives. Both scalable and flexible, the offering is easily customized for specific reporting needs, including Corporate Social Responsibility (CSR) reporting, and it provides a web-based, enterprise-level system to manage GHG emissions and broader sustainability program metrics.

In essence, electric power marketers are replicating many of the services that were previously provided by traditional vertically integrated utilities prior to restructuring.

Utility Distribution Company

A UDC is a monopoly provider of distribution services in a restructured market. Unlike a traditional regulated market distribution company, the UDC may be prohibited by law from selling power (and only be allowed to collect a distribution charge) or may be able to supply power to select customers (typically those not serviced by electric marketers or those who have elected not to be serviced by electric marketers).

The distribution of electric power is an intrastate function under the jurisdiction of state public utility commissions (PUCs). Under the traditional regulatory system, PUCs set the retail rates for electricity, based on the cost of service, which includes the costs of distribution. Retail rates are set by the PUC in rate-making rulings. Rates include the cost to the utility for generated and

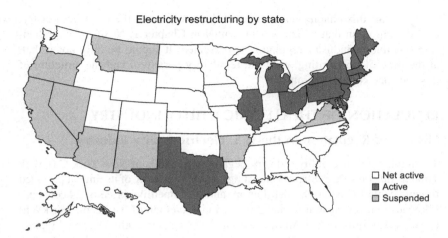

FIGURE 3.4 Status of electric restructuring.

purchased power; capital costs of power, transmission, and distribution plants; all operations and maintenance expenses; and costs to provide programs often mandated by the PUC for consumer protections and energy efficiency, as well as taxes. As the industry restructures, in some states the PUC will eventually no longer regulate retail rates for generated or purchased power. Retail electricity prices will be open to the market forces of competition. PUCs will continue to regulate the rates for distribution of power to the consumer. They also have a say in the siting of distribution lines, substations, and generators. Metering and billing are under jurisdiction of the PUC and, in some states, are becoming competitive functions. As the industry restructures, the responsibilities of the PUC are changing. The goal of each state PUC remains to provide their state's consumers with reliable, reasonably, and fairly priced electric power. More recently, states have taken on the task of promoting renewable energy, which will impact the utility market in sources of generation as well as costs. Figure 3.4 (http://www.eia.doe.gov/cneaf/electricity/page/restructuring/restructure_elect.html) shows the current status of state restructuring.

Energy Service Companies

Energy services companies actually came about in the regulated market to provide services beyond those provided by a regulated provider, such as appliance maintenance, appliance sales, and demand-side management/energy audits. ESCOs are for-profit entities and in a restructured market still continue their traditional roles, as well as other new roles associated with the restructured market. This can include assisting customers in evaluating energy needs and determining the best solutions meeting those needs based on supplies/services from the various market participants.

Thus far, this chapter has examined the structure of the electric industry, expounding upon that which was presented in Chapter 1. Next, a comprehensive timeline of industry regulation is presented. It should be duly noted that, at the time of this writing, both the regulatory paradigm and the structure of this industry continue to unfold.

REGULATION OF THE ELECTRIC UTILITY INDUSTRY

History of Regulation in the U.S. Electric Utility Industry

The modern electric utility industry began in the 1880s in New York City with Thomas Edison's Pearl Street–generating station. A novelty of its time, it provided reliable central generation, distribution, and a competitive price of 24 cents per kilowatt-hour (compared to the 2007 national average of 9.13 cents per kWh) (U.S. Energy Information Administration, 2009). Electric utilities spread rapidly through the rest of the decade, as demand grew from nighttime only to one that required electricity on demand 24 hours a day, every day of the week. The rapid spread of electric utilities characterized the remainder of the century, with most of the electricity supplied by multiservice privately-owned utilities, which competed aggressively for central city markets because of the population density that characterized such areas.

In the early part of the 20th century, it became apparent that the supply of electricity was characterized by growing economies of scale. In addition, technological change precipitated the growth and consolidation of the industry as private suppliers merged into utility-holding companies. Even smaller private and municipally-owned suppliers were forced to merge or to be acquired by these privately-owned multiservice firms. At their peak in the late 1920s, the 16 largest electric power–holding companies controlled over 75% of all U.S. generation. With this came the inevitable: regulation.

The Rise of Regulation

The holding companies that had evolved in the 1920s had significant monopolistic powers over the market. These "natural monopolies" came about due to the need for large capital investments (high entry hurdle) and economies of scale and scope that produced a lower cost of goods and services compared to many smaller firms serving the same market. The potential for excess profits and nondiscriminatory service to all was real and the government felt a need to control these powers. The solution to these natural monopolies was creation of a regulatory structure that granted exclusive service territories and, in exchange, set rates at what was deemed "fair" to investors (IOUs) and to the customer.[3] This paradigm still exists today.

---

[3]Within the boundaries of allowing a fair return on the investment in a utility's stock, regulators also seek to minimize costs to consumers, ensure reliable service, and provide relatively stable rates and an efficient use of resources. Furthermore, ensuring safe practices by utilities is paramount.

Throughout the 1880s and 1890s there was fierce competition among the various utilities in existence at that time. However, by the early 1900s, many local governments perceived a need for change in the market and/or regulation. Reacting to the need to control these natural monopolies, many local governments created municipal utility systems, which effectively eliminated investor-owned utilities. The number of municipally-owned utilities tripled between 1896 and 1906, sending a strong signal to IOUs that they could lose their market to government entities. In 1907 the largest utility associations (e.g., the National Electric Light Association) and business advocacy groups (e.g., the National Civic Federation) started to promote regulation of utilities by the state. States adopted regulation quickly, and by 1916, 33 states had created regulatory agencies. At this time the vertically integrated natural monopoly model was already well established and therefore state and local governments oversaw all aspects of this market.

Regulation by the federal government was also evolving during this time period. Prior to 1905, utilities were allowed to build and operate dams with little or no government oversight. However, in 1905 the federal government began to license dams and charge fees for their construction. In 1920 the Federal Water Power Act and the Federal Power Commission were created to regulate rates, financing, and services of utilities with licensed dams. As a result of the Great Depression, the government started to compete with investor-owned utilities through the creation of federal utility agencies such as the TVA (1933) and numerous dam projects.

During the 1920s, holding companies incurred increasing amounts of debt and the economics of generating and transmitting electricity were changing rapidly. The classic monopoly situation was unfolding in that economies of scale had not been taken advantage of when the marginal costs of new generation addition were less than the average cost of new generation. During the prosperous 1920s, these highly leveraged holding companies managed to remain solvent, but after the stock market crash in 1929 and the resulting lower demand (and hence revenue), these entities could no longer service their debt. As more holding companies went bankrupt, service deteriorated and investors lost millions of dollars. From 1929 to 1936, 53 holding companies went into bankruptcy or receivership and 23 others were forced to default on interest payments.

In 1928, the Federal Trade Commission issued a report warning that the holding company structure was unsound and "frequently a menace to the investor or the consumer or both." As the Great Depression dragged on, the federal government decided that regulatory action was required. The Federal Power Act, which established a federal utility regulatory system, was enacted at the same time as the Public Utility Act of 1935(4); these two acts were intended to work in tandem. Title I of the Public Utility Act of 1935 is known as the Public Utilities Holding Company Act of 1935. PUHCA ultimately broke up interstate holding companies and, between 1935 and 1958, 759 utilities were separated and the number of holding companies dropped from 216 to 18.

Public Utility Holding Company Act (PUHCA) of 1935

The PUHCA, enacted in 1935, was aimed at breaking up the unconstrained and excessively large trusts that then controlled the nation's electric and gas distribution networks. Before PUHCA, almost half of all electricity generated in the United States was controlled by three huge holding companies, and more than 100 other holding companies existed. The size and complexity of these huge trusts made industry regulation and oversight control by the states impossible. Many of these holding companies were looked at as either pyramid schemes (investing a small amount at the top and reaping big rewards) or complex organizations formed solely to avoid state regulations and hide true service costs to the customer. The PUHCA was passed in an era when financial pyramid schemes were extensive through the country and Congress felt a need to act. These pyramids were sometimes 10 organizational layers deep, illustrating the difficulty in regulation as shown in the following discussion from Hyman (1994):

The Insull[4] interests (which operated in 32 states and owned electric companies, textile mills, ice houses, a paper mill, and a hotel) controlled 69 percent of the stock of Corporation Securities and 64 percent of the stock of Insull Utility Investments. Those two companies together owned 28 percent of the voting stock of Middle West Utilities. Middle West Utilities owned eight holding companies, five investment companies, two service companies, two securities companies, and 14 operating companies. It also owned 99 percent of the voting stock of National Electric Power. National, in turn, owned one holding company, one service company, one paper mill, and two operating companies. It also owned 93 percent of the voting stock of National Public Service. National Public Service owned three building companies, three miscellaneous firms, and four operating utilities. It also owned 100 percent of the voting stock of Seaboard Public Service. Seaboard Public Service owned the voting stock of five utility operating companies and one ice company. The utilities, in turn, owned eighteen subsidiaries.

As a result of the Great Depression and the collapse of several large holding companies, the ensuing Federal Trade Commission (FTC) investigations criticized these holding company structures for their many abuses, including higher-cost electricity to consumers. In 1932, presidential hopeful Franklin D. Roosevelt criticized "Insull monstrosity" for inflating the value of its holdings and selling worthless bonds when Middle West Utilities and many of its 284 affiliates were placed in receivership. Under PUHCA, the Securities Exchange Commission (SEC) was charged with administration of the act and regulation of the holding companies. One of the most important features of the act was that the SEC was given the power to break up the massive interstate holding companies by requiring them to divest their holdings until

---

[4]Samuel Insull worked for Thomas Edison and later became the vice-president of Edison General Electric Company. In 1887, Insull established the Chicago Edison Company, and in 1897, Commonwealth Electric was formed. In 1907, Insull consolidated Chicago Edison and Commonwealth Electric to form Commonwealth Edison Company.

each became a single consolidated system serving a circumscribed geographic area. Another feature of the law permitted holding companies to engage only in business that was essential and appropriate for the operation of a single integrated utility. This latter restriction practically eliminated the participation of nonutilities in wholesale electric power sales. The law contained a provision that all holding companies had to register with the SEC, which was authorized to supervise and regulate the holding company system. Through the registration process, the SEC decided whether the holding company would need to be regulated under or exempted from the requirements of the act. The SEC also was charged with regulating the issuance and acquisition of securities by holding companies. Strict limitations on intrasystem transactions and political activities were also imposed.

Aside

The PUHCA effectively reorganized the electric and gas industries and facilitated greater federal and state regulation of both wholesale and retail prices. However, over the years there have been movements to repeal PUHCA and allow holding companies to buy utilities in different parts of the country to provide economies of scale and corresponding lower rates. In 1992, the PUHCA was amended by the Energy Policy Act to except firms engaged exclusively in wholesale sales of electricity. However, during the Enron corporate scandal there was a sense of déjà vu back to the formative years before the PUHCA when investors lost confidence in the basic institutions of capitalism. Many parallels exist between Enron and Insull Investments, including the years it took the FTC to unravel both firms' financial structures, corruption of the chief executive (Insull and Lay), and the ability of both firms to successfully keep the government out of their business dealings. Both firms were even groundbreaking: Enron with EnronOnline, its Internet-based energy-trading system; and Insull, who pioneered "massing production" (later shortened to "mass production") and keeping power plants running 24/7 to defray their high fixed costs. And like Enron, Insull's empire pushed its financial dealings to the brink of manipulating the nation's energy markets. As a result of the Energy Policy Act of 2005, repeal of the PUHCA became effective in 2006, marking the beginning of a new era of holding company regulation, which is discussed further in the following sections.

The Era after PUHCA

Through the 1940s, 1950s, and 1960s the electric industry continued to grow and electric prices dropped significantly due in part from the economies of scale, regulation, and technological advancements. In the early 1960s, the electricity industry created North American Power Systems Interconnection Committee (NAPSIC), an informal voluntary organization, to coordinate the bulk power system in the United States and Canada, resulting in the formation of the largest electricity grid in the world. However, there were unrecognized weaknesses developing in the system that were revealed in the northeast blackout in 1965. The blackout started with the failure of one line and interrupted electric service across 80,000 square miles (eight states), affecting 30 million customers in the northeastern United States and large parts of Canada. In response to this situation, Congress established a regional coordinating body to ensure electricity supply reliability. The North American Electric Reliability Council (NERC) was formed in 1968 as part of the Electric Power Reliability Act of 1967.

Focus on Reliability: The North America Reliability Council

In 1968 nine regional reliability organizations were formalized under the North American Reliability Council, along with regional planning coordination guides as a replacement to the previous voluntary organization, the North American Power Systems Interconnection Committee. The NERC was responsible for promoting reliability efforts and assisting the regional councils by developing common operating policies and procedures. The NERC developed a complex committee structure to bring together volunteer industry experts to consider power integration issues and provide education to support its mission to improve system reliability.

After the northeast blackout of 2003 and passage of the Energy Policy Act of 2005, the FERC was authorized to designate a national Electric Reliability Organization (ERO) to develop and enforce compliance with mandatory reliability standards in the United States. As a nongovernmental body it was designated a "self-regulatory organization" recognizing the interconnected and international nature of the electric grid. In 2006, FERC certified NERC as the ERO for the United States. The North American Electric Reliability Corporation, a nonprofit corporation, was then formed as the successor to the North American Electric Reliability Council (NERC).

Prior to becoming the national ERO, NERC's guidelines for power system operation and accreditation were referred to as policies and, although strongly encouraged, were ultimately only voluntary. As an ERO, the NERC has worked with all stakeholders to revise its policies into standards and now has authority to enforce those standards under financial penalties in the United States as well as several provinces in Canada. U.S. organizations violating the standards can be fined up to $1 million per day per violation. Efforts are currently underway with Canadian and Mexican governments to obtain comparable authority for the NERC.

The NERC currently oversees eight regional reliability entities that control all of the interconnected power systems in the contiguous United States, Canada, and a portion of Baja California in Mexico. NERC's new responsibilities include working with stakeholders to develop standards, monitoring and enforcing compliance, assessing resource adequacy, and providing accredited education and training programs to operators. The NERC also investigates and analyzes the causes of significant power system disturbances to better prevent others from occurring in the future.

The three major and two minor NERC interconnections and the nine NERC Regional Reliability Councils are shown in Figure 3.5. The reliability councils within the Eastern Interconnection are:

- Florida Reliability Coordinating Council (FRCC)
- Midwest Reliability Organization (MRO)
- Northeast Power Coordinating Council (NPCC)
- Reliability First Corporation (RFC)
- SERC Reliability Corporation (SERC)
- Southwest Power Pool, Inc. (SPP)

The Western Interconnection consists of the Western Electricity Coordinating Council (WECC) and the Texas council is the Electric Reliability Council of Texas (ERCOT).

The Northeast Power Coordinating Council (NPCC) covers portions of Canada and is often considered to be part of the Eastern Interconnection. The Alaska Interconnection, Alaska Systems Coordinating Council (ASCC), is not tied to any of the other interconnections and is not generally counted among North America's interconnections, and is an affiliate member of NERC.

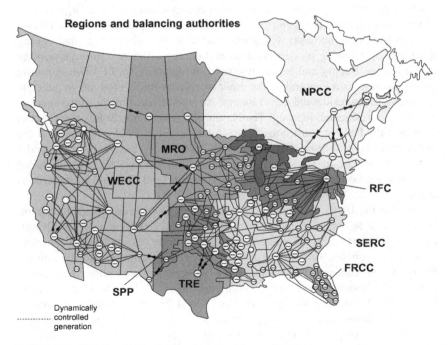

FIGURE 3.5   NERC regional reliability councils. *(Source: NERC.)*

The NERC develops and maintains reliability standards, including regional reliability standards, which must be approved by the FERC and applicable authorities in Canada and Mexico. These are displayed in Figure 3.5.

The NERC relies on regional councils to enforce the NERC standards with bulk power system owners, operators, and users through approved delegation agreements. Regional councils are also responsible for monitoring compliance of the registered entities within their regional boundaries, assuring correction of all violations, and assessing penalties for failure to comply. U.S. law requires that NERC's enforcement actions be filed publicly with the FERC.

The 1970s—A Time of Change

The electric power industry was buffeted by change throughout the 1970s, almost a perfect storm of unforeseen and uncontrollable events. In 1970, Congress passed the Clean Air Act as a response to acid rain caused by power plants as well as other environmental issues. This act significantly reduced allowable emission from power plants, signaling the beginning of massive investments in nongenerating emission control equipment, cleaner technologies, and "compliance" fuels. The Clean Air Act was soon followed by the Water Pollution Control Act of 1972 regulating water emissions and how a power plant used water for importing functions such as cooling and thermal emissions. The year 1973 can be seen as the beginning of the nation's energy crisis when the Organization of Petroleum Exporting Countries' oil embargo hit, resulting in substantial fuel and electric price increases. Congress reacted to the oil embargo and energy shortages by focusing on conservation and energy efficiency through passage of the Energy Supply and Environmental Coordination Act of 1974, requiring utilities to stop using natural gas or other petroleum-based products to generate electricity. Even residential use of natural gas was affected by the curtailment of gas used for residential lighting. The Resource Conservation & Recovery Act of 1976 and the Federal Hazardous Waste Amendment in 1984 gave the EPA the authority to control hazardous waste from "cradle to grave" and established a framework for the management of nonhazardous solid wastes. This included a focus on waste minimization and phasing out land disposal of hazardous waste.

Prior to the 1970s, energy regulation development was controlled by a number of federal agencies, including such cabinet-level departments as the Department of Interior and the Department of Agriculture, as well as independent regulatory agencies, such as the Federal Power Commission and the Atomic Energy Commission. In 1977, a number of government institutions were established to address the energy crisis more centrally, including the Department of Energy (DOE) and the Western and the Federal Energy Regulatory Commissions. The Department of Energy Organization Act in 1977 placed a number of federal energy agencies under the direction of the DOE, including the Federal Energy Administration, Energy Research and Development Administration, and Federal Power Commission, as well as parts of several other agencies and programs, including the nuclear weapons program (with the exception of nuclear energy, which is controlled by the Nuclear Regulatory Commission). The act required the DOE to work with other agencies, including the Environmental Protection Agency, the Bureau of Mines, and the Nuclear Regulatory Commission, in regards to standards and regulations as they pertained to the utility industry. The 1977 act directed the DOE to conduct R&D to support the development of new technologies within the energy field and required the DOE to submit a national energy policy plan biennially.

Federal Energy Regulatory Commission

The FERC was established within the DOE and is an independent agency that regulates the interstate transmission of electricity, natural gas, and oil. The Federal Power Commission (FPC), founded in 1920, was the predecessor to the FERC, created to coordinate federal hydropower development. In 1935, the FPC was transformed into an independent agency to regulate both hydropower and interstate electricity. All FERC decisions are reviewable by the federal courts. In 1983, Congress ended federal regulation of wellhead natural gas prices and, in response, FERC sought greater competition to both natural gas and electric industries. The Energy Policy Act of 2005 expanded FERC's authority to impose mandatory reliability standards on the bulk transmission system and to impose penalties for the manipulation of electricity and natural gas markets. These additional FERC responsibilities as they pertain to the electric industry include:

- Regulation of transmission and wholesale electricity in interstate commerce.
- Review of certain mergers and acquisitions and corporate transactions by electricity companies.
- Review of siting application for electric transmission projects under limited circumstances.
- Licensing and inspection of private, municipal, and state hydroelectric projects.
- Protect the reliability of the high voltage interstate transmission system through mandatory reliability standards.
- Monitor and investigate energy markets.
- Enforce FERC regulatory requirements through imposition of civil penalties and other means.
- Oversee environmental matters related to natural gas and hydroelectricity projects and other matters.
- Administer accounting and financial reporting regulations and conduct of regulated companies.

The FERC regulates approximately 1600 hydroelectric projects in the United States and oversees new mergers and antimarket manipulation regulations.

The FERC was heavily involved in the California electricity crisis and has collected over $6.3 billion from California electric market participants by facilitating settlements. It also investigated allegations of electricity market manipulation by Enron and others in the electric markets. More recently it has promoted voluntary formation of the RTOs and ISOs to facilitate access to the grid and supported the repeal of PUHCA and enactment of the Public Utility Holding Company Act of 2005.

National Energy Conservation Policy Act of 1978

The National Energy Conservation Policy Act of 1978 was a comprehensive energy statute composed of five separate but intertwined public laws dealing with energy conservation (National Energy Conservation Policy Act), coal conversion (Power Plant and Industrial Fuel Use Act), public utility rates (Public Utility Regulatory Policy Act), natural gas pricing (Natural Gas Policy Act), and a series of taxes (Energy Tax Act) designed to discourage energy consumption and to accelerate the transition to alternative fuels. The main purposes of the NEA were to reduce oil imports and promote more efficient use of energy. The Power Plant and Industrial Fuel Use Act specifically restricted the construction of power plants fueled by natural gas or petroleum and promoted the use of coal and alternative fuels supporting the need for greater energy security. However, the most significant of these acts was the Public Utility Regulatory Policy Act.

Public Utility Regulatory Policy Act of 1978

Restructuring of the electric utility industry began with passage of the Public Utility Regulatory Policy Act. In 1978, then President Carter suggested changes to electric pricing structures as he felt that the typical "rate structure" encouraged the use of electricity by charging higher prices for the first increments of electricity used by customers with subsequent increments costing less per unit. As an example, the first 50 kWh of use might cost 5.0 cents/kWh, the next 50 kWh 4.0 cents/kWh, and the next 100 kWh 3.5 cents/kWh, and anything greater than 200 kWh only 3.0 cents/kWh. (This is known as a declining block rate structure.) In an era when utilities were trying to reach an economy of scale and costs were decreasing, this "promotional rate structure" made sense. However, in an era of increasing energy costs, when the country was more concerned about conservation, this declining block rate structure did not support public policy.

The PURPA supported elimination of these declining block rate structures and required state commissions to order utilities to develop new rate structures. However, the most significant part of the PURPA was the requirement of electric utilities to interconnect with and buy whatever amount of capacity and energy was offered from any facility meeting criteria for a Qualifying Facility.

This was a significant departure from traditional regulation, which generally sets the price of electricity on the basis of the cost of production. Utilities were then required to pay for that power at the utility's own incremental or avoided cost of production. An initial interpretation of *avoided* cost was the cost of additional electricity produced by the utility itself. However, under PURPA's requirements, some utilities that already had sufficient supply available to meet demand, either through their own generation or through purchases from other sources, also had to purchase generation from a Qualifying Facility.

To facilitate entry of nonutility companies into the market, Congress exempted most qualifying facilities from rate and accounting regulation by the FERC under the Federal Power Act; from regulation by the Securities and Exchange Commission under the PUHCA; and from the state rate, financial, and organizational regulation of regulated utilities. In addition, by simplifying contracts and the power sales process, it provided increased financial certainty for creditors and equity sponsors and eliminated several barriers into the marketplace for smaller energy producers. Another significant provision of the PURPA was the encouragement of independent power producers and wheeling so that large, industrial firms could contract with other nonutility sources rather than be forced to purchase from the local utility (in fact, the PURPA required that electric utilities contract with certain independent power suppliers for their power).

At the time of the PURPA, energy conservation and the concept of demand-side management were critical components of energy policy but not the only elements. The Carter administration at the time also sought greater production of coal and oil, as well as nuclear energy (President Carter had a nuclear background), with the hope of not affecting the environment adversely. Federal funding through agencies such as the DOE also promoted alternative energy technologies. The PURPA also facilitated the further development of gas-turbine technology and cogeneration. During the military buildup of the 1980s, manufacturers of jet engines received government funding for research and development to advance the efficiency and reliability of this technology. The market responded to these increases in efficiency and reliability by creating aero-derivative jet engines (nominal 10 MW) for use in electric power production (previous to this, industrial turbines dominated the electric generation market). These aero-derivative gas-turbine cogeneration units could be installed at much lower capital costs and obtain thermal efficiencies in the 50% range, which made them much more attractive compared to central station plants. These technological advances were then extended to industrial turbines and included the concept of combining cycles (gas combustion cycle with a steam heat recovery cycle) to produce more efficient gas-turbine technology and reduce the cost-minimizing level of capacity to 400 MW. The reduced cost also made it easier for new players to enter the market; for existing players, the lower cost reduced the risk of investment should market conditions change. The PURPA also supported the development in renewable energy technologies, and new alternative energy producers flourished in states such as California. This situation created a new fledgling industry that would advance the technology to where we are today.

As mentioned earlier, the PURPA also started the movement for the deregulation of the power industry and the breakup of traditional, vertically integrated electric utility firms. With new entrants into the electric market, regulators started to question the need for monopolies and the vertically integrated structure. This act broke the stranglehold that traditional power companies had on the generation function, as any unregulated qualifying facility could now sell electricity into the power grid. However, it should be recognized that the PURPA did not create a competitive market in that qualifying facilities sold/generated power at a premium and not at rates competitive with existing utility generation rates. Rather, the PURPA brought into question the need for a traditional, vertically integrated monopoly to supply future new generation needs as qualifying facilities demonstrated that they could bring new capacity online at the same cost as monopolies.

Prior to the 1970s, utilities could still exploit economies of scale and the increasing thermal efficiencies of steam turbines, generators, and boiler technology. In the 1970s and 1980s, the cost of new generation was now driven more by the economics of cogeneration (the ability to sell both steam and electricity) and later by the lower cost and efficiency of combined-cycle gas turbines, as well as reduced natural prices and greater gas availability (recall earlier that the use of natural gas for electricity generation was actually banned four years earlier in the Energy Supply and Environmental Coordination Act of 1974). In addition to questioning the concept of utilities as natural monopolies, the justification for regulations was even brought into question. As qualifying facilities demonstrated that many generators in a market could exist, the seed for a competitive generation market was planted. Interestingly, there are many similarities between the PURPA and current state renewable portfolio standards (RPSs), including the fact that many cogenerators sold power into the market not at the control of the utility/regional dispatch, but when it just happened to be available, similar to the situation today with wind and solar energy.

The 1970s and passage of the PURPA represented difficult times for the electric utility industry, marked by environmental legislation and concerns, a poor economy, inflation that affected the construction of new plants adversely, occupational safety, and low load growth. However, the 1970s also marked a time when many power plants were still being constructed to supply the forecasted load growth. These primarily coal and nuclear power plants took years to construct and represented significant capital outlays by the industry. Despite the robust regulatory climate and well-intentioned natural monopolies, rapid inflation made the cost of new power plants unpredictable, financing more costly, risks associated with safety concerns and regulation more apparent, and, most importantly, the diminished load growth meant that the industry now had excess generation and reserve margins. These regulatory-approved costs were then passed on to consumers, which resulted in the near tripling of electric energy costs between 1969 and 1985. In 1983 the president of Virginia Electric and Power Company, William Berry, made a profound statement that "[a]s in so many other regulated monopolies, technological developments have overtaken and destroyed the rationale for regulation. Electricity generation is no longer a natural monopoly."

Given all the changes and turmoil of the 1970s, it seemed only fitting that the end of the decade was marked by additional energy turmoil. In January 1979, the Shah of Iran was removed from power, and the resulting oil embargo created worldwide shortages and significantly disrupted the U.S. economy. This was followed by the accident at Three Mile Island in 1979, which increased the cost of nuclear power significantly, initiated regulatory delays, and certainly brought into question the future of nuclear power in this country. In April 1979, responding to growing energy shortages, President Carter announced gradual price controls on oil and proposed a windfall profit tax on oil companies. This was followed quickly by government programs to increase research and development (R&D) funding of renewable energy and to promote commercialization of these technologies through solar development banks. In July 1979, President Carter proposed an $88 billion decade-long effort to support U.S. energy independence through the development of synthetic fuel from the nation's coal and oil shale reserves. In 1980, Congress passed the Energy Security Act to support the creation of a synthetic fuel industry producing two million barrels of oil per day by 1992. This act created the United States Synthetic Fuels Corporation to provide financial assistance and encourage private investment in this new industry.

Industry Restructuring in the 1980s and 1990s

The 1980s represented an era of deregulation for a number of industries in the United States. In 1978 the airline industry was deregulated, followed by the telecommunications industry in 1984. In many areas of the country, electric utilities also distributed natural gas (known as combination utilities), and the movement for deregulation hit this segment of the electric market first when deregulation opened access to transmission pipelines and created a spot market in 1987 (the gas industry has a similar vertical structure to the electric industry in that there are producers, transmission pipelines, and distribution pipelines). Congress, the FERC, and many state regulators believed that electric industry deregulation would lower costs to consumers while increasing supply and improving reliability based on the discussions surrounding the deregulation of other industries at the time.

From its inception, the FERC has encouraged and approved the use of market-based electric rates to support the development of an efficient and competitive market. Between 1985 and 1991 the FERC addressed 31 requests to sell wholesale electric power at market-based rates, although only a few were approved (Notice of Public Conference and Request for Comments on Electricity Issues, Docket

No. PL91-1-000, April 1991). The pace of market-based rate requests picked up substantially after passage of the Natural Gas Utilization Act.

The promotion of market-based rates was a significant step by the FERC in industry restructuring and deregulation. In the traditional regulated culture, wholesale and retail electricity rates are calculated based on a utility's costs plus a negotiated rate of return on the utility's (prudent) investments. This approach ensures investors that the utility will cover its costs of operation but does not encourage that full evaluation of all risks associated with that investment will occur. Should the project be uneconomical (e.g., overruns, demand changes, etc.), the utility could still recover its costs plus the return on the investment by passing along the costs to the customer in the form of higher electric prices.[5] This may shelter the investor from risk and facilitate financing, but does not necessarily promote competitive wholesale power markets. By the mid-1990s the FERC had approved the use of market-based rates for more than 100 power suppliers and a competitive electric power market was emerging.

Other federal legislation in the 1980s that affected the electric power industry included the Pacific Northwest Electric Power Planning and Conservation Act of 1980. This act, in addition to creating the Pacific Northwest Electric Power and Conservation Council, provided for the Bonneville Power Administration to purchase and exchange electric power with northwest utilities at the "average system cost" and gave the agency authority to plan for and acquire additional power to meet its growing load requirements.

The Economic Recovery Tax Act of 1981 introduced a new formula for determining allowable tax depreciation deductions. The Accelerated Cost Recovery System (ACRS) enabled taxpayers to claim generous depreciation deductions based on the system's permitted depreciable life, method, and salvage value assumptions. Generation, transmission, and distribution assets of regulated electric utilities were categorized as public utility property and were assigned relatively long depreciable lives under the ACRS, which influenced new capital investments toward lower-cost technologies.

The Electric Consumers Protection Act of 1986 was the first significant change to the hydropower licensing provisions of the Federal Power Act (FPA) since 1935. Changes included elimination of preferences on relicensing, the importance of environmental considerations in the licensing process associated with an increased role of the state and federal fish and wildlife agencies in reviewing licenses. The act also eliminated PURPA benefits for hydroelectric projects at new dams and diversions unless the projects satisfy stringent environmental conditions. Under this act, FERC's enforcement powers were also increased substantially.

---

[5]However, it is more likely the case that some of these costs will not be recovered in the utility's rate base. This occurred in 1989 when 25% of Louisville Gas & Electric's Trimble County Unit Number 1 was disallowed when anticipated load growth did not materialize. In addition, a $2.5 million refund and an $8.5 million rate reduction were ordered as part of the settlement agreement.

Another important law of the 1980s was the Tax Reform Act of 1986. In this legislation the ACRS method for determining asset depreciation (Economic Recovery Act of 1981) was replaced with the Modified Accelerated Cost Recovery System (MACRS). The MACRS corrected the disparity in treatment of property between regulated and nonregulated utilities. As part of the act, investment tax credits were repealed. [The investment credit of the federal income tax law was a dollar-to-dollar offset and was available for regulated and nonregulated utilities (taxpayers) and was intended to encourage capital investment.]

Natural Gas Utilization Act of 1987

As it affected the electric industry, the Natural Gas Utilization Act amended the Power Plant and Industrial Fuel Use Act of 1978 to repeal prohibitions on the use of natural gas and petroleum as a primary energy source in new power plants and new major fuel-burning installations. This act gave the Secretary of Energy authority to prohibit the use of natural gas in certain boilers and restricted increased use of petroleum by existing power plants. The act also required power plants to have sufficient inherent design characteristics to permit the addition of equipment, such as pollution control devices, necessary to allow the plant capable of using coal or another alternate fuel as the primary energy source. In addition, this act stipulated that no new power plant could be constructed or operated as a base-load power plant without the inherent design capability of being able to be converted to coal from natural gas or oil in the event market conditions warranted. Exempt from this requirement were peak-load and intermediate-load power plants, which opened the door for IPPs as often such plants were constructed to meet peak or intermediate loads using integrated gasification combined-cycle gas turbines fueled by natural gas. Supporting the use of natural gas to generate power was the act's repeal of the incremental natural gas pricing provisions in the Natural Gas Policy Act of 1978.

The 1990s—The Pace of Industry Restructuring Accelerates

In January 1990 energy prices were relatively stable. However, this changed dramatically in August 1990 when Iraq invaded Kuwait, marking the beginning of the Gulf War and another era of energy concern. Another important event affecting the electric utility industry was the passage of the Clean Air Act Amendments of 1990 (CAAA). These amendments impacted the industry significantly, both in the need to reduce emissions and in changing fuel-buying practices. One of the major objectives of the CAAA was to reduce annual sulfur dioxide emissions by 10 million tons and annual nitrogen oxide emissions by 2 million tons from 1980 levels, understanding that electric generators were responsible for a large portion of the proposed reductions. The program instituted under the CAAA established a market-based approach to sulfur dioxide emission reductions (cap and trade) while relying on more traditional technological methods for nitrogen oxide reductions.

Despite PURPA's objective of providing market access to qualifying facilities, many new players in the market accused vertically integrated electric utilities of favoring their own generation and of control area operators giving preference to their company's resources. Both the FERC and Congress believed that without open access to the transmission system the end-use customer would not realize all the benefits of market-based rates and new generation technologies. The primary intent of the Energy Policy Act of 1992 was to create open access to the transmission system for qualifying facilities, other utility-generating companies, and independent power producers.

Energy Policy Act (EPACT) of 1992

The EPACT created a structure of competition in the wholesale electric generation market and defined a new category of electric generator, the exempt wholesale generator. EWGs were not constrained by PUHCA-imposed limitations, which made it easier for them to enter the wholesale electricity market. EWGs differ from PURPA-qualifying facilities in that they were not required to meet PURPA's cogeneration or renewable fuels provisions and utilities were not required to purchase power from EWGs. Like qualifying facilities, EWGs did not sell to retail customers nor did they own transmission facilities. Unlike qualifying facilities, however, EWGs were not regulated and were able to charge market-based rates. The EPACT also mandated that the FERC provide transmission system access to wholesale suppliers on a case-by-case basis. This access provision eliminated a major barrier for utility-affiliated and nonaffiliated power producers to compete for new nonrate-based power plants. At the time of the EPACT, many of the new nonrate-based generating units were expected to be gas turbines due to the lower capital costs compared relative to coal-fired plants.

The EPACT has been considered the most significant piece of legislation in the history of the industry. The act had a significant impact on municipal and rural cooperatives in that it provided access to new generators (EWGs and qualifying facilities) in distant wholesale markets, freeing them from their dependency on surrounding investor-owned utilities for their wholesale power requirements. This has led to a nationwide open-access electric power transmission grid supporting the wholesale market (the EPACT prohibits the FERC from ordering retail wheeling to end-use customers). Anyone selling power at wholesale now has the ability to gain access to transmission at "just and reasonable" rates as defined by the FERC. The EPACT directs the FERC when it issues a transmission order to approve rates that permit the utility to recover all legitimate, verifiable economic costs incurred in connection with the transmission services. Such costs include "an appropriate share, if any, [of] necessary associated services, including, but not limited to, an appropriate share of any enlargement of transmission facilities." The language also says that the FERC "shall ensure, to the extent practicable," that costs incurred by the wheeling utility are recovered from the transmission customer rather than "from a transmitting utility's existing wholesale, retail, and transmission customers."

The EPACT also provided reforms to the PUHCA, including the expansion of FERC's authority and the creation of players in the market now exempt from SEC regulation. PUHCA reform was seen as critical by many of the market players. Nonutility groups argued that revising the PUHCA without revising transmission-access rules would reinforce the utility monopolistic structure, while regulated public utilities expressed concern that the increased access to transmission would jeopardize the reliability of the grid. Prior to the EPACT, utilities had no obligation to provide access to their transmission lines, except under the PURPA they were required to interconnect with and purchase power from qualifying facilities. Under the Federal Power Act, as amended by the PURPA, the FERC also appeared to have authority to require wheeling under limited circumstances. Wheeling is defined when a transmission utility owner allows another utility or independent power producer to move (or wheel) power over its transmission lines. However, the FERC and federal courts later ruled that PURPA authority was limited and did not allow the FERC to require a utility to wheel power to its wholesale customers or to encourage competition in bulk power markets. It should be noted that in addition to the FERC, federal courts can also require wheeling, but only when the Sherman Antitrust Act has been violated. This would include circumstances where a regulated public utility refuses to wheel power in an attempt to monopolize a particular market. The Atomic Energy Act of 1954, the Nuclear Regulatory Commission, and the U.S. attorney general may require wheeling access as a condition for issuing a construction permit for a nuclear plant. The EPACT broadened available exceptions substantially by giving the FERC new authority to order utilities to provide wheeling over their transmission systems to utilities and nonutilities alike.

In addition to granting greater access to the transmission grid and defining EWGs, the EPACT also encouraged utilities to make investments in conservation and energy efficiency as amendments to the PURPA. Utilities were now required on a regular basis to perform integrated resource planning, file those plans with the state regulatory authority, allow for public participation and comment, and implement the plan. The law also stipulated that state-regulated utilities invest in energy conservation, energy efficiency, and demand-side management programs, and be allowed a rate of return to be (EPACT, 1992)

at least as profitable, giving appropriate consideration to income lost from reduced sales due to investments in and expenditures for conservation and efficiency, as its investments in and expenditures for the construction of new generation, transmission, and distribution equipment.

Energy efficiency under the EPACT also extended to the utilities' own assets. The act provided for rate charges to be sufficient to encourage energy-efficiency investments in cost-effective improvements associated with power generation, transmission, and distribution. State regulatory authorities and nonregulated electric utilities under EPACT were (EPACT, 1992)

required to consider the disincentives caused by existing ratemaking policies, and practices, and consider incentives that would encourage better maintenance, and investment in more efficient power generation, transmission, and distribution equipment.

The EPACT also repealed the alternative minimum tax for some smaller producers. In addition to other regulatory measures, such as the Natural Gas Utilization Act of 1987, the EPACT was intended, in part, to expand the use of natural gas and contributed to the rise of gas-fired nonutility generators as the fastest growing source of electric generation capacity. Electric generation from natural gas grew from 17% of the market in 1996 to 22% of the market in 2003.

It quickly became clear after the passage of the EPACT that additional work was still needed to support industry restructuring. In 1993 the FERC issued a policy statement regarding regional transmission groups (RTGs) to clarify the provision of transmission services and facilitate the resolution of disputes. The FERC believed that RTGs would encourage negotiated agreements among transmission providers and minimize litigation before FERC (Policy Statement Regarding Regional Transmission Groups, RM93-3-000, July 30, 1993).

During the mid-1990s the FERC also established guidelines for "comparable transmission access" for third parties. The concept of comparable access is based on the assumption that owners of the transmission grid should offer third parties access to the grid on the same or comparable basis and under the same or comparable terms and conditions as the transmission owner's use of the system. Comparable access is one of the key provisions in the open-access transmission tariff specified in Order 888 (67FERC61, 168). The FERC also issued in 1995 its transmission pricing policy statement expanding the prior "postage-stamp" and contract path pricing mechanisms to a variety of other pricing methods more suitable for competitive wholesale power markets (Transmission Policy Statement, RM93-19-001, October 1994, Final Rule Order on reconsideration and clarifying the policy statement, May 22, 1995).

Despite the FERC's authority and rulings regarding wheeling, disparities still existed in the comprehensiveness and quality of transmission services provided by transmission owners to other users. To further encourage open access to transmission grids the FERC applied the comparability standard during a utility requested as a condition for approval for market-based rates or approval to merge with another utility. Despite these FERC efforts, open nondiscriminatory transmission access still did not exist universally at the end of 1995. In 1996 the FERC issued Order 888 to correct the lack of universal access, which was considered at the time to be the most ambitious and far-reaching ruling by the FERC to eliminate impediments to competition in the electric power industry.

FERC Orders 888 and 889 (1996)

FERC Order 888 required all public utilities that own, control, or operate transmission facilities to have an open-access nondiscriminatory transmission tariff on file. The order also allows public utilities to seek recovery of stranded costs associated with providing open access (Order 888, Final Rule, RM95-8-000, and RM94-7-001, April 24, 1996). The FERC issued Order 889 establishing the Open-Access Same-Time Information System (OASIS). By eliminating anticompetitive practices through a universally applied open-access transmission tariff, the FERC also recognized that regulated utilities could have stranded costs and provided for the recovery of those costs as part of the transition to competitive markets. The FERC nondiscriminatory transmission tariff specified that by July 9, 1996, utilities that own or control transmission must have filed a single pro forma open-access tariff specifying minimum conditions that offer load-based and point-to-point network services, contract-based services, and ancillary services to eligible customers comparable to the service they provide themselves at the wholesale level. The universal transmission tariff eliminated FERC's time-consuming case-by-case evaluation of wheeling requests. Including the rights, terms, and conditions to wheel power in the tariff meant that a company could respond immediately to opportunities in short-term markets that were previously not available in a timely manner and facilitated the proper function of a competitive short-term power market.

FERC Order 888 also required transmission owners to unbundle their activities, which meant that they were now under the same tariff as other transmission users (comparability standard) and had to rely on the same electronic information network that its customers relied on to obtain information about prices and available capacity of the transmission system. Transmission owners were also required to separate their rates for wholesale generation and ancillary services. Functional unbundling essentially eliminated the vertically integrated utility by separating its transmission services functions from other business activities in the company. Six ancillary services were defined in Order 888 as part of the open-access tariff, including

- Scheduling, system control, and dispatch
- Reactive supply and voltage control from generation sources
- Regulation and frequency response
- Energy imbalance
- Operating reserve—spinning reserve
- Operating reserve—supplemental reserve

The transmission customer must purchase the first two services from the transmission provider.

As mentioned earlier, Order 888 was included in the provision, enabling electric utilities to recover their stranded costs in the new competitive marketplace. These transition costs (stranded costs) represent a utility's capital investments that are unrecoverable due to the transition to competition. Because regulations allowed the cost recovery of prudent investments and potential billions of dollars in the industry could be impacted, the recovery of these stranded costs was critical to the restructuring process and creation of a competitive market, as well as in gaining support and cooperation from industry participants. The FERC also acknowledged that the recovery of these stranded costs could delay some of the benefits of competition. FERC Order 888 specified that cost recovery was limited to the loss of wholesale power customers and FERC required the open-access transmission tariff and that wholesale stranded costs should be assigned to the departing wholesale customer (typically the regulated utility). At the retail level, states still retained primary jurisdiction over cost recovery resulting from retail competition, although the FERC indicated that it would entertain requests to recover costs resulting from retail competition when a state did not have the authority. Since Order 888 was issued, the FERC has had relatively few stranded costs cases, with most being in states that have implemented retail competition.

FERC Order 889 also required all IOUs to participate in the OASIS, which provides unrestricted timely and accurate day-to-day information about transmission and is accessible to all transmission users. The OASIS is an interactive Internet-based database containing information on available transmission capacity, capacity reservations, ancillary services, and transmission prices. The underlying idea of the OASIS is to create an interactive computerized market for transmission-related products and services that is accessible by all qualified users of the transmission system. In that role, the OASIS facilitates the functioning of competitive power markets. OASIS "nodes" are Internet-based interfaces to each transmission system's market offerings and transmission availability announcements. Power marketers that sign an open-access tariff agreement are allowed complete access to view existing transmission and service availability as well as existing service requests made by other parties.

Transmission facilities have power transfer limits that must be maintained to allow for reliable operation of the power grid. Transmission operators perform system studies in various future timeframes to determine how much transfer capacity is required to serve their own "native load" and how much capacity must remain as a buffer to prevent unscheduled or accidental overflows, which can damage high-voltage equipment. The difference between the capacity needed to serve the load and to maintain safe flow margins can be made available for purchase on the OASIS node. The OASIS also provides for "firm" and "nonfirm" transmission rates to allow for unplanned outages and other conditions limiting capacity. These different transmission rates carry different cost structures.

Open access has caused much higher loads on the transmission systems. In reality, power flows along the paths of least resistance, resulting in "loop flows" of energy flowing on alternate paths, creating overload stress on the system, which requires curtailments. To address this issue, the NERC developed a tagging method designed to capture the entire transaction from beginning to end, which could then be used to determine scheduling and possible curtailments. The NERC also assumed control of the Transmission System Information Network database, a comprehensive listing of generation points, transmission facilities, and delivery points, as well as transmission and generation priority definitions that support the various OASIS nodes and NERC tagging application.

FERC Order 888 also encouraged formation of the independent system operators. Although the FERC in 1993 issued a policy statement recommending that transmission owners, transmission customers, and other interested parties form regional transmission groups to coordinate transmission planning and expansion on a regional and interregional basis, few RTGs were established. Order 888 now encouraged the formation of ISOs to facilitate the transfer of a utility operating control of their transmission assets to the ISO. Ownership would remain with the utility, and participation in an ISO was voluntary. An unbundled utility and control of the grid by an ISO with no economic interest in marketing and selling power now ensured fair and open-access transmission tariffs and eliminated discriminatory practices while achieving an efficient marketplace and regional control of the grid. By the end of 1998, the FERC had conditionally approved five independent system operators: California ISO, ISO New England, New York ISO, Pennsylvania, New Jersey, Maryland (PJM Interconnect) ISO, and the Midwest ISO.

Despite the creation of these ISOs, the development of wholesale power markets after FERC Orders 888/889 was slow, and obstacles to competition still remained. Four major obstacles were identified, including continued complaints of transmission owners' discriminating against independent power companies. The increased number of market participants and transactions made detecting the discriminatory behavior more difficult to identify. The second issue was that functional unbundling under the new orders did not produce sufficient separation between operating the transmission system and marketing and selling power, contributing further to discriminatory behavior. The third element was the fact that voluntary ISO formation had yet to occur in some areas of the country, despite the expectation that more regions would seek grid regionalization to support a competitive market. Finally, the increase in trading and movement of electricity in new directions within the grids made it more difficult to manage the grids, and the concern about grid reliability and its capacity to deal with these new loads was brought into question.

Due to the fact that transmission congestion increased, the existing procedures at the time were found to be inadequate. As the FERC itself pointed out, "current transmission loading relief (TLR) procedures [for relieving congestion] are cumbersome, inefficient, and disruptive to power markets because they rely exclusively on physical measures of [electricity] flows with no attempt to assess the relative costs and benefits of alternative congestion management techniques." Furthermore, due to uncertainties in predicting load growth, responsibilities for transmission expansion were not always clear and financial motivations for the construction of new facilities appeared to carry a greater risk. Pancake pricing, the additive charge customers paid every time power crosses the boundary of a transmission owner, had the effect of increasing transmission costs and reducing the geographic size of competitive power markets. Shortly after FERC Orders 888/889 it became clear that additional changes were needed.

FERC Order 2000 and Grid Regionalization

FERC Order 2000, issued in December 1999, asked all transmission-owning utilities, including nonpublic utilities, to voluntarily place their transmission facilities under the control of a regional transmission organization and defined the characteristics and minimum functions required by the RTO. Order 2000 was designed to take the transmission system from one that was owned and controlled mostly by vertically integrated electric utilities to a system owned and/or controlled by unaffiliated RTOs. The FERC believed that a voluntary approach would be successful, as many vertically integrated utilities would recognize the benefits and clear rules and guidance were established for the function of the RTO. Furthermore, the FERC established a collaborative process for RTO formation and provided rate-making incentives for utilities that assume the risks of a transition to a new corporate structure.

FERC Order 2000 required that each public utility that owns, operates, or controls interstate transmission facilities (except those already participating in an approved regional transmission entity) file a proposal to participate in a regional transmission organization or must file a description of efforts to participate in an RTO, obstacles to participation, and plans and a time table for future efforts. Each public utility that was already a member of an existing transmission entity that conformed with the ISO principles contained in Order 888 was required to file a description that explained the extent to which the transmission entity in which it participates meets the minimum characteristics and functions of an RTO and how it proposes to modify the entity to become an RTO or a description of efforts, obstacles, and plans to conform to an RTO's minimum characteristics and functions. All RTOs were given a time table to implement their minimum functions, including congestion management, parallel-path flow coordination, and transmission planning and expansion.

FERC Order 2000 also required that the RTO must be independent of market participants. Independence was defined as a set of conditions that included that the RTO, its employees, and any nonstakeholder director must not have any financial interest in any market participants; the RTO must have a decision-making process independent of control by any market participant; and the RTO must have exclusive authority under Section 205 of the Federal Power Act to file changes to its transmission tariff. The RTOs were also required to be of sufficient scope and configuration to perform effectively their required function and to support efficient and nondiscriminatory power markets. The FERC provided guidance on the term "sufficient scope" through the following nine criteria:

- Facilitates performing essential RTO functions
- Encompasses one contiguous geographic area
- Encompasses a highly interconnected portion of the grid
- Deters the exercise of market power
- Recognizes existing trading patterns
- Takes into account existing regional boundaries (e.g., NERC regions)
- Encompasses existing regional transmission entities
- Encompasses existing control areas
- Takes into account international boundaries

In addition, the RTO must have operational authority for all transmission facilities under its control, including security aspects to ensure real-time operating reliability. The RTO was also tasked with exclusive authority for maintaining short-term reliability of the transmission grid. Additional RTO minimum functions included

- Tariff administration and design
- Congestion management
- Parallel-path flow management
- Promote or provide ancillary services
- Provide OASIS and capability calculations
- Market monitoring
- Planning and expansion
- Interregional coordination

FERC Order 2000 was designed to create effective transmission rates to promote economic efficiency in the generation and transmission sectors and support the success of RTOs as a standalone transmission business. Under Order 2000, the FERC is responsible for an RTO's transmission rate schedule, and rates were required to address the following issues:

- Eliminate pancake pricing
- Reciprocal waiving of access charges between RTOs
- Uniform access charges
- Congestion pricing
- Service to transmission-owning utilities that do not participate in an RTO
- Performance-based regulation

- Other RTO transmission rate reforms
- Additional rate-making issues
- Filing procedures for innovative rate proposals

In Order 2000, the FERC identified eight issues, other than the ones discussed earlier, that may have an impact on the structure, completeness, regulation, and design of RTOs. They are as follows.

1. *Public power and cooperative participation in RTOs.* The FERC expects public power entities to participate in the formation of RTOs, but it is aware that public power entities face several obstacles. Internal Revenue Service codes may prevent facilities financed by tax-exempt debt from wheeling privately-owned power or may prevent transfer of operational control of transmission facilities financed by tax-exempt debt to a for-profit transmission company. State and local government laws may prevent public power entities from participating in RTOs. The lack of participation of public power entities may negate some of the effectiveness and expected benefits of RTOs.

2. *Participation by Canadian and Mexican entities.* The FERC opined that Mexican and Canadian participation in an RTO would be beneficial.

3. *Existing transmission contracts.* The FERC indicated that it will examine, case by case, how to handle existing contractual arrangements when forming an RTO. For example, one issue may involve how to handle pancaked rates in existing contracts for others when transmission-owning utilities design a nonpancaked rate for their own transactions.

4. *Power exchanges.* The FERC will leave it to each region to determine a need for a power exchange, and if the RTO should operate the exchange should there be a need.

5. *Effects on retail markets and retail access.* The FERC opined that formation of an RTO will not affect the ability of states to implement retail markets and competition. In Order 2000, the FERC noted that experience with ISOs indicates that an RTO could be a benefit to states implementing retail competition.

6. *Effects on states with low-cost generation.* Some states are concerned that an RTO would result in local utilities selling their low-cost power to other states. The FERC asserted that an RTO will provide access to future low-cost generation plants and that new low-cost generation plants will be attracted to regions with an RTO because of dependable and nondiscriminatory access to the transmission system.

7. *States' role with regard to RTOs.* The FERC believes that states have an important role to play, but they chose not to specify what role in Order 2000.

8. *Accounting issues.* The FERC will require that RTOs conform to the Uniform System of Accounts, but they also indicated that changes in the industry require them to reexamine existing accounting and related reporting requirements.

The FERC envisions that bid-based markets for wholesale electric power will be a central feature in many RTO proposals. Although bid-based markets for electric power do not now represent the dominant method for buying and selling electricity, this method is expected to grow. In Order 2000, the FERC summarized lessons learned from its analysis and approval of bid-based markets for four independent system operators. As these and other power markets mature, additional information on how to design and operate power markets will develop. The lessons learned include the following.

- Multiple-product markets: Efficiency of a multiproduct market operating in the same time period is maximized when arbitrage opportunities reflected in the bids are exhausted. That is, it is efficient when, after the RTO's market has cleared, no market participant would have preferred to be in another RTO's market.
- Physical feasibility: Transaction in the market should be physically feasible.
- Access to real-time balancing market: Real-time balancing refers to the moment-to-moment matching of loads and generation on a system-wide basis. A real-time balancing market should be available to all grid users for purposes of settling their individual imbalances.
- Market participation: Markets are more efficient with a broad participation.
- Demand-side bidding: Current wholesale power markets do not offer customer demand-side bidding, only power suppliers bid into the markets. However, demand-side bidding, to the extent it is practical, is desirable to make electricity supply and prices more responsive to competitive markets.
- Bidding rules: The market should allow generators to make bids that approximate their costs.
- Transaction costs and risks: Transaction costs should be low and participation in the market should involve no unnecessary risk.
- Price recalculations: Market clearing prices should minimize electricity price recalculations.
- Multisettlement markets: These markets may involve a day-ahead market and a real-time market. If day-ahead market bids are needed for reliability, these bids need to be physically binding and may be subject to penalties for failing to adhere to the bid.
- Preventing abusive market power: The FERC highlights three items that will help lessen the potential for market power: (1) have fewer restrictions on importing power into the region, (2) have less segmentation of geographic markets for the same product, and (3) stop allowing market participants to change bids before they complete the financial settlement. Bid changing can be used as signaling to facilitate collusive behavior.
- Market information and marketing monitoring: Market clearing prices and quantities should be transparent so that market participants can assess the market and plan their business efficiently.

One of the significant issues not addressed in FERC Order 2000 was that of federally-owned and other public power and cooperative utilities as defined as "FERC nonjurisdictional utilities." In essence, they have no filing requirements under Order 2000 and the FERC has no leverage in obtaining their participation. Because these utilities own approximately 30% of the nation's power grid, the potential exists for substantial gaps in regional coverage. In Order 2000, the FERC encourages nonjurisdictional utility participation, but also recognizes that municipally-owned utilities face numerous regulatory and legal obstacles. The Internal Revenue code has private use restrictions on the transmission facilities of municipally-owned utilities financed by tax-exempt bonds. State and local government limitations, such as prohibitions on participating in stock-owning entities and other restrictions, may also impede full participation.

Energy Policy Act (EPACT) of 2005

The EPACT of 2005 contained several provisions affecting the electric industry structure. The EPACT required the Secretary of Energy to identify critically constrained transmission corridors that cross the borders of two or more states. Proposed transmission projects in these corridors may petition the FERC, under certain conditions, to exercise federal eminent domain authority to allow the acquisition of rights of way to construct new transmission facilities. Historically, transmission siting and eminent domain authority have been left to state governmental authorities. The FERC was granted this authority to resolve impediments to construction of multistate transmission. The act also authorized the FERC to approve incentive rates for the construction of transmission facilities to enhance reliability and expand the system to increase the efficiency of the supply of generation in wholesale power markets. This was followed quickly by FERC-issued Order No. 679, which set forth criteria for a new transmission infrastructure to qualify for incentive rate treatment. The act also resulted in the establishment of a loan guarantee program within the DOE for advanced generation technologies, including nuclear, coal, and renewables, as well as other technologies enhancing the efficient delivery and use of electricity.

Regarding renewables, the EPACT extended and modified the Renewable Electricity Production Tax Credit (PTC), which is a per-kilowatt-hour tax credit for electricity generated by qualified energy sources. Originally part of the Energy Policy Act of 1992, it is a corporate tax credit to owners or operators of electric generation facilities that produce electricity from qualified energy resources, including wind, biomass, geothermal, solar, and hydropower over the first 10 years of operation. The Renewable Energy Production Incentive also originated with the Energy Policy Act of 1992 but expired and appropriations were reauthorized by EPACT 2005 to extend through 2026. An interesting twist to this is that it is intended for the following:

- Not-for-profit electric cooperatives
- Public utilities

- State governments, commonwealths, and U.S. possessions
- Indian tribal governments and native corporations

In essence, this applies to nonvertically integrated, noninvestor-owned utilities, as the production payment applies only to electricity sold to another entity.

As mentioned earlier, EPACT 2005 repealed the PUHCA, which significantly limited the merger of electric utilities and subjected holding companies to SEC regulation and allowed holding companies to only acquire or merge with interconnected utilities that would operate as a single integrated system. Under the EPACT, utilities will no longer need to apply to the SEC to determine their compliance with the PUHCA or find "transmission paths" by which to connect their utility with others they wish to acquire or with which to merge. In exchange, the new act requires utilities to provide additional data from their "books and records" so that states and the U.S. Federal Energy Regulatory Commission can mitigate a utility's potential to exercise undue market power or to cross-subsidize between utility and nonutility activities. Also, the FERC is given responsibility for reviewing the loans and other utility encumbrances to assess financial risks that utilities or their holding companies may assume by virtue of an acquisition.

Energy Independence and Security Act (EISA) of 2007

In December 2007, the EISA provided a legislative framework for transmission system modernization, including initiating "smart grid" expansion, providing tax incentives for investment, creating federal "smart grid" committees, and assigning federal funding for research and development. "Smart grids" would present consumers with real-time electricity prices, thereby encouraging efficient consumption and possibly reducing demand.

The EISA also established four standards under Section 111(d) of the Public Utility Regulatory Policies Act. The EISA requires that each state regulatory authority and nonregulated electric utility must consider adopting the new PURPA standards. This includes consideration by the Tennessee Valley Authority as a nonregulated electric utility with respect to its own operations and retail sales to directly served customers and in its separate capacity as the designated state regulatory authority under the PURPA for the distributors of TVA power. The EISA further requires that consideration of the new PURPA standards should be addressed in proceedings concluded by December 19, 2009.

Also in 2007 the FERC issued an Advance Notice of Proposed Rulemaking (ANOPR) identifying four specific issues in organized market regions that were not being addressed adequately. These included the greater use of market prices to elicit demand response during periods of operating reserve shortage, increasing opportunities for long-term power contracting, stronger market monitoring, and enhancing the responsiveness of RTOs and ISOs to customers and other stakeholders. Based on comments received from the ANOPR, the FERC issued Order 719.

FERC Order 719

Order 719 issued in October 2008 addressed reforms to improve the operation and competitiveness of organized wholesale electric power markets. These reforms represented an incremental improvement to the operation of organized wholesale electric markets in the areas of demand response, long-term power contracting, market monitoring policies, and RTO and ISO responsiveness. Regarding demand response, the FERC ordered the use of market prices to elicit demand response and required RTOs and ISOs to

1. Accept bids from demand response resources on a basis comparable to other resources.
2. During a system emergency, elimination of a charge to a buyer that takes less electric energy in the real-time market than it purchased in the day-ahead market.
3. In select situations, permit an aggregator of retail customers to bid demand response on behalf of retail customers directly into the organized energy market.
4. As necessary, modify market rules to allow the market-clearing price to reach a level that rebalances supply and demand so as to maintain reliability while providing sufficient provisions for mitigating market power during periods of operating reserve shortage.

Order 719 also required that qualifying-demand response resources in RTOs and ISOs be eligible to bid to supply energy imbalance, spinning reserves, supplemental reserves, reactive and voltage control, and regulation and frequency response. The FERC also required that RTOs and ISOs establish web sites to allow market participants to place bids to buy or sell power on a long-term basis to promote long-term contracts and transparency among market participants. Market monitoring units of RTOs/ISOs were also required to identify ineffective market rules, reporting, and notification of market participant's behavior that may require investigation.

Consolidated Appropriations Act of 2008

The Consolidated Appropriations Act of 2008 passed in December 2007 directed the U.S. EPA to develop a mandatory reporting rule for greenhouse gases (GHGs). The rule requires that emitters of GHGs from 31 different source categories report their emissions to the EPA. Approximately 80–85% of total U.S. GHG emissions from 10,000 facilities are covered by the rule and reporting and monitoring began by January 1, 2010 and the first annual emissions reports were due in 2011.

FERC Order 980

In March 2009, FERC Order 890 was issued to prevent undue discrimination and preference in transmission service. This included strengthening the

Open-Access Transmission Tariff pro forma to reduce undue discrimination, provide greater details in the tariff to facilitate enforcement efforts, and increase transparency in rules applicable to planning and use of the transmission system. Order 890 also included changes to the terms and conditions of point-to-point and network transmission services and the information required to be posted on OASIS. To address pricing, Order 890 provided for changes to the pricing of energy and generator imbalances and requirements to provide conditional firm service and planning redispatch associated with point-to-point service. Transmission providers were also required to implement an open and transparent transmission planning process.

American Recovery and Reinvestment Act (ARRA) of 2009

The ARRA was an economic stimulus package intended to mitigate the effects of the U.S. recession and followed other such emergency acts passed by Congress in 2008. Regarding the electric industry, the act allocated funding for a number of initiatives affecting the industry in the areas of renewable energy, energy efficiency, modernization of the nation's electrical grid, smart metering, energy R&D, green-job training, and carbon capture demonstrations. The act also extended tax credits for renewable energy production (until 2014) and provided for renewable energy and electric transmission technology loan guarantees and grants. The long-term impact of these stimulus measures have yet to be seen in the industry.

American Energy and Security Act (ACES) of 2009

Also known as H.R. 2454, the ACES passed the House of Representatives in June 2009 but not the Senate. A nice summary is provided by the congressional budget office:

H.R. 2454 would make a number of changes in energy and environmental policies largely aimed at reducing emissions of gases that contribute to global warming. The bill would limit or cap the quantity of certain greenhouse gases (GHGs) emitted from facilities that generate electricity and from other industrial activities over the 2012–2050 period. The Environmental Protection Agency (EPA) would establish two separate regulatory initiatives known as cap-and-trade programs—one covering emissions of most types of GHGs and one covering hydrofluorocarbons (HFCs). EPA would issue allowances to emit those gases under the cap-and-trade programs. Some of those allowances would be auctioned by the federal government, and the remainder would be distributed at no charge.

Other major provisions of the legislation would:

- Provide energy tax credits or energy rebates to certain low-income families to offset the impact of higher energy-related prices from the cap-and-trade programs.

- Require certain retail electricity suppliers to satisfy a minimum percentage of their electricity sales with electricity generated by facilities that use qualifying renewable fuels or energy sources.
- Establish a Carbon Storage Research Corporation to support research and development of technologies related to carbon capture and sequestration.
- Increase, by $25 billion, the aggregate amount of loans the DOE is authorized to make to automobile manufacturers and component suppliers under the existing Advanced Technology Vehicle Manufacturing Loan Program.
- Establish a Clean Energy Deployment Administration within the Department of Energy, which would be authorized to provide direct loans, loan guarantees, and letters of credit for clean energy projects.
- Authorize the Department of Transportation to provide individuals with vouchers to acquire new vehicles that achieve greater fuel efficiency than the existing qualifying vehicles owned by the individuals.
- Authorize appropriations for various programs under the EPA, DOE, and other agencies.

American Clean Energy Leadership Act of 2009

The U.S. Senate had its own bill, which is known as the American Clean Energy Leadership Act of 2009. This too is defunct. Nonetheless, the provisions of this bill were as follows:

- Set up a new Clean Energy Deployment Administration to facilitate tens of billions of dollars in new financing to get breakthrough clean energy technologies introduced into U.S. markets and expanded as quickly as possible.
- Require electric utilities nationwide to meet 15% of their electricity sales through renewable sources of energy (e.g., sun, wind, biomass, geothermal energy, hydropower) or energy efficiency by 2021.
- Establish an "interstate highway system" for electricity by creating a new bottoms-up planning system for a national transmission grid based on regional, state, and local planning and input; allowing states to take the initial lead in deciding where to build high-priority national transmission projects; ensuring that if an impasse develops over high-priority projects that have been identified in the consensus planning process that they can proceed with federal authority as a backstop; and making sure that the costs of "interstate highway system" transmission projects are shared fairly.
- Promote distributed generation by harmonizing and streamlining the current patchwork of interconnection standards and processes. It directs the Federal Energy Regulatory Commission to establish a national interconnection standard for small power production facilities (15 kW or less), which would cover nearly all residential-sized distributed generation.
- Revitalize America's manufacturing industries by boosting their use of clean energy and energy efficiency so that they remain competitive, preventing American jobs from being lost overseas as energy costs rise in the future.

- Improve efficiency in buildings, homes, equipment, appliances, and the federal government to cut costs to consumers and stop energy waste.
- Ensure that the U.S. electrical grid is protected from cyber vulnerabilities, threats, and attacks by giving the Secretary of Energy and the Federal Energy Regulatory Commission the authority and responsibility to respond quickly to threats and attacks that might emerge.
- Modernize the Strategic Petroleum Reserve through creation of a 30-million-barrel petroleum product reserve so that U.S. supplies of gasoline and diesel fuel will not face sudden shortfalls and price spikes due to the shutdown of refineries by hurricanes and other natural disasters, as occurred in 2008.
- Open the eastern Gulf of Mexico to leasing and exploration for oil and gas, making over 3.8 billion barrels of new oil resources and 21.5 trillion cubic feet of new natural gas resources available.
- Lay out a four-year integrated plan to double U.S. investment in energy innovation and technology to a total of almost $6.6 billion, with a complementary set of programs to enhance energy jobs training and workforce development. The bill also facilitates the large-scale demonstration and early deployment of carbon dioxide capture and storage technologies by providing a legal and regulatory framework for the first 10 "early-mover" projects.
- Protect U.S. energy consumers and businesses from energy price manipulation and volatility by increasing the transparency of what is happening in oil markets in the United States and around the world, including the role of financial markets in driving oil prices, and by giving U.S. energy regulators the same strong enforcement authorities against market tampering and manipulation now available in financial markets.
- Reform the federal energy planning process by requiring a new comprehensive energy plan one year into each new presidential term and by providing a baseline of specific studies of resources and international climate and energy policies.

FERC Order 1000

FERC Order 1000 is a Final Rule that reforms the commission's electric transmission planning and cost allocation requirements for public utility transmission providers. The rule builds on the reforms of Order 890 and corrects remaining deficiencies with respect to transmission planning processes and cost allocation methods.

Background

On June 17, 2010, the FERC issued a Notice of Proposed Rulemaking seeking comment on potential changes to its transmission planning and cost allocation requirements. Industry participants and other stakeholders provided extensive comment in response to the Notice of Proposed Rulemaking.

The rule establishes three requirements for transmission planning:

- Each public utility transmission provider must participate in a regional transmission planning process that satisfies the transmission planning principles of Order 890 and produces a regional transmission plan.
- Local and regional transmission planning processes must consider transmission needs driven by public policy requirements established by state or federal laws or regulations. Each public utility transmission provider must establish procedures to identify transmission needs driven by public policy requirements and evaluate proposed solutions to those transmission needs.
- Public utility transmission providers in each pair of neighboring transmission planning regions must coordinate to determine if there are more efficient or cost-effective solutions to their mutual transmission needs.

Cost Allocation Reforms

The rule establishes three requirements for transmission cost allocation:

- Each public utility transmission provider must participate in a regional transmission planning process that has a regional cost allocation method for new transmission facilities selected in the regional transmission plan for purposes of cost allocation. The method must satisfy six regional cost allocation principles.
- Public utility transmission providers in neighboring transmission planning regions must have a common interregional cost allocation method for new interregional transmission facilities that the regions determine to be efficient or cost effective. The method must satisfy six similar interregional cost allocation principles.
- Participant funding of new transmission facilities is permitted, but is not allowed as the regional or interregional cost allocation method.

This wraps up the discussion at the federal level. Now let us move onto state-level regulations and legislation.

State Regulations

Thus far, the focus has been at the federal level. However, states play an integral role in developing and implementing energy policy as well. As stated previously, the distribution of electric power is an intrastate function under the jurisdiction of state public utility commissions. Under the traditional regulatory system, PUCs set the retail rates for electricity, based on the cost of service, which includes the costs of distribution. Retail rates are set by the PUC in rate-making rulings. Rates include the cost to the utility for generated and purchased power; capital costs of power, transmission, and distribution plants; all operations and maintenance expenses; and costs to provide programs often mandated by the PUC for consumer protection and energy efficiency. Prior to

1992, essentially all vertically integrated IOUs were regulated, but much has transpired since then.

Despite whether retail choice is offered, many states have generally chosen to restructure by transitioning existing power pools to ISO/RTOs, creating new state or regional RTO/ISOs, or implementing restructuring without establishing new competitive wholesale markets. In reality, no two states have implemented restructuring in the same way and each state is a unique example. In terms of unbundling vertically structured utilities, states have taken a number of approaches, including requiring utilities to create new and separate departments, moving of assets into separate subsidiary companies, or divesting generation and, in some cases, transmission assets as well. Clearly, federal acts and FERC rulings as discussed earlier have played a role in how state PUCs regulate.

Allowing retail access has meant allowing all customers to choose, no customer choice, or defining classes of customers granted choices. To maintain the competitiveness of local business, the most accepted approach has been to allow commercial and industrial customers choices. Less popular is to allow the customer choices but still regulate the default supply service from the utility to provide competition and a regulated price point.

As the industry restructures, in some states the PUC will eventually no longer regulate the retail rates for generated or purchased power. Retail electricity prices will be open to the market forces of competition. The PUCs will continue to regulate the rates for distribution of power to the consumer and have a say in the siting of distribution lines, substations, and generators. Metering and billing are under jurisdiction of the PUC and, in some states, are becoming competitive functions. As the industry restructures, PUCs' responsibilities are changing. The goal of each state PUC remains to provide their state's consumers with reliable and reasonably and fairly priced electric power.

LOOKING FORWARD: RENEWABLE RESOURCES AND GENERATING TECHNOLOGIES

Renewable Portfolio Standard

More recently, states have taken on the task of promoting renewable energy that will impact the utility market in sources of generation as well as costs. Figure 3.6 displays the 29 states that have adopted a renewable portfolio standard as of August 2011. Another 8 states have renewable portfolio goals, including North Dakota, South Dakota, Utah, Oklahoma, Indiana, Vermont, Virginia, and West Virginia.

A renewable portfolio standard requires utilities to own or acquire renewable energy or renewable energy certificates corresponding to percentage goals of sales or generating capacity within certain timeframes as set forth

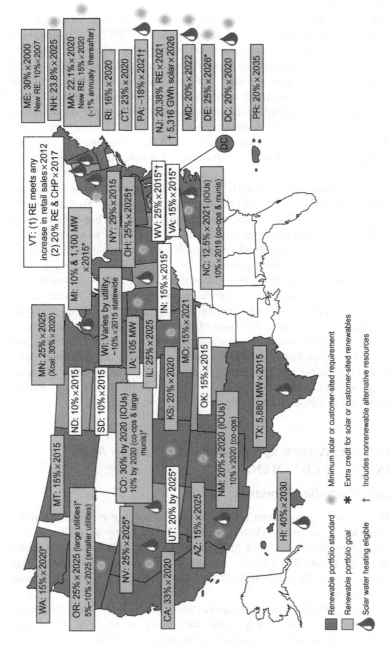

ME: 30% × 2000
New RE: 10% × 2007

NH: 23.8% × 2025

MA: 22.1% × 2020
New RE: 15% × 2020
(~1% annualy thereafter)

RI: 16% × 2020

CT: 23% × 2020

PA: ~18% × 2021†

NJ: 20.38% RE × 2021
† 5,316 GWh solar × 2026

MD: 20% × 2022

DE: 25% × 2026*

DC: 20% × 2020

PR: 20% × 2035

VT: (1) RE meets any
increase in retail sales × 2012
(2) 20% RE & CHP × 2017

MI: 10% & 1,100 MW
× 2015*

NY: 29% × 2015

OH: 25% × 2025†

WV: 25% × 2015*†

VA: 15% × 2015*

NC: 12.5% × 2021 (IOUs)
10% × 2018 (co-ops & munis)

DC

MN: 25% × 2025
(Xcel: 30% × 2020)

WI: Varies by utility;
~10% × 2015 statewide

IA: 105 MW

IL: 25% × 2025

IN: 15% × 2015*

MO: 15% × 2021

ND: 10% × 2015

SD: 10% × 2015

CO: 30% by 2020 (IOUs)
10% by 2020 (co-ops & large
munis)*

KS: 20% × 2020

OK: 15% × 2015

MT: 15% × 2015

NV: 25% × 2025*

UT: 20% by 2025*

AZ: 15% × 2025

NM: 20% × 2020 (IOUs)
10% × 2020 (co-ops)

TX: 5,880 MW × 2015

WA: 15% × 2020*

OR: 25% × 2025 (large utilities)*
5%–10% × 2025 (smaller utilities)

CA: 33% × 2020

HI: 40% × 2030

Renewable portfolio standard

Renewable portfolio goal

Solar water heating eligible

Minimum solar or customer-sited requirement

* Extra credit for solar or customer-sited renewables

† Includes nonrenewable alternative resources

FIGURE 3.6 States with renewable portfolio standards. *(Source: Database of State Incentives for Renewables and Efficiency (DSIRE) (http://dsireusa.org/ summarymaps/index.cfm?ee=1&RE=1).)*

by the state. Renewable portfolio goals differ only in that they are not legally binding. In addition to broad percentage goals, many state RPSs have set-asides that are renewable resource specific, calling for a certain percentage of retail sales or a certain amount of generating capacity from these specific resources. A renewable portfolio standard may have the most sweeping impact on the utility industry, as goals and timeframes are relatively aggressive given the historical evolution of the industry. Renewable energy targets differ by state but generally range from 10 to as much as 27% of a state's retail generation sales, and most programs require that these goals be met within the 2015–2025 timeframe.

States also offer incentives (grant programs) to support the development of renewable energy technologies. Renewable technologies supported by state grants often focus on local renewable sources. These grants are generally available only to nonresidential customers.

The following discussions briefly cover the various types of state initiatives and their impact on the utility industry.

Green Power Purchasing and Aggregation Policies

In 2009, nine states (primarily in the midwest and northeast) implemented programs to purchase electricity from renewable sources or the purchase of renewable energy credits. These programs are designed to support the growth of renewable energy–based electrical generation through voluntary means, as frequently the cost of green power is more expensive than that from fossil fuels. These programs include commitments and percentage targets by state and local governments. Some states even allow local governments to aggregate the electric load of an entire community (community choice programs) to purchase green power. Green power is typically purchased directly from project developers or power marketers and facilitates the development of renewable projects that may not otherwise compete in the market.

Interconnection Standards

As of July 2009, 39 states have implemented interconnection standards. These standards specify the technical, contractual, rate, and metering rules regarding the connection of a customer's electric-generating system to the grid (see Network for New Energy Choices, 2008). Interconnection standards for local distribution standards have typically been adopted by state PUCs, while the FERC has adopted standards for systems interconnected at the transmission level. These interconnection standards have clearly introduced competition to incumbent generation utilities, even in states with regulated IOUs. Furthermore, these interconnection standards support the robustness of the RTO/ISO regions by providing access to the grid and increasing the number of participants, which is critical to an effective marketplace.

Utility Green Power Consumer Option

By mid-2009, nine states required specific classes of electric utilities to offer their customers the option of buying electricity generated from renewable resources by either the utility or purchased elsewhere under contract. Some states also allow for the purchase of renewable energy credits from renewable energy providers certified by a state PUC (DSIRE, http://www.dsireusa.org/).

Net Metering

As shown in Figure 3.7, by mid-2009, 42 states and the District of Columbia had adopted net metering practices. Net metering allows electric customers who generate their own electricity to meter the flow of electricity both to and from the customer's premise. When a customer is generating electricity in excess of the customer's needs, excess electricity can flow back into the distribution system, offsetting electricity consumed by the customer from the grid. Indirectly, the customer is "storing" excess power on the grid and using it later when needed to avoid buying power at the full retail price.

Public Benefit Funds and System Benefit Charges

Public benefit funds (PBFs) or system benefit charges represent programs run by the state typically used to support renewable energy generation, energy efficiency programs, and low-income energy support. Most often funds are raised

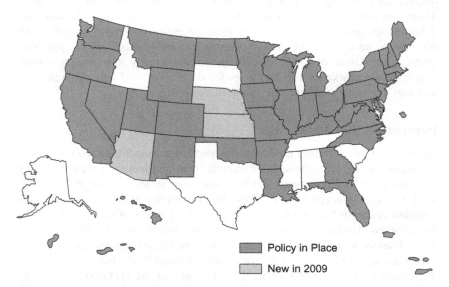

FIGURE 3.7   State adoption of net metering. *(Source: Database of State Incentives for Renewables and Efficiency (DSIRE), National Renewable Energy Laboratory (NREL).)*

as a surcharge tied to electricity consumption, but other models, such as Maine's voluntary funding program and Pennsylvania's self-sustaining program supported by load repayments and other returns on investments, exist. PBFs typically support rebate programs for renewable energy systems, loan programs, grants, and energy education programs. Loans and equity have been used by some states to support the development of clean energy projects, while business development grants, marketing support, R&D grants, and other forms of technical assistance have been used to promote clean energy industry development. By mid-2009, 17 states and the District of Columbia had implemented PBFs.

Rebate Programs

Eighteen states offer rebate programs to promote the installation of renewable energy systems such as solar water heaters and photovoltaic systems, primarily at the residential and commercial level. The bulk of these programs are administered by states, municipal utilities, and electric cooperatives. Rebate amounts can vary considerably depending on the technology and size of investment.

Renewable Energy Access Laws

Nearly 35 states have recognized that in addition to financial issues, many barriers exist related to the installation of renewable energy technologies, such as solar and wind. Wind and solar access laws protect a consumer's right to install and operate these types of energy systems at a home or business. In some states, laws exist to protect an owner's access to sunlight and prohibit homeowners associations, neighborhood covenants, or local ordinances from restricting a homeowner's right to use solar or wind energy. The most prevalent type of access law is the granting of easements to allow for access to a renewable resource and prevent adjoining property owners from restricting that resource.

Renewable Energy Production Incentives

This type of program typically provides cash payments based on renewable energy generated to support the creation of new energy projects. Many states have found that payments based on actual production (performance) are more effective in generating actual kilowatt hours compared to payments based on project capacity ratings. By mid-2009, six states had enacted renewable energy production incentives.

Tax Incentives

In addition to the numerous federal tax incentives enacted as part of energy legislation and other acts, states have created multiple types of tax incentives to promote renewable energy. These include corporate, industry recruitment and support, personal, property, and sales tax incentives. Generally, state tax incentives are not renewable-energy resource specific, to provide developers the opportunity to select the technology most suited for their needs and locally available renewable resources. Corporate tax incentives offered by states include credits, deductions, and exemptions. Twenty-four states provide a corporate tax incentive to promote renewable energy. In addition to electrical generation technology, these corporate tax incentives may also pertain to energy-efficiency equipment or green building construction. For production technologies, incentives may be a function of energy produced. Minimum investment amounts in an eligible project are also called for by some states and typically there is a maximum limit on the dollar amount of the credit or deduction.

To create new jobs, many states offer financial incentives to promote the manufacturing and development of renewable energy resources and technologies. In 2009, 15 states had industry recruitment and support incentives. These tax credits, tax exemptions, and grants may be based on the quantity of electricity produced or the resources used. In most cases these tax incentives are established as temporary measures to encourage growth in their early years and are reduced or eliminated as the industry becomes self-sufficient.

Personal tax incentives include personal income tax credits and deductions to reduce the expense of installing renewable energy systems. Although credit and deductions vary by state, in most cases there is a maximum limit on the dollar amount of the credit or deduction. Twenty-two states provided a personal tax incentive in 2009.

Property tax incentives are also used by states to promote renewable resources; these can include exemptions, exclusions, and credits, and generally only pertain to the added value of the renewable technology and not the entire property value. As property taxes are collected locally, some states have granted local taxing authorities the option of allowing a property tax incentive for renewable energy systems. Thirty-five states in 2009 provided a property tax incentive to promote renewable energy development.

State sales taxes are also being used to promote renewable energy sources and technologies. Sales tax incentives may include exemptions from the sales tax, or sales and use tax, for the acquisition of renewable energy technology. Thirty states in 2009 provided a sales tax incentive to promote renewable energy development.

Feed-in Tariffs

A feed-in tariff generally requires a utility to purchase electricity from an eligible renewable energy generator. The tariff provides a guarantee of payment associated with each unit of energy produced for the full capacity output of the system over a guaranteed period of time, usually spanning 15–20 years. This payment guarantee often includes access to the grid, and payments are frequently structured as a function of the technology type, project size, quality of renewable resources, and other local variables that may affect project economics. Feed-in tariff payments can be based on the levelized cost of service plus a specified return or the value of generation to the utility and/or society. The advantage of the levelized cost-of-service approach is that feed-in tariff payments can be designed to be more conducive to market growth by setting favorable returns. In the second approach, value to the utility and/or society, the value can be defined by the utility's avoided costs or by attempting to internalize the "externality" costs associated with fossil fuel–powered generation. These externality costs can include the value of climate mitigation, health and air quality impacts, and/or effects on energy security to name a few. Value-based approaches must be tested to ensure that value is set higher than the actual generation cost to ensure that payments are sufficient to promote market development. Based on European feed-in tariff policies, payments structured to cover the renewable project costs plus an estimated profit have proven to be the most successful (Klein et al., 2008). However, U.S. states typically use value-based cost approaches, which have so far been unsuccessful (Grace, 2008; Jacobsson and Lauber, 2005).

Feed-in tariff and RPS policies differ, as RPS mandates prescribe how much customer demand must be met with renewable energy by a utility, while properly structured feed-in tariff policies support new renewable energy supply projects by providing investor certainty regarding rates of return and long-term contract structures. As of early 2009, only a few states have passed feed-in tariff policies, including California and Washington. Several utilities in the United States have also created fixed-price, production-based incentive policies that can be considered feed-in tariffs. Approximately 14 states are currently considering feed-in tariff legislation and a federal feed-in tariff has been proposed.

FUTURE OF THE ELECTRIC INDUSTRY

Both federal and state regulations and initiatives continue to shape the future of the electric utility industry at an ever-increasing rate. Today, no one business model can describe the structure of the industry, and regulated vertically integrated utilities, traditional government, municipal and cooperative utilities, and unregulated utility and service organizations exist in various areas of the United States. Even in regulated states with RTOs, the true sense of a vertically integrated utility has been broken as the transmission grid is open to competition and controlled by an unrelated third party.

Rate-making mechanisms are also being tested, as utilities are increasingly being asked to address and account for externalities such as global warming and national energy security without a clear path to investment recovery. Historically, a utility's rates are a function of the estimated costs of providing service (including an allowed rate of return) divided by a forecasted amount of unit sales over a given period. The allowed rate of return is achieved when sales equal forecasts. If forecasts are exceeded, the utility would earn additional profits; conversely, not meeting forecast sales reduces profit. To increase profits, the utility must increase demand, which is contrary to conservation, emission reductions, and national security goals. To change this approach to rate making, the concept of decoupling profits from electrical demand is gaining acceptance. In the simplest form, *decoupling* refers to rate adjustment mechanisms that "decouple" the utility's ability to recover its agreed-upon fixed costs and earnings from electrical consumption. Various rate adjustment mechanisms exist, but essentially most are based on a "true-up" mechanism once actual sales levels are known. The "true-up" mechanism accounts for both lower than and higher than forecast sales, compensating the utility when demand decreases and penalizing the utility for higher demand with the "true-up" mechanism incentivizing the utility to meet the stated objectives (externalities). Typically these "adjustments" are small and "caps" are established to limit risk from both the ratepayer's and the utility's viewpoint. By reducing cost recovery risks to the utility and its investors, society benefits by encouraging energy-efficiency programs. This has been proven in California and Oregon where decoupling has produced some of the highest levels of utility funding for energy efficiency.

In addition to changes in rate making, a number of legislative, environmental, and energy conservation initiatives are being debated that may change the structure of the industry dramatically.

In December 2009 the Environmental Protection Agency formally declared that carbon dioxide from the burning of fossil fuels poses a threat to human health and welfare. This designation established the path for the federal government to regulate greenhouse gas emissions from power plants and other major sources. This action was anticipated based on the 2007 Supreme Court decision that declared carbon dioxide and five other greenhouse gases pollutants covered by the Clean Air Act. However, in January 2010 a Senate "disapproval resolution" was filed. This is a rarely used procedural move that prohibits rules written by executive branch agencies from taking effect. The disapproval resolution would essentially throw out the process by which the EPA found that greenhouse gases endanger public health.

In addition to federal initiatives, states and regions of the country have become involved with greenhouse emission issues and cap-and-trade programs. The Regional Greenhouse Gas Initiative (RGGI), composed of 10 northeastern and midatlantic states, is the first mandatory, market-based effort in the United States to reduce greenhouse gas emissions with the goal of reducing CO_2 emissions from the electric utility sector 10% by 2018. In the RGGI, states sell emission allowances through auctions and use the proceeds to invest in energy efficiency, renewable energy, and other clean energy technologies. The states of Connecticut, Delaware, Maine, Maryland, Massachusetts, New Hampshire, New Jersey, New York, Rhode Island, and Vermont are signatory states to the RGGI agreement. The RGGI conducts individual CO_2 budget trading for each of the 10 participating states. State regulations, based on a RGGI model rule, govern activities and are linked across the states through CO_2 allowance reciprocity. Regulated power plants (>25 MWe) can use allowances issued by any of the participating states to show compliance with their own state programs. In effect, the 10 state programs function as a single regional compliance market.

The RGGI market-based cap-and-trade approach establishes a multistate CO_2 emissions budget (cap) that decreases gradually over time. Electric-generating utilities are required to acquire allowances equal to their CO_2 emissions over a specified period. Allowances are purchased through a market-based emissions auction and trading system where CO_2 allowances can be bought and sold. The RGGI also employs offsets, which are greenhouse gas emission reduction projects outside the power generation sector, to further facilitate compliance. To minimize dramatic changes in electricity prices, the RGGI uses a gradual phased-in approach, which is more conducive to generation planning and provides a certain degree of regulatory certainty. However, it should be noted that H.R. 2454, if enacted, would prohibit state and regional cap-and-trade programs from operating between 2012 and 2017. The first three of four auctions conducted by the RGGI in 2009 involved approximately 90 million in allowances, generating more than $432.8 million.

Other important trends influencing the electric industry are energy conservation and the concept of demand management. Reducing energy consumption has been demonstrated in states such as California where annual per capita electric consumption has remained relatively flat since 1980 (about 7200 kWh) while the rest of the nation's per capita electrical consumption has increased by 50% over that period (California Energy Commission, 2007). Increases in per capita electrical consumption have historically been encouraged through rate models charging customers a flat rate per kilowatt hour of electricity consumed.

This "average pricing" structure does not accurately reflect the true cost of producing, transmitting, and delivering power, especially during peak demand periods. Average pricing has encouraged electrical consumption, made markets inefficient, increased wholesale price volatility, and underutilizes utility assets during nonpeak periods.

Average pricing has been set to make energy seem more affordable and thus provided the customer with no incentive to reduce energy consumption. In addition to average pricing, some utilities used pricing structures that actually encouraged more energy consumption, such as declining block rates (the price per unit of energy decreased as consumption increased). Higher consumption encourages more capital investment, resulting in more revenue and profits (i.e., the Averch–Johnson effect). This inefficient model has been replaced in some areas of the country by decoupling mechanisms that have separate utility revenues from the sale of electricity. Due to this lack of incentive to change ratepayer consumption behavior, utilities have been forced to overinvest in generation and transmission capacity to meet peak demand. Average pricing and certainly declining block rates have therefore created an inefficient market with poorly utilized assets, volatile wholesale prices, and reduced grid reliability.

Demand-side management and the concept of dynamic pricing are designed to send the consumer price signals to support appropriate consumer response such as during peak demand hours. Instead of a flat rate and a monthly bill showing total energy consumed, dynamic pricing incentivizes customers to shift their load from peak to off-peak hours and exposes customers to more realistic energy prices based on generation costs and overall market supply and demand.

It enables customers to gain a better understanding of how they use energy and how the cost of energy changes during the day. The Energy Independence and Security Act of 2007 took the first step in this direction by supporting the need for a "smart grid" and advanced meters that allow utilities to monitor energy consumption in real time. In addition to smart meters, the "smart grid" must also include enabling elements, such as two-way communication networks, data storage systems, automated device control, and advanced billing and cost modeling to make dynamic pricing possible. Nearly 60 utilities in the United States have participated in voluntary real-time pricing tariffs, either through pilots or permanent programs (see Barbose and Goldman, 2004).

A number of dynamic pricing models exist. Time-of-use rates are the most widely implemented and differentiate between peak and off-peak periods throughout the day. A day is typically divided into sections, such as off-peak, midpeak, and peak time periods. Predetermined prices are assigned to those periods with peak demand consumption producing a higher energy bill, which should provide an incentive to reduce energy consumption. Conversely, customers shifting their load to off-peak times will experience lower energy bills. Pricing reflects the utility's cost of generating electricity and/or purchasing power in the wholesale markets to meet customer demand. Clearly, cost models play an important role in determining pricing under dynamic conditions.

Time-of-use rates have demonstrated their effectiveness in bringing down ratepayer energy bills, and by exposing customers to price signals for marginal demand costs, the utility can improve its economic efficiency. These rates also encourage conservation and eliminate cross-subsidies between customer classes, producing fairer rate structures. Customers are feeling empowered as they now have choices regarding their management of energy consumption and costs.

Other forms of dynamic pricing include critical peak pricing (CPP), which is a form of time-of-use pricing except that very high electric prices are set for critical peak times during the year to discourage consumption during these limited number of hours per year. These critical peak periods are typically identified the day before, allowing price signals to be set to traditional peak-use customers. Critical peak pricing can be based on a maximum number of days with defined time and duration (fixed-period CPP), no predetermined duration or time with little advanced warning (variable-period CPP), off-peak and midpeak period prices set in advance for a designated length of time (variable-peak CPP), and critical peak rebates where customers remain on fixed rates but receive rebates for load reductions during critical peak periods. By making customers more responsive to increases in energy demand and market prices, as well as supply shortages, CPP programs provide several benefits, including reducing the use of peak power and stress on the transmission/distribution systems. By curtailing peak demand, utilities also avoid the use of expensive generation assets and avoid high transmission marginal costs to produce overall lower costs to the ratepayer.

Real-time pricing (RTP) programs are another form of dynamic pricing that provides the ratepayer with hourly retail prices that reflect hourly changes in the cost of production or purchases from the grid. Unlike time of use and CPP, real-time pricing is not based on preset prices for specified periods. Real-time pricing sends actual market prices to the customer, exposing them to sudden and some-times unexpected changes in the price of electricity caused by events such as unexpected high demand, supply interruptions, weather, and other factors. In RTP programs, the cost per kilowatt hour changes hourly and is based on either the marginal cost to produce electricity or the current market price. Variations in RTP also include two-part pricing. In this model, a base usage level is priced at fixed rates and real-time prices are used for demand above the base usage level. The advantage of this approach is that low-usage customers see no change in rates and high-usage customers are encouraged to reduce usage.

Because real-time prices reflect the real-time marginal costs of producing electricity more accurately, they tend to be of more benefit than time-of-use rates. Dynamic pricing and the use of robust cost modeling can provide numer-ous benefits to both the customer and the utility. By reducing peak demand, uti-lities can defer the construction of new transmission or distribution systems and offer new programs, such as giving customers the option to purchase risk man-agement tools, including caps and other hedging contracts. Dynamic pricing also enables utilities to implement demand response, conservation, and energy-efficiency programs that can impact the grid positively. Utilities involved in dynamic pricing programs have experienced improved operational efficiency, lower capital costs, and greater quality and reliability of service.

From a utility standpoint, dynamic pricing and demand-side management can reduce or eliminate in the near term the construction of more generation assets. Utilities that have implemented dynamic pricing programs have experi-enced improved operation efficiency, lower capital costs, and greater power quality and reliability. Dynamic pricing has also lowered the cost of service over time and allocates resources more efficiently. By eliminating cross-subsidies that occur with average pricing or in the implementation of distributed generation and net metering programs without dynamic pricing, distributed generators that reduce grid and distribution load can purchase electricity when prices are lower than their cost of generation and the distributed generation can then be used to meet peak loads more cost effectively.

Customers who adjust their demand behavior will receive the most direct benefits in reduced energy costs. For those ratepayers who either choose to or cannot adjust their demand behavior, dynamic pricing can still have benefits in that reduced peak demand from others will still reduce transmission and distribution losses and dispatch will not call upon the most expensive generation assets as frequently. Dynamic pricing also shifts market power to the customer when they respond to price signals actively, limiting the ability to increase prices. Over time, dynamic pricing has been shown to decrease wholesale prices. Changes in electric pricing structures and the use of dynamic pricing will also encourage innovations, such as the development of energy use monitors, remote device control, Internet-connected appliance, and certainly more robust cost models as presented in this book. These concepts are discussed in more detail in subsequent chapters.

Chapter 4

The Economics (and Econometrics) of Cost Modeling

GENERAL COST MODEL

In general, a cost function describes the relationship among inputs, outputs, and other factors on total cost, that is,

$$C = f(Y, P, O) + e, \tag{4.1}$$

where C is total cost, which is equal to total operating expenses plus the cost of capital; Y is a vector of total outputs; P is a vector of input prices; and O is a vector of other factors.

It is critical to note that not every such relationship describes a proper cost function, which follows from the hypothesis of cost minimization. Thus, a proper cost function is characterized by the following.

1. **Monotonic in Y.** Given a cost function of the general form above, an increase (decrease) in output should always increase (decrease) total cost. Mathematically, this is given by

$$\partial C / \partial Y > 0, \tag{4.2}$$

   where $\partial C / \partial Y$ is marginal cost. A second-order condition would be that cost increases with output at a decreasing rate, that is,

$$\partial^2 C / \partial Y^2 < 0, \tag{4.3}$$

   which yields the cost curve displayed in Figure 2.1 (Chapter 2).
2. **Homogeneous of degree one in input prices** (also known as linear homogeneity). Formally, this implies that, for $t > 0$,

$$C(tp, Y) = t\,C(p, Y). \tag{4.4}$$

   Simply put, if $t = 2$, then a doubling of input prices will double total cost. Furthermore, derivatives of this cost function, factor demands (or inputs), are homogeneous of degree zero.

3. Nondecreasing in input prices. Let p and p' denote vectors of input prices such that $p' \geq p$. Nondecreasing in input prices implies that

$$C(p', Y) \geq C(p, Y), \tag{4.5}$$

that is, an increase in the price of an input cannot cause total production cost to fall. In addition, because the derivatives of this cost function result in optimal factor demands, this restriction is further warranted; that is, it is the result of Shephard's lemma, which states that the optimal factor demands, x_i, are derived from

$$\partial C(p, Y)/\partial p_i = x_i(p, Y) \geq 0 \tag{4.6}$$

(see also footnote 9).

4. Concavity in input prices. This is not an obvious outcome of cost minimization and has several implications. First, the cross-price effects are symmetric, that is, by Young's theorem,

$$\partial x_i(p, Y)/\partial p_j = \partial^2 C(p, Y)/\partial p_j \, \partial p_i = \partial^2 C(p, Y)/\partial p_i \, \partial p_j = \partial x_j(p, Y)/\partial p_i. \tag{4.7}$$

Second, own price effects are nonpositive, that is,

$$\partial x_i(p, Y)/\partial p_i = \partial^2 C(p, Y)/\partial p_i^2 \leq 0. \tag{4.8}$$

That is, cost minimization requires that the matrix of first derivatives of the factor demand equations be negative semidefinite, which requires that the diagonal terms of the Hessian matrix be nonpositive (these are the own-price coefficients) (see the Appendix to this chapter for details).

5. For multiproduct firms, the cost function should be able to accommodate output vectors in which some goods take on a value of zero, which is certainly the case throughout the cost function literature for electric utilities. As will be shown, two of the more commonly used specifications in the literature violate this criterion: the Cobb–Douglas and the translogarithmic functional forms. Each is explored in more detail, but for now it will suffice to review the general form of each as well as its salient properties.

Cobb–Douglas Cost Function

The (two-output) Cobb–Douglas cost function (input prices are omitted for simplicity) is

$$C(Y, p) = A(Y_1)^{\alpha_1} (Y_2)^{\alpha_2}, \tag{4.9}$$

where A is a technology parameter, Y_1, Y_2 are outputs, and α_1, α_2 are parameters to be estimated.

It is easy to see that if one of the outputs is zero, then total cost equals zero, which is not likely the case for a firm producing positive amounts of the other output. Note: Because the parameters to be estimated enter nonlinearly in the equation, ordinary least squares (OLS) cannot be used.

Translogarithmic Cost Function

The (two-output) translogarithmic cost function (with input prices, p_k) is

$$\ln C(Y,p) = \alpha_0 + \sum_i \alpha_i \ln Y_i + \sum_i \sum_j \alpha_{ij} \ln Y_i\, Y_j + \sum_k \beta_k \ln p_k \qquad (4.10)$$

for $i, j = 1, 2$.

Because the natural log of zero is undefined, any output vector in which Y takes on a value of zero renders total cost of production undefined. Despite its popularity, this form is simply not appropriate for modeling cost in multiple-output industries.

6. The functional form of the cost function should not influence the analysis and resulting conclusions. Because cost functions are often used to determine the optimal industry structure (i.e., number of firms, etc.), via concepts such as economies of scale and scope, of horizontal and vertical integration, and subadditivity of the cost function. That is, the form itself must be flexible enough to allow for the emergence of these important cost concepts, which can then be used to shape industry structure and optimal policy making.

Again, despite its popularity, the translogarithmic functional form precludes a finding of economies of scope or subadditivity of the cost function due to its inability to deal with outputs that take on a value of zero.

The Cobb–Douglas form is also excluded. Recalling from Chapter 2, cost complementarity, which is a sufficient condition for subadditivity, is given by

$$\partial^2 C(Y)/\partial Y_i \partial Y_j < 0, \qquad (4.11)$$

where $Y = \sum y_i$ for $i, j = 1, \ldots, n$ and $i \neq j$.

In the case of the Cobb–Douglas functional form, this is equivalent to

$$\partial^2 C(Y)/\partial Y_i\, \partial Y_j = \alpha_i\, \alpha_j\, C/Y_i\, Y_j, \qquad (4.12)$$

where $C = A(Y_i)^{\alpha i}\, (Y_j)^{\alpha j}$ and $i \neq j$.

Clearly, as long as marginal costs are positive and all outputs are positive, then

$$\partial^2 C(Y)/\partial Y_i \partial Y_j > 0, \qquad (4.13)$$

which precludes a finding of cost complementarity and hence subadditivity of the cost function.

Thus, it should be data, not the functional form, that determine the findings and support the conclusions reached.

7. Finally, whenever possible, parsimony is key. Include only variables that are theoretically relevant for which accurate data can be obtained. Be certain to review data carefully, looking for outliers and other oddities.

THE ECONOMETRICS OF COST MODELING: AN OVERVIEW

Ordinary Least-Squares Estimation

You may recall that there are certain assumptions that must be met for OLS estimators to be the "best" (a.k.a. BLUE—best linear unbiased estimators). Known as classic assumptions, when met, OLS estimators are unbiased, efficient (i.e., minimum variance), consistent, and distributed normally. These assumptions are:

1. The model is linear in coefficients, correctly specified, and the error term (ε_i) is additive; that is,

$$Y = \beta X + \varepsilon_i. \tag{4.14}$$

2. The error term has a population mean $= 0$.
3. Explanatory variables are not correlated with the error term.
4. Error terms are not correlated serially (i.e., no autocorrelation).
5. The error term has a constant variance (i.e., no heteroscedasticity).
6. No explanatory variable is a perfect linear function of another explanatory variable (i.e., no multicollinearity).
7. The error term, ε_i, is distributed normally; that is, $\varepsilon_i \sim N(0, \sigma^2)$, or

$$E(\varepsilon_i) = 0 \text{ and } \mathrm{Var}(\varepsilon_i) = \sigma^2. \tag{4.15}$$

Regression Analysis and Cost Modeling

Basically put, regression analysis is a statistical technique that attempts to quantify the effect of a change in an independent variable on a dependent variable. In the case of cost modeling, the dependent variable, cost, is a function of other explanatory variables, such as prices of inputs and, of course, output. That is, at the very least, a cost function is of the general form:

$$C = f(Y, p), \tag{4.16}$$

where C is cost, Y is output, and p is (a vector of) input prices.

In its simplest form, a single-output linear cost model is of the form:

$$C = \alpha_0 + \alpha_y Y + \sum_i \beta_i p_i. \tag{4.17}$$

In this case, the parameters to be estimated via regression are coefficients on output (α_y), input prices (β_i), and the constant (α_0).

On the surface, this particular specification appears to conform to the classic assumptions mentioned earlier, which indicates that the parameters can be estimated via OLS. In upcoming chapters we will see much more complex models that cannot be estimated via this technique but for now we will continue along this vein. At this juncture, some examples are probably in order.

Examples: Examining Data—An Illustration of Salient Points

The data set RUS97_Basic contains data on the cost of distributing electricity for 707 rural electric cooperatives that distributed electricity in 1997. To get familiar with these data (we will be using them extensively in upcoming chapters), the variables of which are defined in Table 4.1, a review of the summary

TABLE 4.1 Variables Included in RUS97_Basic Data Set

Variable	Definition
State FIPS	State indicator
Coop	Coop indicator
Borrower	RUS borrower (distribution coop)
Supplier	Type of supplier: G&T cooperative, IOU, federal, other
Cost	Total cost of distributing electricity (thousands $)
RES	Electricity distributed to residential customers (MWhs)
CISmall	Electricity distributed to small commercial/industrial customers (MWhs)
Y_1	Electricity distributed to "small" (i.e., low-voltage) customers (MWhs)
Y_2	Electricity distributed to "large" (i.e., high-voltage) customers (MWhs)
Y	Total electricity distributed ($Y_1 + Y_2$) in MWhs
P_k	Price of capital (interest on LTD/LTD)
P_p	Price of purchased power (cost of purchased power/MWh purchased)
P_l	Price of labor (total payroll expense/number hours worked)
T_R	Miles of transmission lines
DensAll	Customer density (customers/mile distribution line)

TABLE 4.2 Summary Statistics—RUS97_Basic Data Set: 1997 Rural Electric Cooperatives

Variable	Mean	Std Dev	Minimum	Maximum
Cost (000)	19,397	23,255	429	186,796
RES	172	214	2	1677
CISmall	53	77	–	694
Y_1	225	280	2	2371
Y_2	63	230	–	4040
Y	288	400	2	4573
P_k	5	1	–	7
P_p	43	10	18	90
P_l	16	4	–	59
T_R	28	58	–	348
DensAll	6	3	–	24
ResCust	13,422	15,621	282	132,780
CISmallCust	1365	1788	–	12,327
CILargeCust	11	57	–	1336

statistics and variable plots can prove to be helpful in the determination of outliers and other data irregularities. Table 4.2 shows summary statistics of the relevant variables. Note in particular the composition of the load: mostly residential customers and low density, which has implications in terms of optimal industry structure and public policy.

Example 4.1 Check for Outliers

One of the first things that one should always do is review data. Plots of the variables can yield important clues as to appropriate specification and whether there are errors in data. Because purchased power is one of the largest costs faced by a distribution utility, it is informative to plot this variable, which is displayed in Figure 4.1.

Figure 4.1 displays the frequency distribution of the price of purchased power (in $/MWh) for the coops in the sample. Note the wide range of values and that it does not appear to be distributed normally. This is not surprising; in the case of electricity, power prices tend to vary with more observations at higher levels, especially given the types of suppliers and the various sources of generation fuel sources.

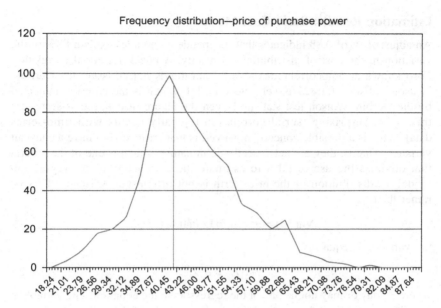

FIGURE 4.1　Distribution of the price of purchased power for 1997 coops.

Example 4.2 Suppliers—Cost of Purchased Power by Supplier Type

Despite the fact that many distribution coops are members of generation and transmission (G&T) cooperatives, some are not. In fact, 46 purchase power from an investor-owned utility and over 10% ($n = 72$) purchase power from a federal power supplier. As an exercise, you will be asked to review the descriptive statistics associated with the distribution coops that purchase power from each of these types of supplier.

Example 4.3 Estimating a Basic Cost Function

Using data contained in RUS97_Basic, estimate the parameters of a single-output, three-input regression equation of the form

$$C = \alpha_0 + \alpha_y Y + \beta_k p_k + \beta_l p_l + \beta_p p_p, \qquad (4.18)$$

where p_k is price of capital, p_l is price of labor, and p_p is price of purchased power. Using OLS, the estimated equation is (t stats in parentheses):

$$\begin{aligned} C = -17{,}635 + 53.11^*Y + 938.80^*p_k + 123.30^*p_l + 351.07^*p_p \\ (-6.86) \quad (68.14) \quad (2.46) \quad\quad (1.51) \quad\quad (11.52) \end{aligned} \qquad (4.18')$$

Estimation Results: Basic Cost Model

An adjusted R^2 of 0.88 indicates that independent variables explain 88% of the variation in the cost of distributing electricity. A check for serial correlation (also known as autocorrelation) reveals that this is not an issue; the Durbin–Watson statistic of 1.827 is very close to 2.0, which is the criterion indicated by the Durbin–Watson test statistic [given the nature of data (cross-sectional) this is not surprising (serial correlation typically occurs with time-series data]. What is a possible concern, however, is that errors do not have a constant variance. That is, they are heteroscedastic in nature, which is one of the criteria that obviates the use of OLS to estimate the parameters of the regression model, as distribution of the error term is not of constant variance. That is, rather than

$$\text{Var}(\varepsilon_i) = \sigma^2 \ [\text{from Equation (4.2)}],$$

the variance is equal to

$$\text{Var}(\varepsilon_i) = \sigma^2 Z_i^2, \tag{4.2'}$$

where Z_i may or may not be one of the regressors in the equation. The variable Z is called a proportionality factor because the variance of the error term changes proportionally to the square of Z_i (Studenmund, 1997). White's[1] test reveals that errors are indeed heteroscedastic. The heteroscedasticity-corrected standard errors generated the t statistics displayed in parentheses in the following equation:

$$C = -17,635 + 53.11^*Y + 938.80^*p_k + 123.30^*p_l + 351.07^*p_p \tag{4.18''}$$
$$(-4.99) \quad (10.99) \quad (2.01) \quad (1.91) \quad (6.89)$$

Evaluating Equation (4.18″) at various levels of output yields the average and marginal cost curves displayed in Figure 4.2.

The average and marginal cost curves displayed in Figure 4.2 are not consistent with economic theory. Due to the linear form of the basic cost equation, average cost declines with output and marginal cost is constant, unlike the appropriate U-shapes that each should have due to the law of diminishing returns in production (see Chapter 2, Figure 2.2).

---

[1]White's test detects heteroscedasticity by running a regression of the squared residuals (from the original regression, e_i) as the dependent variable on the original explanatory variables, their squares, and their cross-products. White's test statistic is computed as

$$nR^2, \tag{4.18a}$$

where n is the sample size and R^2 is the (unadjusted) coefficient of determination of the equation that contains the original explanatory variables, their squares, and their cross-products. The test statistic has a χ^2 distribution with degrees of freedom equal to the number of slope coefficients in the equation containing the original explanatory variables, their squares, and their cross-products.

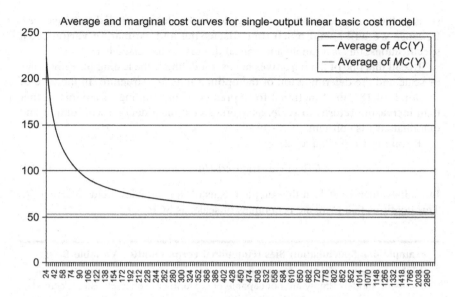

FIGURE 4.2 Average and marginal costs generated by basic cost model.

Consequences of Heteroscedasticity

1. Pure (as opposed to impure) heteroscedasticity does not cause bias in the parameter estimates. Note that these did not change, which indicates that there is no bias in the estimates themselves in the presence of heteroscedasticity.
2. It does, however, increase the variance of the distribution around the estimated coefficients so that they are no longer minimum variance.
3. In addition, it causes OLS to underestimate standard errors of the coefficients, which means that the t statistics of the coefficients will be inflated. This, in turn, can increase the probability of committing a type I error, which is rejection of a true null hypothesis.

A comparison of the two equations yields the following: While the estimates themselves have not changed (again, OLS estimates are unbiased even in the presence of heteroscedasticity), standard errors of the estimates have clearly changed; however, in some cases, we see that they have not changed in the same direction. This is an indication that there is a deeper, more complex issue at hand that must be addressed.

Impure Heteroscedasticity

Impure heteroscedasticity results from an error in specification, such as an omitted variable or an incorrect functional form. Either can result in incorrect signs on the estimated parameters in addition to the aforementioned consequences. Because we will spend much time discussing and estimating various cost specifications, this is a timely and critical discussion to pursue here.

Recall that one of the objectives in cost modeling is the testing for economies of scale and the determination of the optimal industry structure. In the case of Equation (4.18), the functional form precludes any finding of anything other than increasing returns to scale, as average cost must decline with output and marginal cost is constant.

Recalling from Chapter 2,

$$SCE(y) = \text{average cost/marginal cost.} \tag{4.19}$$

Indeed, evaluating SCE at the sample means of the variables yields SCE = 1.26. This confirms that there are economies of scale, as SCE > 1.

Example 4.4 Specification Bias (Functional Form, Omitted Variable Bias)

While parsimony might be an appealing property, it is often the case that very basic regression models simply are not capable of capturing the characteristics deemed important and necessary for their intended purposes. In the example given earlier, constant marginal cost and an ever-declining average cost curve do not accord with economic theory and can provide no assistance in determination of the optimal industry structure and appropriate public policy.

The specification bias that must be addressed here is the possible omission of a relevant variable and/or the incorrect functional form of the equation. For the issue at hand, a fairly straightforward solution could be to allow output to enter both linearly and as a quadratic; that is, to estimate an equation of the following form:

$$C = \alpha_0 + \alpha_y\, Y + 1/2 \,{}^*\alpha_{yy}\, Y^2 + \Sigma_i\, \beta_i\, p_i. \tag{4.20}$$

This particular form allows for both marginal and average cost to be a function of output, which is certainly appealing and accords to economic theory. (The coefficient of the quadratic term in output is multiplied by 1/2 for simplicity.) Furthermore, specifying that output enters as a quadratic "allows for the unconstrained emergence of economies or diseconomies of scale and scope as well as subadditivity" (Kwoka, 1996, p. 59).

Prior to estimating any equation, it is important to determine the expected signs of the estimated coefficients, which should accord with economic theory. In this case, a priori expectations are that the coefficients of the input price variables (β_i) are all positive (nondecreasing in factor prices) and that α_y, too, is greater than zero (monotonic in output). To generate a region of economies of scale, it is necessary that α_{yy}, the coefficient of output squared, be negative in sign and statistically different from zero; however, because marginal cost is a function of output, one must ensure that its magnitude does not yield a negative marginal cost, which is given by

$$\partial C(Y, \boldsymbol{p})/\partial Y = \alpha_y + \alpha_{yy} Y. \tag{4.21}$$

Estimation Results—Quadratic Cost Model

Estimation results for the quadratic cost model are displayed in Equation (4.20′):

$$C = -14{,}417 + 73.02^*Y - 0.017^*Ysq - 143.91^*p_k + 110.12^*p_l + 319.48^*p_p$$

$$(-4.92) \quad (28.94) \; (-8.59) \quad (-0.55) \quad (1.86) \quad (6.96)$$

$$\tag{4.20′}$$

The adjusted R^2 of 0.92 indicates that the model fit has improved over the previous specification. Indeed, the signs of the estimated coefficients on the output variables are of the expected sign and statistical significance. However, a review of the other estimated coefficients gives cause for concern, particularly that of capital price, which is now negative (although not statistically different from zero). Furthermore, White's test[2] indicates that heteroscedasticity is still an issue. These examples illustrate some of the issues involved in modeling costs. Exercises at the end of this chapter provide hands-on experience in dealing with some of these practical, real-world issues. Subsequent chapters investigate other, more sophisticated cost specifications, including the Cobb–Douglas and translogarithmic, as they have been employed throughout the economics literature to model costs for electric utilities. In addition, the quadratic form developed by Greer (2003) and the Cubic form also by Greer (and introduced in this manuscript) will be explored in more detail. The latter is the subject of a case study presented in Chapter 6.

A BRIEF HISTORY OF COST MODELS AND APPLICATIONS TO THE ELECTRIC INDUSTRY

Cobb–Douglas Functional Form

One of the first to estimate cost models in the electric industry was Marc Nerlove (1963), who employed the dual to the Cobb–Douglas production

---

[2] For Equation (4.20′), the regression that generated the White test statistic is given by

$$(e_i)^2 = \alpha_0 + \alpha_y Y + 1/2^* \alpha_{yy} Y^2 + \beta_l p_l + \beta_k p_k + \beta_p p_p$$
$$+ \beta_{ll}(p_l)^2 + \beta_{kk}(p_k)^2 + \beta_{pp}(p_p)^2 + \beta_{lk} p_l p_k + \beta_{kp} p_k p_p + \beta_{pl} p_p p_l \tag{4.21a}$$
$$+ \alpha_{yl} Y p_l + \alpha_{yk} Y p_k + \alpha_{yp} Y p_p + 1/2^*(\alpha_{yyl} Y^2 p_l + \alpha_{yyk} Y^2 p_k + \alpha_{yyp} Y^2 p_p + 2^* \alpha_{yyy} Y^2 Y),$$

which yields a White's test statistic $= nR^2 = 539.4$. The null hypothesis (homoscedasticity) is rejected and the presence of heteroscedasticity is confirmed. Note: Most econometric software packages will calculate the White statistic and indicate whether the null hypothesis of homoscedasticity can be rejected.

function, which was introduced in the seminal paper by Charles Cobb and Paul Douglas in 1928. In general, a production function summarizes the relationship among inputs and output. More specifically, the production function (f) indicates the maximum possible output (y) given any combination of inputs (x_i, $i = 1,..., n$); that is,

$$y = f(x_1, x_2, ..., x_n; A),\tag{4.22}$$

where A is a technical knowledge variable, which reflects improvements in technology and human capital.

Succinctly put, "the fundamental principle of duality in production: the cost function of a firm summarizes all of the economically relevant aspects of its technology" (Varian, 1992, p. 84). In other words, a production function can be recovered from a cost function, and vice versa.

One of the most commonly used Cobb–Douglas production functions is given by

$$Y = A^* K^\alpha L^{1-\alpha},\tag{4.23}$$

where Y is value-added output, A is technological knowledge, K is capital (input), L is labor (input), and α is a parameter to be estimated.

Returns to Scale

In the case of this particular function, it is clearly the case that there are constant returns to scale in the production technology (since $1 - \alpha + \alpha = 1$). Mathematically, in the two-input case [such as Equation (4.23)]:

$$ty = tf(x_1, x_2) = f(tx_1, tx_2).\tag{4.24}$$

In general, if all inputs are scaled up by some amount t (a scalar), then output would increase by t times. Similarly, increasing (decreasing) returns to scale exist if

$$ty = tf(x_1, x_2) > (<) f(tx_1, tx_2).\tag{4.25}$$

Nerlove's Cobb–Douglas Function

As the basis for his study, Nerlove employed a three-input (capital, labor, and fuel) cost model, which is the dual to the production function of the form

$$Y = A x_1^{\alpha 1} x_2^{\alpha 2} x_3^{\alpha 3}.\tag{4.26}$$

For the Cobb–Douglas production function, returns to scale (r) are equal to the sum of the exponents, or

$$r = \alpha_1 + \alpha_2 + \alpha_3.\tag{4.27}$$

After much algebra, the Cobb–Douglas cost function employed by Nerlove is given by (in natural logs)[3]

$$\ln C = \ln k + (1/r) \ln y + \sum_i (\alpha_i/r) \ln p_i \qquad (4.28)$$

for all $i = 1, 2, 3$, where $\ln C$ is the natural log of cost, $\ln y$ is the natural log of output, and $\ln p_i$ is the natural log of input prices, and

$$k = r \left[A \sum_i (\alpha_i^{\alpha i}) \right]^{-1/r} \qquad (4.29)$$

$$r = \sum_i \alpha_i \text{ (parameters to be estimated via regression)}. \qquad (4.30)$$

Substituting and arranging terms yields

$$\ln C^* = \beta_0 + \beta_y \ln Y + \beta_1 \ln p_1^* + \beta_2 \ln p_2^*, \qquad (4.31)$$

where

$$\ln C^* = \ln C - \ln p_3 \qquad (4.32)$$

$$\ln p_1^* = \ln p_1 - \ln p_3 \qquad (4.33)$$

$$\ln p_2^* = \ln p_2 - \ln p_3 \qquad (4.34)$$

$$\beta_0 = \ln k \qquad (4.35)$$

$$\beta_y = 1/r \qquad (4.36)$$

$$\beta_1 = \alpha_1/r \qquad (4.37)$$

$$\beta_2 = \alpha_2/r, \qquad (4.38)$$

which imply that

$$\alpha_1 = \beta_1 r = = > \alpha_1 = \beta_1/\beta_y \qquad (4.39)$$

and

$$\alpha_2 = \beta_2 r = = > \alpha_2 = \beta_2/\beta_y. \qquad (4.40)$$

From Berndt (1991), linear homogeneity in input prices (recall that this is one of the conditions for a properly specified cost model) implies that the constraint on the underlying parameters is given by

$$(\alpha_1 + \alpha_2 + \alpha_3)/r = 1 \qquad (4.41)$$

so that

$$\alpha_3 = (1 - \beta_1 - \beta_2)/\beta_y \qquad (4.42)$$

---

[3]See Berndt (1991, pp. 69–71) for details.

(Berndt, 1991, p. 71). It is left as an exercise to check for linear homogeneity of the underlying input price parameters (α_1, α_2, α_3).

Economies of Scale

Analogous to the concept of returns to scale in production is that of economies of scale. You may recall from Chapter 2 that this is defined as the ratio of average cost to marginal cost, and economies of scale indicate the situation in which the cost of producing an additional unit of output (i.e., the marginal cost) of a product decreases as the volume of output increases. That is, an $x\%$ increase in all inputs yields a more than $x\%$ increase in output. For the purposes here, it is often more useful to define the degree-of-scale economies, which is given by

$$SCE(y) = C(y)/y\,C'(y), \tag{4.43}$$

which is equivalent to the ratio of average cost to marginal cost. Returns to scale are said to be increasing, constant, or decreasing as SCE is, respectively, greater than, equal to, or less than unity.

The degree-of-scale economies at y, SCE, is the elasticity of output at y with respect to the cost of producing y. Alternately, it is also the elasticity of output (at y) with respect to the magnitude of a proportionate (or any efficient) expansion in input levels [for more details, see Baumol et al. (1982)]. This concept will be extremely important (and relevant) in upcoming chapters.

Minimum Efficient Scale

Geometrically, Figure 4.3 displays the relevant regions of an average cost curve. As output is expanded, cost increases at a decreasing rate until average cost is at its minimum. Known as the minimum-efficient scale, this point indicates the optimal level of output for a firm (or firms) to produce. After this, diminishing marginal returns set in (i.e., marginal cost begins to rise, thus causing average cost to increase at an increasing rate). This is displayed in Figure 4.3.

This concept is extremely important because it is an important factor in determining the optimal size and number firms in an industry. As such, it can have major implications for public policy, and regulatory considerations.

Nerlove's Results

Nerlove estimated the parameters of this Cobb–Douglas cost model [Equation (4.31)] via ordinary least squares, which work well under certain conditions.

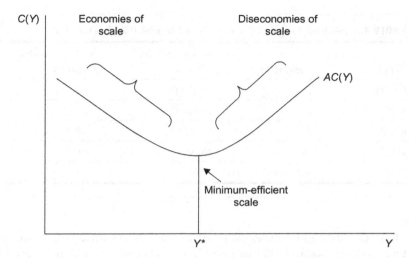

FIGURE 4.3 Minimum efficient scale and optimal industry output. Y^* indicates that output at which average costs are minimized, also known as the minimum efficient scale.

Before reviewing his results, let us briefly discuss the a priori expectations of the estimation results.

A Priori Expections

First and foremost, the estimated parameter (or coefficient) of the output variable, β_y, should be positive, as an increase in output should always increase total cost (i.e., monotonicity in output). Second, and also critical, is that the estimated parameters of the input price variables (β_1, β_2) should also be positive, as an increase in the price of an input should also increase total cost (i.e., nondecreasing in factor prices).

Turning now to Table 4.3, which contains the estimation results, you will note that the coefficient of the output variable (ln Y) is indeed positive and statistically significant (given the t statistic of 41.334). The interpretation is straighforward: A 1% increase in output yields a 0.72% increase in total cost, ceteris paribus. Similarly, the coefficient of ln p_1^* (β_1) is also positive in sign and statistically significant (keep in mind that the true parameter estimate of the price of the labor variable must be recovered [it is actually α_1, which is given by Equation (4.39)]). Finally, the estimated coefficient of the "other" input price ln p_2^* (β_2) is negative in sign (but not statistically different from zero). Again, one must recover the actual input price coefficients to determine whether this negatively signed coefficient is truly problematic (as it could be the case that α_2, $\alpha_3 > 0$ but $\alpha_2 < \alpha_3$ so $\beta_2 < 0$.). Using Equations (4.39)–(4.42), this is done in Exercise 4.5.

TABLE 4.3 Nerlove Original Data and Cost Model (Cobb–Douglas)

Variable	Parameter	Nerlove Model: Estimate*	Nerlove Model: t Statistic
Constant	β_0	−0.6908	−5.301
ln Y	β_y	0.72069	41.334
ln p_1*	β_1	0.59291	2.898
ln p_2*	β_2	−0.00738	−0.039
	Adjusted R^2	0.93	

Exercises at the end of the chapter use Nerlove's data to recover the parameter estimates of the original production function. In addition, you will examine data to look for outliers, which may be the cause of any econometric issues that may be present.

In Chapter 5 (cost models), there is an end-of-chapter exercise in which you will use data supplied by the Rural Utilities Service on the 711 cooperatives that distributed electricity in the United States in 1997 to estimate the parameters of a Cobb–Douglas cost function and calculate the returns to scale implied by the estimation results.

Elasticities

Some of the drawbacks of the Cobb–Douglas functional form were discussed earlier in the chapter. Because input prices were omitted from that discussion, the concept of substitution elasticities among inputs was avoided but will now be broached because we have included input prices into the cost specification. With this said, another concern about the Cobb–Douglas production function was that substitution elasticities among inputs are always equal to unity. That is, in Nerlove's model, substitution elasticities between capital and labor, capital and fuel, and labor and fuel always equal unity. Formally, the elasticity of substitution between inputs, which measures the degree of substitutability between inputs (x_i, x_j), is given by

$$\sigma_{ij} = \partial \ln (x_i / x_j) / \partial \ln (P_i / P_j), \qquad (4.44)$$

where P_i, P_j are the marginal products of x_i, x_j.[4]

---

[4]The marginal product (P_i) of an input is equal to the partial derivative of y with respect to that input (x_i):

$$\partial y / \partial x_i = P_i. \qquad (4.44a)$$

Constant Elasticity of Substitution Functional Form

To get around this limitation, an important paper by Arrow et al. (1961) showed that by solving for $\partial \ln(x_i/x_j)$ and then integrating Equation (4.44) yields

$$\ln(x_i/x_j) = \text{constant} + \sigma \ln(F_i/F_j),$$ (4.45)

where F_i/F_j is the marginal rate of technical substitution (MRTS) between x_i and x_j.

The integral of the MRTS yielded the implied production function, which is known as the constant elasticity of substitution (CES) production function. Imposing constant returns to scale, this is given by [for details, see Berndt (1991), p. 452)]

$$Y = A^*[\delta x_i^{-\rho} + (1-\delta)x_j^{-\rho}]^{-1/\rho},$$ (4.46)

where $\sigma = 1/(1 + \rho)$. The Cobb–Douglas is a special case of the CES function in the limiting case in which $\rho \to 0$, $\sigma \to 1$.

This particular form is not well suited for empirical analyses due to its nonlinear form. There is only one instance of which the author is aware that a CES production function was used to estimate economies of scale in the distribution of electricity. In his 1987 study, Claggett employed it to estimate economies of scale for 50 TVA electric distribution cooperatives. Specifically, he estimated a two-input equation of the form

$$Y = \tau^*[\delta K^{-\rho} + (1-\delta)L^{-\rho}]^{-1/\rho},$$ (4.47)

where τ is the efficiency parameter, δ is the intensity parameter, and ρ are parameters to be estimated.

As stated, this form is somewhat problematic in that it is highly nonlinear in the parameter ρ, which must be estimated to calculate the elasticity of substitution between capital and labor. As such, it cannot be estimated via ordinary least squares.[5] Nonetheless, Claggett found that there were increasing economies of scale in distribution and concluded that the cooperatives in his data set were too small (in terms of the quantity of electricity distributed) to be truly efficient (i.e., attain the minimum-efficient scale) from a cost-minimization perspective. This conclusion has been reached by others, including Greer (2003).

Generalized Leontief Cost Function

It was not until 1971 that Erwin Diewert employed Shephard's duality theory to estimate a general Leontief cost function, a flexible cost function associated with the Leontief form of production technology. In his classic article, Diewert

---

[5]However, more recently developed econometric software programs are capable of estimating such equations that use more sophisticated modeling techniques (Proc Model in SAS™, for example).

provided to researchers functional forms that placed no a priori restrictions on substitution elasticities but were consistent with economic theory. Among the first to implement the generalized Leontief cost model empirically were Berndt and Wood (1975) in a study of the U.S. manufacturing industry from 1947 to 1971. In this study, a four-input, multiple-equation system was estimated, which allowed both price and substitution elasticities to vary among the inputs. The form of the cost equation is described in detail later, but first an overview of the Leontief production technology and its underlying assumptions is provided.

ASIDE: Leontief Production Technology[6]

The implied L-shaped isoquants[7] of the Leontief production function are shown in Figure 4.4. Such technology is referred to alternatively as "fixed proportions," "no substitution," or "input–output" technology (or some iteration thereof). At any particular output level (Y^*) there is a necessary level of capital (K^*) and labor (L^*) that cannot be substituted. Note that these levels are determined purely technologically. Increasing only labor inputs (from L^* to L' for instance) will *not* result in any higher output. Rather, the extra labor, without the extra capital to work with, will be entirely wasted. The implication is that fixed proportions technology is "no less than a formal rejection of the marginal productivity theory. The marginal productivity of any [factor] ... is zero" (Leontief, 1941, p. 38).

The production function for a no-substitute case can be written as

$$Y = \min(K/v, \, L/u), \tag{4.48}$$

---

[6]Source: http://cepa.newschool.edu/het/essays/product/technol.htm.

[7]An *isoquant* (derived from quantity and the Greek word iso, meaning equal) is a contour line drawn through the set of points at which the same quantity of output is produced while changing the quantities of two or more inputs. While an indifference curve helps answer the utility-maximizing problem of consumers, the isoquant deals with the cost-minimization problem of producers. Isoquants are typically drawn on capital-labor graphs, showing the trade-off between capital and labor in the production function, and the decreasing marginal returns of both inputs. Adding one input while holding the other constant eventually leads to decreasing marginal output, which is reflected in the shape of the isoquant. A family of isoquants can be represented by an *isoquant map*, a graph combining a number of isoquants, each representing a different quantity of output. An isoquant shows that the firm in question has the ability to substitute between the two different inputs at will to produce the same level of output. An isoquant map can also indicate decreasing or increasing returns to scale based on increasing or decreasing distances between the isoquants on the map as output is increased. If the distance between isoquants increases as output increases, the firm's production function is exhibiting decreasing returns to scale; doubling both inputs will result in placement on an isoquant with less than double the output of the previous isoquant. Conversely, if the distance decreases as output increases, the firm is experiencing increasing returns to scale; doubling both inputs results in placement on an isoquant with more than twice the output of the original isoquant.

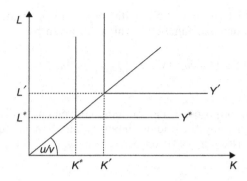

FIGURE 4.4   Leontief production technology and Isoquants.

which is also referred to as a *Leontief production function*—as this form was introduced by Wassily Leontief in 1941. Note that if K is at K^* and L is at L', then

$$K^*/v < L'/u. \text{ Thus, } Y = K^*/v. \tag{4.49}$$

If so, then the technically efficient level of labor would, by definition, be where

$$K^*/v = L/u \quad \text{or} \quad L = (u/v)K^*, \tag{4.50}$$

which, as is obvious in Figure 4.4, is at L^*. As a result, then it is the case that the following holds all along the ray from the origin:

$$Y/L = (1/v)K/L. \tag{4.51}$$

Leontief Cost Function

Again via duality, the generalized Leontief cost function is given by

$$C = Y^* \left[\sum_i \sum_j \beta_{ij} (p_i p_j)^{1/2} \right], \tag{4.52}$$

where $\beta_{ij} = \beta_{ji}$ (by Young's theorem—or the symmetry of second derivatives)[8] and $i, j = 1, \ldots, n$.

---

[8]Formally, Young's theorem states the following (source: http://www.economics.utoronto.ca/osborne/MathTutorial/CLN.HTM): Let f be a differentiable function of n variables. If each of the cross-partials f''_{ij} and f''_{ji} exists and is continuous at all points in some open set S of values of $(x_1, \ldots, x_n)$, then

$$f''_{ij}(x_1, \ldots, x_n) = f''_{ji}(x_1, \ldots, x_n) \text{ for all } (x_1, \ldots, x_n) \text{ in } S. \tag{4.52a}$$

Differentiating Equation (4.52) with respect to p_i yields optimal, cost-minimizing input demand functions, which are given by

$$\partial C/\partial p_i = x_i = Y^* \left[\sum_j \beta_{ij} \left(p_j/p_i \right)^{1/2} \right] \qquad (4.53)$$

for $i, j = 1, \ldots, n$.

For estimation purposes, it may be more convenient to divide through by output (Y), thus yielding the optimal input–output demand equations, which are denoted by a_i (Berndt, 1991) and are given by

$$a_i = x_i/Y = \sum_j \beta_{ij} \left(p_j/p_i \right)^{1/2} \qquad (4.54)$$

for $j = 1, \ldots, n$.

Hicks–Allen Partial Elasticities of Substitution

As stated previously, one of the attractive features of flexible functional forms like this one is that they place no a priori restrictions on the substitution elasticities between inputs. The Hicks–Allen partial elasticities of substitution for a general dual-cost function with n inputs are computed as

$$\sigma_{ij} = C^* C_{ij} / C_i^* C_j, \qquad (4.55)$$

where i, j are the first and second partial derivatives of the cost function with respect to input prices (p_i, p_j) and $i, j = 1, \ldots, n$.

We will revisit substitution elasticities in more detail in Chapter 5. For now, we proceed onto one of the more popular flexible functional forms, the translogarithmic cost specification.

Translogarithmic Cost Function

Around the same time, Lauritis Christiansen, Dale Jorgenson, and Lawrence Lau presented a paper introducing the translogarithmic (translog) specification for production and cost functions, which placed no a priori restrictions on the substitution elasticities. The translog function is a second-order Taylor's series approximation to any arbitrary cost function (in logarithms).[9] Christensen and Greene (1976) employed this cost specification in their seminal paper, "Economies of Scale in U.S. Electric Power Generation." Their study used Nerlove's 1955 and 1970 data for those same firms to estimate economies of scale in the U.S. electricity

---

[9]To expand a function $y = f(x)$ around a point x_0, means to transform that function into a polynomial form in which the coefficients of the various terms are expressed in terms of the derivative values $f'(x_0)$, $f''(x_0)$, etc., all evaluated at the point of expansion, x_0 (Chiang, 1984, p. 254).

market. This study employed a single-output, three-input translog cost function of the form

$$\ln C = \alpha_0 + \alpha_y \ln y + \sum_i \beta_i \ln p_i + (1/2)\alpha_{yy}(\ln y)^2$$
$$+ (1/2)\sum_i \sum_j \phi_{ij} \ln p_i \ln p_j + \sum_i \omega_i \ln y \ln p_i, \tag{4.56}$$

where output ($\ln y$) and input prices ($\ln p_i$) enter linearly, as quadratics and as cross-products.

At the time, Nerlove recognized that the Cobb–Douglas model did not adequately account for the relationship between output and average cost [witness the negatively signed coefficient on capital price, which was derived in Equation (4.40′)]. Christensen and Greene augmented the data set with cost-share data to estimate the complete demand system, including the cost-minimizing factor demand equations (i.e., the optimal level of inputs—capital, labor, and fuel in this case), which are given by Shephard's lemma[10]:

$$x_i^* = \partial C(Y, \boldsymbol{p})/\partial p_i, \tag{4.57}$$

where $\boldsymbol{p}$ is a vector of input prices.

By differentiating logarithmically, the cost-minimizing factor cost-share equations are obtained, which are given by

$$s_i = \partial \ln C(Y, \boldsymbol{p})/\partial \ln p_i, \tag{4.58}$$

that is,

$$s_i = \partial \ln C/\partial \ln p_i = p_i x_i/C = \beta_i + \beta_{ii} \ln p_i + \omega_i \ln Y + (1/2)\sum_j \beta_{ij} \ln p_j \tag{4.59}$$

for $i \neq j$.

Defining cost shares, s_i, as the proportion of total cost allocated to each input (again fuel, capital, or labor), or

$$s_i = p_i x_i/C, \tag{4.60}$$

it follows that

$$\sum_i s_i = 1. \tag{4.61}$$

Thus, for the three inputs employed in Christensen and Greene's translogarithmic cost model (again fuel, labor, and capital), the respective cost-share equations are given by the following:

Fuel

$$s_f = \beta_f + \beta_{ff}\ln p_f + \omega_P \ln Y + (1/2)\sum_j \beta_{fj} \ln p_j, \tag{4.62}$$

where $j = k$ (capital), l (labor).

---

[10]Shephard's lemma states that the optimal, cost-minimizing demand for an input (or factor) can be derived by differentiating the cost function with respect to the price of that input.

Labor

$$s_l = \beta_l + \beta_{ll} \ln p_l + \omega_l \ln Y + (1/2) \sum_j \beta_{lj} \ln p_j, \tag{4.63}$$

where $j = f$ (fuel), k (capital).

Capital

$$s_k = \beta_k + \beta_{kk} \ln p_k + \omega_K \ln Y + (1/2) \sum_j \beta_{kj} \ln p_j, \tag{4.64}$$

where $j = l$ (labor), f (fuel).

As stated, a well-behaved cost function must be homogeneous of degree one in input prices, which means that, for example, doubling the price of an input doubles total cost. This implies that the following restrictions can be imposed on Equation (4.56):

$$\sum_i \beta_i = 1, \text{ and } \sum_i \beta_{ij} = \sum_j \beta_{ji} = \sum_i \omega_{iy} = 0.$$

Finally, symmetry, which implies that (by Young's theorem)

$$\beta_{ij} = \beta_{ji}, \tag{4.65}$$

can also be imposed on the cost model, which reduces the number of parameters to be estimated and results in the (single-output) final form to be estimated. This is given by

$$\ln C = \alpha_0 + \alpha_Y \ln Y + \sum_i \beta_i \ln p_i + \alpha_{YY} \ln Y^2$$
$$+ \sum_i \omega_{iy} \ln Y \ln p_i + \sum_i \sum_j \beta_{ij} \ln p_i \ln p_j. \tag{4.66}$$

Once the restrictions are imposed, the system of equations (4.56 and 4.62–4.64) can then be estimated simultaneously by Zellner's seemingly unrelated regression method, which is described here.

Note: Because the cost shares (s_i) sum to unity, only $n-1$ of the share equations are linearly independent, which implies that the covariance matrix is singular and nondiagonal (i.e., does not have an inverse). As such, the parameters of the equations cannot be estimated. The solution is to divide through by one of the input prices (thus deleting one of the share equations).

Dividing through by (p_k), the capital price input variable, yields the share equations for fuel and labor:

$$s_f = \beta_f + \beta_{ff} \ln (p_f/p_k) + \omega_{fy} \ln Y + \beta_{fl} \ln (p_l/p_k) \tag{4.67}$$

$$s_l = \beta_l + \beta_{ll} \ln (p_l/p_k) + \omega_{ly} \ln Y + \beta_{lf} \ln (p_f/p_k). \tag{4.68}$$

This particular form has been widely used and still remains a chosen specification for numerous studies in the electric industry, particularly those testing for scale economies and the appropriate structure of the industry. [Ramos-Real (2004) provides a nice overview of these studies.] As such, Chapter 5 (cost models) contains a series of examples/exercises in which you will employ data on the requisite variables to estimate various translogarithmic cost models for the rural electric cooperatives that distributed electricity in 1997. In addition, the concepts of economies of scale and scope described in Chapter 2 (natural monopoly) are examined in much more detail. Examples and exercises are provided for hands-on experience working with cost models of this form.

Digression: Use of Zellner's Method (Seemingly Unrelated Regressions Method)

In the translog cost function estimation literature, the most popular estimation technique is that of Zellner's iterated seemingly unrelated regression. One nice feature of the translog specification is that, via Shephard's lemma, the optimal input demand equations can be derived and then estimated simultaneously with the cost equation via this method, which yields estimates that are more efficient than by the ordinary least squares equation. This method requires an estimate of the cross-equation covariance matrix, which increases the sampling variability of the estimator and yields estimates that are numerically equivalent to the maximum likelihood estimator (Berndt, 1991, p. 463). Before estimation can proceed, however, the following precautions must be made:

1. Because the shares always sum to unity and only $n - 1$ of the share equations are linearly independent, for each observation the sum of the disturbances across equations must always equal zero. This implies that the disturbance covariance matrix is singular and nondiagonal. Thus, one of the equations must be deleted and its parameters inferred from the homogeneity condition. This raises the question of whether the parameter estimates are invariant to the choice of the equation to be dropped. However, as long as either maximum likelihood or Zellner's method (one step or iterated seemingly unrelated regressions) estimation is performed, then the estimates are invariant to the choice of the equations to be estimated.
2. To preserve the linear homogeneity of the system, both cost-share equations must be normalized by dividing each input price by the input price that corresponds to the deleted cost-share equation (in this case the price of capital, p_k). Thus, the remaining cost-share equations take the form of Equations (4.67) and (4.68).

Quadratic Cost Models

Flexibility notwithstanding, the translogarithmic functional form does have its limitations. As described earlier, the inability of the translog to deal with zero levels of outputs (or input prices) has been considered a serious flaw by many. The quadratic model specification offers a nice alternative, exhibiting the flexibility of the translog while conforming to the properties of economic theory. In fact, for multiple-output markets, it is far superior to the translog. This too is discussed in much more detail in Chapter 5. But for now, let it suffice to introduce the general form (single output with input prices) and a discussion of its salient properties.

In general, the quadratic cost function is given by

$$C = \alpha_0 + \alpha_y Y + (1/2)\alpha_{yy} Y^2 + \sum_i \beta_i p_i, \tag{4.69}$$

where Y is output (in this case, electricity), Y^2 is output squared, and p_i is input prices.

Cost models often include other variables (known as cost-shift variables). In the case of electricity, the cost of distribution often includes miles of transmission, miles of distribution, and/or the number of customers per mile (also known as density). The quadratic model specification is the subject of much detail, examples, and exercises in Chapter 5 (cost models). An extension to multiple-output models is also included in this chapter.

Digression: Reasons That the Quadratic Cost Functional Form Is Well Suited for Modeling Industry Structure

In the econometric literature, the quadratic form has been gaining popularity in recent years because of its favorable properties, especially where multiple outputs are concerned. Let us now examine these properties in more detail.

Multiple-Output Quadratic Cost Function

In general, a multiple-output quadratic cost function is given by

$$C = \alpha_0 + \sum_i \alpha_i Y_i + (1/2)\sum_i \sum_j \alpha_{ij} Y_i Y_j + \sum_i \beta_k p_k, \tag{4.70}$$

where i, j, and $k = 1, \ldots, n$.

In the case of the quadratic form, marginal cost is given by

$$\partial C / \partial Y_i = \alpha_i + \sum_j \alpha_{ij} Y_j \tag{4.71}$$

and average cost by

$$C / Y_i = \left(\alpha_0 + \sum_i \alpha_i Y_i + (1/2)\sum_i \sum_j \alpha_{ij} Y_i Y_j + \sum_i \beta_k p_k\right) / Y_i. \tag{4.72}$$

Degree-of-Scale Economies

Recalling that the degree-of-scale economies, S_N, is equal to the ratio of average cost to marginal cost, in the multiple-output case we have

$$S_N(Y) = C(Y)/Y_i C_i(Y), \tag{4.73}$$

where $C_i(Y)$ is the marginal cost with respect to Y_i (Baumol et al., 1982, p. 50).

Substituting Equations (4.71) and (4.72) into Equation (4.73) and ignoring input prices yields

$$S_N(Y) = \left(\alpha_0 + \sum_i \alpha_i Y_i + (1/2) \sum_i \sum_j \alpha_{ij} Y_i Y_j \right) \Big/ \left(\sum_i \alpha_i Y_i + (1/2) \sum_i \sum_j \alpha_{ij} Y_i Y_j \right) \tag{4.74}$$

so that

$$S_N > 1 \text{ (increasing returns to scale) as } \alpha_0 > 1/2 \sum_i \sum_j \alpha_{ij} Y_i Y_j. \tag{4.75}$$

Ray Average Cost

Relating this to the concept of ray average cost (RAC) allows us to envision this concept geometrically. Figure 4.5 displays the case in which ray average cost is U-shaped for any point Y^0, which is a point along a ray emanating from the origin and is a composite commodity (i.e., in the two-output case, a function of both Y_1 and Y_2). This U-shaped ray average cost occurs when

$$\alpha_0 > 0 \quad \text{and} \quad \sum_i \sum_j \alpha_{ij} Y_i^0 Y_j^0 > 0 \tag{4.76}$$

for any point Y^0 on the ray.

For economies of scale to hold throughout the ray, it is necessary that

$$\sum_i \sum_j \alpha_{ij} Y_i^0 Y_j^0 < 0, \tag{4.77}$$

which, of course, implies that

$$\alpha_{ij} < 0. \tag{4.78}$$

Product-Specific Returns to Scale

You may recall from Chapter 2 the discussion on product-specific returns to scale and its relation to average incremental costs, which are relevant for multiple-output markets. For the quadratic form given in Equation (4.70), the degree of product-specific returns to scale is given by

$$S_i(y) = (\alpha_i Y_i + (1/2)\alpha_{ii} Y_i^2 + \sum_{j \neq i} \alpha_{ij} Y_i Y_j)/(\alpha_i Y_i + \alpha_{ii} Y_i^2 + \sum_{j \neq i} \alpha_{ij} Y_i Y_j), \tag{4.79}$$

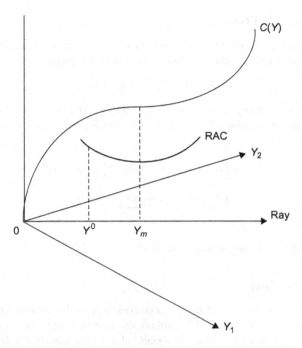

FIGURE 4.5   U-shaped RAC, which describes the behavior of cost along a ray emanating from the origin. In this case, the minimum occurs at the point of inflection of the total cost curve, $C(Y)$.

which implies that

$$S_i > 1 \text{ as } 0 > \alpha_{ii}. \tag{4.80}$$

Likewise,

$$S_i < 1 \text{ as } 0 < \alpha_{ii}. \tag{4.81}$$

As such, the average incremental cost of product i is globally declining, constant, or rising as α_{ii} is negative, zero, or positive, respectively (Baumol et al., 1982, p. 454). The interpretation is straightforward. Recalling Equation (2.22) from Chapter 2:

$$S_i(y) = \text{AIC}(y_i)/(\partial C/\partial y_i). \tag{4.82}$$

When $S_i < 1$, then the average incremental cost of product i is less than its marginal cost of production. Under the assumption that the price of product i is *at least* equal to its marginal cost, then the revenue from the production of product i exceeds its average cost, which improves the financial viability of the firm.

Economies of Scope

Recalling from Chapter 2 that economies of scope (or joint production) are integral to discussions regarding efficient industry structure (recall that economies of scope are a necessary condition for natural monopoly in a multiple-output firm). For Equation (4.70), the degree of economies of scope is given by

$$S_c = [C(Y_1, 0) + C(0, Y_2) - C(Y_1, Y_2)]/C(Y_1, Y_2). \tag{4.83}$$

The quadratic form is well represented in the literature in the estimation of economies of scale and scope, vertical integration, and subadditivity in the electric utility industry. Notable is a paper by John Mayo (1984), who employed multiple-output quadratic cost models to test whether there were economies of scope in the distribution of electricity and natural gas. To accomplish this, he estimated an equation of the form:

$$C = \alpha_0 + \sum_i \alpha_i Y_i + (1/2)\sum_i \sum_j \alpha_{ij} Y_i Y_j + \sum_k \beta_k p_k, \tag{4.84}$$

which is not homogeneous of degree one in input prices. To remedy this, Mayo multiplicatively appended input prices in the following fashion:

$$C = (\alpha_0 + \sum_i \alpha_i Y_i + (1/2)\sum_i \sum_j \alpha_{ij} Y_i Y_j)\prod_k \beta_k p_k. \tag{4.85}$$

Mayo imposed strict input–output separability[11] and linear homogeneity in input prices (recall that separability was rejected by Karlson). According to Baumol and coworkers (1982, p. 458),

Such cost functions (with p and y multiplicatively separable) require all input demands to vary with outputs in the same fashion. In fact, input ratios and cost shares are thus assumed to be independent of output levels, and input demand elasticities with respect to each output become equal to independent of p. Consequently, multiplicatively separable cost functions are not well suited for investigation of those properties of input demand functions that relate to the effects of input prices on industry structure.

Thus, we have another example of the tradeoff between tractability in empirical analysis of the functional form chosen for the cost relationship and its usefulness in testing the many properties that theory suggests are important for industry analysis.

---

[11]Chambers (1988) gives the necessary conditions for input–output separability for the profit-maximizing producer as

$$\partial(x_i/x_j)/\partial p = 0 \tag{4.85a}$$

$$\partial(y_i/y_j)/\partial r = 0. \tag{4.85b}$$

The first condition implies that a change in output prices, p, does not influence the composition of inputs x_i and x_j, while the second condition implies that a change in input prices, r, does not influence the composition of outputs y_i and y_j. Rejecting input–output separability means that a change in input (output) price alters the composition of output (input) quantities.

Greer (2003) offers an improvement over Mayo's quadratic cost model in that it is strictly concave in input prices (not equal to zero like the cost model presented by Mayo in his 1984 paper "The Multiproduct Monopoly, Regulation, and Firm Costs"). In this cost model, input prices (and other variables) enter multiplicatively and are nonlinear in parameters, which allows second derivatives with respect to input prices to take on nonzero values. (A proof is offered in the appendix to this chapter.) As such, Greer's is a properly specified cost model in that it conforms to all properties of economic theory. The cost model is given by

$$C = (\alpha_0 + \sum_i \alpha_i Y_i + 1/2 * \sum_i \sum_j \alpha_{ij} Y_i Y_j) \prod p_m^{\beta_m} e^\varepsilon \qquad (4.86)$$

for $i, j = 1, 2$ and $m = 1, 2, 3$.

In this case, input price parameters (β_m) enter nonlinearly, which allows for the concavity in input price criterion to be satisfied (as long as the parameter estimates are of the expected sign. (A proof is offered in the Appendix to this chapter.)

This model is monotonic (increasing) in output, increasing and concave in input prices, and capable of estimating the parameters despite a variable taking on a value of zero (recall the translogarithmic cost function's inability to do so). As such, it is a properly specified multiple-output cost function unique to the literature. (Additional proofs are offered in the appendix to Chapter 5.)

Cubic Cost Models

It is a well-established fact that total cost functions are cubic in nature; that is, there is a region of increasing returns, constant returns, and decreasing returns to scale, which yield the classic "hook-shaped" marginal cost and U-shaped average cost curves displayed in Figure 4.6.

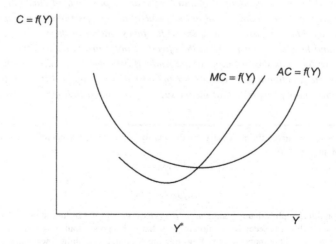

FIGURE 4.6 Average and marginal cost curves for cubic total cost function.

Also displayed in Chapter 2, the cubic cost function generates the average (*AC*) and marginal (*MC*) cost curves displayed in Figure 4.6. For $Y < Y^*$, marginal cost declines and pulls average cost down with it; this is the region of the total cost curve in which cost is rising at a decreasing rate, which generates increasing returns to scale. Once diminishing returns set in, marginal costs rise and eventually cause average cost to rise as well, which occurs at Y^* when total costs begin to increase at an increasing rate.

That is, total cost increases at a decreasing rate (marginal cost declines with output) and then increases at an increasing rate (marginal cost increases with output), thus causing average cost to rise. This is also analogous to Figure 4.3.

These are the classic "textbook" examples that all principles of economics students are taught: marginal cost declines in the increasing returns-to-scale portion of the total cost curve, which causes average cost to decline; when decreasing returns to scale set in, marginal cost begins to rise, thus causing average cost to rise (marginal cost cuts the average cost curve at its lowest point), which results in the classic U-shaped average cost curve.

As stated in Chapter 2, the cubic form of the cost equation is the appropriate specification for generating average and marginal cost curves that accord to economic theory. The author has began experimenting with the cubic form of her quadratic cost model, which is given by

$$
\begin{aligned}
C = [\alpha_0 + \alpha_1 Y_1 + \alpha_2 Y_2 + 1/2(\alpha_{11}Y_1{}^2 + \alpha_{22}Y_2{}^2 + 2^*\alpha_{12}Y_1 Y_2) \\
+ 1/3(\alpha_{111}Y_1{}^3 + \alpha_{222}Y_2{}^3) + \alpha_{112}Y_1{}^2 Y_2 + \alpha_{221}Y_2{}^2 Y_1]\prod w_i{}^{\beta i}\prod O_m{}^{\theta m}e^{\varepsilon}
\end{aligned}
\tag{4.87}
$$

for outputs Y_1 and Y_2. This cost model is the subject of a case study presented in Chapter 6.

CONCLUSION

This chapter provided a rather comprehensive overview of cost models, concepts, and some of the econometric issues that one faces in the estimation of such models. It is a natural precursor to the next chapter, which is devoted solely to cost models and their estimation. Several exercises, along with the appropriate data sets, are provided to give the interested reader a hands-on experience with estimating these cost models and the author encourages you to try them.

APPENDIX

This is a proof that Equation (4.86) is concave in input prices, that is, own prices are nonpositive.

By Shephard's lemma, the optimal factor demand equation for the *i*th input price is given by

$$
x_i^* = \partial C(Y, p)/\partial p_i,
\tag{4.88}
$$

where $i = 1, 2$.

The Hessian matrix is the matrix of second derivatives of the cost function with respect to input prices, which is equivalent to the matrix of first derivatives of the factor demand equations, x_i^*, with respect to input prices.

Because cost minimization requires that the matrix of first derivatives of the factor demand equations be negative semidefinite, which requires that the diagonal terms of the Hessian matrix be nonpositive, define:

$$C_{ii} = \partial^2 C(p, Y) / \partial p_i^2 \qquad (4.89)$$

for the ith input and

$$C_{jj} = \partial^2 C(p, Y) / \partial p_j^2 \qquad (4.90)$$

for the jth input, and that

$$C_{ij} = \partial^2 C(p, Y) / \partial p_i \partial p_j, \qquad (4.91)$$

where i, j represent the cross-price effects (the nondiagonal elements of the matrix).

Then the (Bordered) Hessian matrix, H, is given by

$$H = \begin{vmatrix} C_{ii} & C_{ij} & H_1 \\ C_{ji} & C_{jj} & H_2 \\ H_1 & H_2 & 0 \end{vmatrix}, \qquad (4.92)$$

where H_1, H_2 are constraints (on output) that result from the firm's optimization problem, which is to minimize cost subject to satisfying the demand for its output, that is,

$$\text{minimize cost} \, (Y, p) \qquad (4.93)$$

subject to $Y \geq$ the demand for the firm's output.

EXERCISES

1. One of the causes of heteroscedasticity is that there are large changes in the explanatory variables from one observation to another, which likely contributes to the large cost differences among the coops in the sample. For example, the price of purchased power exhibited large differences, from a minimum of $18.00/MWh to a maximum of $90.00/MWh with a mean value of $43.00 (Figure 4.1).
   a. Create a data set named "RUS97_low" where Cost < $10,000 and use this data set to estimate the parameters of Equation (4.20). Do the results accord with economic theory in terms of expected signs and statistical significance? Why or why not?
   b. Calculate the White test statistic. Is heteroscedasticity still an issue?

c. In the example that used the original data set to estimate the quadratic equation [Equation (4.20)] there was a potential issue with the coefficient of the price of capital, p_k. Has the new data set corrected this issue?

d. Examine the summary statistics in this newly created data set. Are there any other variables that could be causing the heteroscedasticity?

2. You will note that there are three types of power suppliers in the data set entitled "RUS97_Basic": generation and transmission cooperatives (G&Ts), federally owned entities (Federal), and investor-owned utilities (IOUs).

   a. Create a data set for each type of supplier and review the summary statistics. Do you see the same variation in the variables across the observations in each data set? Would you expect to see more or less variation in the price of purchased power among the three supplier types? Why or why not?

   b. Create a data set from "RUS97_low" in which supplier = "G&T." Reestimate Equation (4.20) and review the results.

      i. Calculate the White statistic—is the presence of heteroscedasticity still confirmed?

      ii. What about the other coefficient estimates—are they as expected (in terms of sign and statistical significance)? Why or why not?

3. Unlike investor-owned utilities, rates charged by distribution cooperatives are not regulated in every state. In fact, fewer than 20 state utility commissions have jurisdiction over rates charged to various end users by cooperatively owned entities.

   a. Create two new data sets from "RUS97_low" with supplier = "G&T": one for regulated entities (REG = 1) and one for nonregulated entities (REG = 0). Examine the summary statistics of the relevant variables. Are they as expected? That is, is the price of purchased power generally lower or higher for regulated entities than for nonregulated entities?

   b. Again, estimate Equation (4.20) using the newly created data set in which firms are regulated. What does White's test statistic indicate now about the presence or absence of heteroscedasticity?

   c. What about the signs and statistical significance of the coefficients on the explanatory variables—do they accord with a priori expectations?

4. Open and examine the data set "Nerlove.xls."

   a. Do there appear to be any outliers?

   b. Does the price of fuel for firms in the sample exhibit the same degree of volatility as prices paid for power by the coops in "RUS97_Basic"?

   c. Is it necessarily the case that the total cost increases with output? Why or why not?

5. Using results obtained in Table 4.3, obtain the original parameters [i.e., those from the underlying production function, which is displayed in Equation

(4.26)]. Use these to estimate returns to scale for firms in Nerlove's data set. Are they increasing, constant, or decreasing?

6. In the appendix to this chapter there is proof that Equation (4.86) is concave in input prices. Using a similar approach, prove that Equation (4.87) is also concave in input prices.

Chapter 5

Cost Models

INTRODUCTION

This chapter delves more deeply into cost models and the concepts applicable to the efficient structure of the electric industry for both single- and multiple-output markets. A review of the literature is included, and data on rural electric cooperatives are used to estimate various functional forms, including translogarithmic, quadratic, and cubic cost specifications. Examples and exercises provide the interested reader with hands-on experience in understanding the somewhat complex concepts put forth in this chapter, and the appendix offers proofs that the quadratic and cubic models developed by Greer (2003, 2011) are appropriately specified and, hence, proper cost models from which robust results can be gleaned.

DETERMINATION OF AN APPROPRIATE OBJECTIVE FUNCTION: A BRIEF OVERVIEW OF THE LITERATURE

While numerous studies have attempted to measure economies of scale and scope, vertical integration, and subadditivity in the electric utility industry, they are not all conducted in the same manner nor do they consider the same type of firms. Some employ a production function, taking input quantities as exogenous and output as endogenous. Others have estimated the dual-cost function, where input prices and the level of output are exogenous. Whether to estimate a cost function or a production function typically depends on what type of data is available. Because much data in the electric utility industry are disaggregated, firm level data, it is preferable to employ a cost function in which input prices are regressors rather than a production function in which input quantities are the right-hand variables. Also, as Nerlove aptly noted, because electricity rates (in the United States) were set by regulators, they were exogenous so that cost rather than production functions were appropriate (Berndt, 1991). As such, the optimization problem is to choose inputs so that the cost of production is minimized given the input prices and the level of output, which is also exogenous, as regulated utilities have an obligation to serve native load as part of their franchise agreement with the state regulatory commission.

Some of the earlier studies employed relatively simple functional forms, such as the Cobb–Douglas (detailed in Chapter 4) and constant elasticity of substitution (CES) (also detailed in Chapter 4), to model cost or production technology. Unfortunately, these functional forms are rather limited in that they place a priori restrictions on the elasticities of substitution among the factors of production and the returns to scale inherent within this industry. To get around such restrictions, the translog functional form was introduced circa 1961 (Heady, 1961), which added quadratic and cross-product terms to a second-degree polynomial in logarithms and placed no a priori restrictions on substitution elasticities, a major breakthrough for empirical analysis.

It was not until 1971 that W. Erwin Diewert employed Shephard's duality theory to estimate a general Leontief cost function, a flexible cost function associated with the Leontief form of production technology. Among the first to implement this was a study of the U.S. manufacturing industry from 1947 to 1971 by Berndt and Wood (1975), who were among the first to model the interrelated demands in the energy industry (see Chapter 7).

Although most of the studies thus far have concentrated on privately-owned firms (i.e., investor-owned), there are a few that have examined firms that are publicly-owned (usually municipally) and attempted to quantify some of the differences between public and private ownership. Among the first were studies by Moore (1970) and Alchian and Demsetz (1972), whose property rights theory of the firm has been cited as the primary motivation in the differences between each ownership type. This theory was extended to electric utilities by both Pelzman (1971) and De Alessi (1974), who observed that, on average, the price of power supplied by publicly-owned utilities is lower than power from firms that are privately-owned. They concluded that price concessions are the chosen method of conferring political benefits.

Meyer (1975) employed a cost function in which output enters as a cubic function to examine scale economies in production, transmission, and distribution of public electric utilities versus private electric utilities. What was disconcerting about his study was that fixed costs did not seem to play much of a role in either transmission or distribution costs, despite the fact that both are capital intensive and remain the naturally monopolistic aspects of the industry (the constant is not statistically significant in either regression). This was reconciled by Neuberg (1977), who pointed out that Meyer only considered distribution operating costs—he did not include costs associated with maintenance and capital (nor did he include factor prices so the model likely suffered from omitted variable bias); as such, it is not surprising that he did not find any fixed cost effects. In addition, his model is of a linear additive form, which leads to the "a priori expectation of zero values for the constant term and shift parameter, as Meyer himself concedes."

Among the first to employ a multiproduct cost model was Neuberg (1977). In his study, there were four interdependent outputs: number of customers served, number of megawatt hours sold, size of distribution territory, and miles of overhead distribution line. He found evidence of increasing returns in distribution and that investor-owned firms were not more cost efficient than municipally-owned firms; in fact, he found that the opposite occurred in most of the various regressions he specified.

The next wave of studies focused primarily on power generation only; for example, studies by Pescatrice and Trapani (1980), Fare et al. (1985), and Hayashi et al. (1985) each found that publicly-owned systems do indeed have lower costs. Berry (1994) compared the costs of producing various three-product output bundles between cooperatives and investor-owned firms and found that investor-owned firms do so more efficiently.

As stated previously, most studies focused on investor-owned utilities (IOUs). In addition to the seminal paper by Christensen and Greene (1976), Mayo (1984) specified a multiproduct cost function of the quadratic nature for firms that produce both electricity and natural gas. He found that economies of scope prevailed for smaller firms but not for larger firms. His model is revisited later in this chapter. More recently, Kwoka (1996) examined, among other things, deregulation of the industry on both privately and municipally-owned firms. In a comprehensive study, he estimated a fully simultaneous system of pricing, cost, and demand equations to examine such issues as public versus private ownership, economies of scale and of vertical integration, and monopoly versus competition in the distribution of electricity. He found that,

Clearly, publicly-owned utilities have lower costs than comparable IOUs—5.5 percent lower overall. Moreover, their lower costs appear to arise in the distribution function, attesting to the "comparative advantage" of public systems in end-user tasks.

As comprehensive as Kwoka's study is, it ignored a rather large faction of the market for electric utility distribution systems: rural electric cooperatives. In fact, very few industry studies have even considered rural electric cooperatives, which is one way that Greer (2003, 2008, 2011) has distinguished her work from others.

RURAL ELECTRIC COOPERATIVES

Cooperative ownership is quite different from investor or public ownership. Some of these differences were discussed in *Electricity Cost Modeling Calculations* (Greer, 2011) and are repeated here for convenience.

Reasons That Cooperatively-Owned Utilities Are Different

In the United States, electric cooperatives (coops) are organized as either generation and transmission (G&Ts) or distribution only (member coops), with long-term, full-requirement contracts in place to render them quasivertically integrated. However, unlike IOUs, not all states regulate electric cooperatives. In fact, fewer than 20 states have jurisdiction over coops, and the degree of regulation is not consistent among these states. For example, in Florida, Indiana, Maine, and Mississippi, public regulatory commissions do not regulate the rates charged by coops. This inconsistency creates a special challenge for federal policy makers in the United States. Other differences between investor-owned utilities and coops include:

- Urban differences versus rural differences
- Institutional differences
- Regulatory differences
- Philosophical differences

Each of these is discussed in turn.

DIFFERENCES BETWEEN COOPS AND IOUs

Urban versus Rural

Electric cooperatives were created in 1936 by the Rural Electrification Act to serve those areas that profit-driven investor-owned utilities were unwilling to serve. Prior to the creation of this act, fewer than 12% of rural farm homes were electrified. By 1941, over 35% of these homes had been electrified (U.S. Bureau of the Census, 1975).

Rural areas are quite different from urban areas, which tend to be served by investor-owned utilities. As a result, the effects of electric restructuring on those customers served by rural electric cooperatives are different than those served in urban areas by IOUs. Given the special properties that characterize electricity (e.g., the fact that it is essentially nonstorable and that it follows the path of least resistance), coupled with the yet-to-be discussed differences between rural and urban areas, a standard, federally mandated cookie-cutter approach to deregulation of this industry will not work. These urban rural disparities are discussed here.

First, rural areas tend to be far less populated, with a terrain that is more rugged compared to urban areas, which render them more costly to serve. In addition, coops in rural areas serve mostly residential loads, which tend to be much smaller and more volatile in nature, often demanding power during peak times. Thus, on a per-customer basis, distribution coops face far higher costs than their investor-owned counterparts.[1]

In addition, rural incomes tend to be lower than those in urban areas. According to the Economic Research Service, U.S. nonmetropolitan incomes were 26% below those in metropolitan areas in 1996. Not surprising, rural adults tend to have less education than their urban counterparts and often face local labor markets that offer few opportunities to advance beyond low-paying entry-level jobs (Economic Research Service, Briefing Room, Rural Labor and Education).

---

[1]In 1996, the average rural distribution coop served fewer than 6 customers per mile of distribution line and received about $7000 per mile, while their investor-owned counterparts in urban areas served almost 40 customers per mile and received over $60,000 per mile of line. Furthermore, according to the National Rural Electric Cooperative Association, the average investment in distribution plant per consumer was $1975, which implies that coops invested $11,850 per mile versus $1535 for IOUs.

Institutional

Institutional differences exist between coops and IOUs in the United States. The first is the 85% rule, which applies to both G&T and distribution cooperatives. In the United States, both G&Ts and distribution cooperatives must obtain 85% of their revenues from their members or they will lose their tax-exempt status. This limitation has two significant effects. First, it effectively limits (and possibly even precludes, depending on capacity and native load) G&Ts from participating in the wholesale power market, even though their generation costs tend to be lower than those of investor-owned firms (due, in part, to their access to less expensive federally-produced power). Second, it severely limits the extent of retail competition that can take place because distribution cooperatives are subject to the same restriction for tax-exempt status; that is, 85% of revenues must come from their member end-use customers. As such, there would be a high cost penalty to rural areas for coops to participate in retail choice. Moreover, because the increase in cost due to the tax would be passed directly through to members, on a per-member basis, this could result in significant increases in rates given the low densities that tend to prevail in rural areas.

In addition to the tax implications of restructuring, a potential stranded cost issue must be examined. For G&Ts to obtain low-interest financing from the Rural Utilities Service, they are required to have long-term (30 years), full requirement contracts with their member distribution cooperatives. Because of deregulation, these contracts could become stranded as retail choice is made available to rural customers. These contracts will not be forgiven; they are a debt that must be repaid. As owners of the G&T, member cooperatives are responsible for any debt incurred by the G&T on their behalf; in other words, residential customers could end up bearing most of this burden, especially if larger users choose to leave the cooperative system and purchase power elsewhere. As they leave the cooperative system, the G&T's member load will be reduced, as will its revenues, and the possibility of not being able to repay this debt will become a reality (unless rates of the remaining customers increase enough to offset the lost revenues).

Furthermore, it is well documented that industrial users subsidize residential ones; despite contributing to the peak demand, demand charges are typically not assessed to residential customers but are imposed upon large commercial and industrial users. As retail choice is made available, these larger users may choose an alternative supplier, thus imposing the full cost of service onto those small users who may or may not have a choice of suppliers.[2] As a result, rural residential users are doubly harmed; because of the lack of density, these higher costs would be spread across even fewer customers, thus resulting in rather substantial increases in rural residential rates.

---

[2]According to Brockway (1997, p. 14), "in a competitive market, low volume, low income customers likely will be the last served....In fact, in a deregulated environment, firms uniformly may refuse to serve all customers with low incomes." Furthermore, recent experience from the telecommunications industry tends to confirm this.

Regulatory

As stated previously, not all states in the United States regulate cooperatives. In some of the states that do, the state's restructuring legislation includes an opt-out clause or other special protections for cooperatively-owned firms specifically included (e.g., the state of Texas). For states that do not regulate coops, opt-in clauses were included in for those coops that wanted to participate in retail choice (e.g., the state of Iowa). In other cases, cooperatives were required to participate but were on a delayed schedule (in the state of Delaware, for example, where coops were on a six-month delay from investor-owned utilities in the offering of retail access). The state of Maine provided an interesting twist: All consumers had the right to choose their electric supplier but the coops are prohibited from selling generation service outside their service territory. Finally, there are several states that regulate coops but offer neither opt-out clauses nor any special language written into any restructuring legislation that had been passed (e.g., in the state of Michigan, the coops were required to implement retail access on the same schedule as investor-owned utilities).

Philosophical

Cooperative ownership implies different incentives than those of either investor or even municipal ownership. Due to the coincidence of buyers and sellers, presumably cooperatives follow a welfare-maximizing strategy, entailing both producer and consumer surplus maximization. As a result, certain principal agent problems tend to arise that cannot be mitigated as easily as those in investor-owned firms. For example, in the case of the latter, in the United States there is often incentive pay (i.e., bonuses) or profit sharing and the ownership of stock. However, nonperformance of the firm may lead to the transfer of property rights (i.e., the selling of one's stock), a depressed stock price, and the loss of one's job. These motivating schemes are typically not offered to the managers of coops (the agents), which are nonprofit entities operating on a one-vote-per-member system. Hence, the "stock" of the firm confers no real property rights upon the principals (the consumer owners), and while there is the possibility of accruing capital credits, these credits cannot be received upon demand.

For now, suffice it to say that they have not been the subject of much research but provide an interesting alternative to studies performed on investor ownership. A brief review of the literature is provided here.

LITERATURE REVIEW: COST STUDIES ON RURAL ELECTRIC COOPERATIVES

Few studies of the electric utility industry include cooperatively-owned firms in their data sets. There is a group of authors, however, whose work tends to focus on the coops and on cooperatively-owned distribution systems in the continental

United States. Claggett (1987) uses a constant elasticity of substitution (CES) production function to estimate economies of scale for 50 cooperative distributors of the Tennessee Valley Authority (TVA), finds that there are increasing economies of scale in distribution, and concludes that the individual cooperatives in this group should be increased in size.[3]

Studies on property rights include Hollas and Stansell (1988) and Claggett et al. (1995), who employ a profit function to test for absolute and relative price efficiency for proprietary, cooperative, and municipal electric distributors in the United States. Not surprising, they find little evidence to support profit-maximizing behavior on the part of either cooperatives or municipals. Hollas and Stansell (1991) employ a similar methodology but limit the study group to rural electric distribution cooperatives. They include a binary variable to differentiate among those states that regulate coops and those that do not. As in their previous study, they find that coops are not absolute price efficient (i.e., do not maximize profits). The effect of property rights is examined further in Claggett (1994), who estimates a translogarithmic cost function and found that TVA cooperatives distribute power at a lower cost than do the municipal distributors of TVA power. Hollas et al. (1994) also limited their study to municipal and cooperative distributors of TVA power. Assuming welfare maximization, the authors reject the null hypothesis that there is no difference between rates charged by municipals and cooperatives, finding that municipally-owned firms charge lower rates to residential and commercial users but higher rates to industrial ones.

Berry's 1994 paper is a product of his dissertation, the purpose of which is to show that cooperatively-owned electric utilities are far less efficient than their privately-owned counterparts. However, his results are suspect since, to compare the results at various levels of output, Berry essentially adds the G&T's total cost to that of the distribution cooperatives in an attempt to provide a common structure to that of investor-owned firms in the sample. The problem herein is that the G&Ts and the distribution cooperatives in the sample are not necessarily related (in the sense that the distribution cooperatives in the sample are not necessarily member coops of the G&Ts in the sample), thus rendering this type of aggregation invalid.

Greer (2003) used data on distribution coops to estimate economies of scale and scope and determined that the average coop was too small (in terms of delivered electricity) to take advantage of the inherent economies that exist in distribution. Greer (2008) used data on coops (both G&T and distribution) to determine lost vertical economies due to the fact that coops are not truly vertically integrated. Both of these studies are included in *Electricity Cost Modeling Calculations* (Greer, 2011).

---

[3]His results confirm those of Gallop and Roberts (1981) and Neuberg (1977), who found that significant economies exist in both transmission and distribution.

DATA

Data employed to perform the exercises and examples in several chapters of this book come exclusively from the Rural Utility Service (RUS). Given the presence of almost 900 rural electric distribution cooperatives and 66 G&T cooperatives across the United States, such studies are warranted. Cooperatives are interesting entities, because while they are privately-owned and have some similarity to their investor-owned counterparts, they are not profit maximizing but rather follow a strategy of welfare maximization, which was defined in Chapter 1.

Given the nature of data, cost functions rather than production functions are appropriate for estimating the cost models that were reviewed in Chapter 4. For comparison purposes, several different models have been specified, and results are then used to calculate efficiency measures, which were detailed in Chapter 2.

REVIEW OF THE LITERATURE: COST FUNCTION ESTIMATION IN THE ELECTRIC UTILITY INDUSTRY

Economies of Scale

As detailed in Chapter 2, in the electric utility industry, numerous studies have employed single-output cost models in determination of the efficient structure of the industry. However, they pertain mostly to the generation of electricity only. Among those studies that estimate economies of scale in generation are Nerlove (1963), who employed a Cobb–Douglas cost model (defined in Chapter 4) and found that, in 1955, all but the very largest utilities experienced increasing returns to scale. In their seminal paper, Christensen and Greene (1976), using both Nerlove's 1955 data and also 1970 data, found that by 1970, most firms were generating electricity at (and some even beyond) that point at which economies of scale had been exhausted. They were among the first to employ the translogarithmic cost function, the properties of which are discussed in an upcoming section. In a later study, Huettner and Landon (1977), using an ad hoc, semilog quadratic cost function, confirmed the Christensen and Greene results, although they found that scale economies are exhausted at an even lower level of output. Finally, Atkinson and Halvorsen (1984) employed a shadow cost function (whereby firms base their decisions on shadow prices that reflect the effects of regulation on the effective prices of inputs). Upon reestimation of the Christensen and Greene model, they obtained a different result: They found that firms were not operating in the upward-sloping portion of the long-run average cost curve.

While these studies focus on the generation component, a few studies focus on either transmission or distribution alone, two of the three components, and all three components. Of those that focus on some combination of the components, most do so to study the economies associated with vertical integration, which is discussed later in this chapter. In virtually all of these studies, the consensus is that distributed electricity is not a homogeneous good. This was discussed further in the empirical chapter (Chapter 4), but for now suffice it to say that different end users have different elasticities of demand and that some users are more costly to serve than others.

NERLOVE'S COBB–DOUGLAS COST MODEL

As stated previously, among the first to estimate a cost function for the electric utility industry was Marc Nerlove, who in 1963 employed a Cobb–Douglas cost function to assess returns to scale in the generation of electricity. The form of the equation he estimated is given by

$$\ln C = \beta_0 + \beta_y \ln Y + \beta_1 \ln p_1^* + \beta_2 \ln p_2^*, \tag{5.1}$$

where p_i^* denotes transformed input prices. [See Chapter 4, Equations (4.26)–(4.40), for derivation.]

Using 1955 data, Nerlove found that there were significant scale economies in the generation of electricity for nearly all firms in the sample. Years later, Christensen and Greene used a more flexible functional form to reestimate Nerlove's model using both 1955 and 1970 data and found that, by 1970, most of the scale economies had been exhausted. The models by Nerlove and Christensen and Greene are the subject of the following examples and exercises.

Example 5.1 Estimating a Basic Cost Model

It was stated that, at the very least, a cost model is a function of output and input prices. In other words,

$$C = f(Y, \boldsymbol{p}). \tag{5.2}$$

The data set GT97 contains data on the 43 G&T cooperatives that were RUS borrowers and provided electricity in 1997. The summary statistics of the relevant variables are described in Table 5.1. (The reader should verify these.)

Employing the most basic model, estimate the parameters of the following equation:

$$\ln C = \beta_0 + \beta_y \ln Y + \beta_1 \ln p_1 + \beta_2 \ln p_2 + \beta_3 \ln p_3. \tag{5.3}$$

Next, use Nerlove's specification to estimate the parameters of the log-linear Cobb–Douglas specification given by

$$\ln C^* = \beta_0 + \beta_y \ln Y + \beta_1 \ln p_1^* + \beta_2 \ln p_2^*. \tag{5.4}$$

(Hint: You must create the variables $\ln C^*$, $\ln p_1$, and $\ln p_2^*$, which are derived in Chapter 4.) Table 5.2 displays the estimation results.

Using results from the basic cost model given by Equation (5.3), it is straightforward to calculate r (returns to scale) and determine the degree-of-scale economies for coops in the data set. Recalling from Chapter 4,

$$\beta_y = 1/r, \tag{5.5}$$

TABLE 5.1 Summary Statistics for 1997 RUS Borrowers—G&T Cooperatives

Variable	Name	Mean	Standard Deviation	Minimum	Maximum
Total cost (millions)	TC	214,145	216,549	5791	1,073,260
Y (Electricity in MWh)	GTY	4158	3823	24.00	15,250
Price of fuel ($/MWh)	gt_pf	20.26	36.12	0.00	222.32
Price of purchased power ($/MWh)	gt_pp	27.34	11.92	0.00	82.85
Average price of power (weighted average)	AvgPP_gt	28.73	8.37	0.00	44.15
Price of capital (%)	gt_pk	6.26	2.63	0.00	19.12
Price of labor ($/hour)	gt_pl	19.49	7.36	0.00	38.46

TABLE 5.2 Basic Cost Model versus Nerlove Cost Model Specification

Parameter (Variable)	Basic Cost Model: Estimate	Basic Cost Model: t Statistic	Nerlove Model: Estimate	Nerlove Model: t Statistic
β_0	11.3498	6.92	7.58348	8.88
$\beta_y(\ln Y)$		12.85	1.00729	11.96
$\beta_1(\ln pl)$	0.9575	−0.4	0.53291	2.52
$\beta_2(\ln pk)$	−0.1619	−0.65	0.2236	0.96
$\beta_3(\ln pf)$	0.06002	0.49		
Adjusted R^2	0.918		0.94	

which implies that

$$r = 1/\beta_y = 1/0.9575$$

or

$$r = 1.044.$$

Because $r > 1$, results of the basic cost model indicate that firms in the sample were operating in the increasing returns-to-scale portion of the average cost curve.

FURTHER CONSIDERATIONS

It was stated previously that a proper cost model is monotonic (increasing) and homogeneous of degree one in input prices, which implies that, for a given level of output, a doubling of all input prices results in a doubling of total cost. What this means is that the estimated parameters of the input price variables should be positive in sign and sum to unity. Reviewing the results of the basic cost model, it is clear that neither condition holds; as such, the basic cost model does not represent an appropriately specified cost function.

At first blush, it may appear that the Cobb–Douglas specification employed by Nerlove does not conform either. However, you may recall that the estimated model is derived from the underlying production function and that the parameter estimates are actually functions of other parameters and variables, which was detailed in Chapter 4 and reviewed here. More specifically, we have

$$\beta_1 = \alpha_1/r \tag{5.6}$$

and

$$\beta_2 = \alpha_2/r, \tag{5.7}$$

which imply that

$$\alpha_1 = \beta_1 * r \Longrightarrow \alpha_1 = \beta_1/\beta_y \tag{5.8}$$

and

$$\alpha_2 = \beta_2 * r \Longrightarrow \alpha_2 = \beta_2/\beta_y \tag{5.9}$$

From Berndt (1991), linear homogeneity implies that the constraint on the underlying parameters is given by

$$(\alpha_1 + \alpha_2 + \alpha_3)/r = 1 \tag{5.10}$$

so that

$$\alpha_3 = (1 - \beta_1 - \beta_2)/\beta_y \tag{5.11}$$

(Berndt, 1991). It is left as an exercise to check for linear homogeneity of the underlying input price parameters (α_1, α_2, α_3).

END OF SECTION EXERCISES: BASIC COST MODEL VERSUS NERLOVE COST MODEL

1. Refer to Table 5.2.
   a. What do you notice about the estimated parameters of the basic cost model given in Equation (5.3); in other words, do they seem reasonable (i.e., do they accord to economic theory in terms of sign and statistical significance)? Why or why not?

TABLE 5.3 Nerlove Original Data and Cost Model (Cobb–Douglas)

Variable	Parameter	Nerlove Model: Estimate[a]	Nerlove Model: t Statistic
Constant	β_0	−4.6908	−5.301
(ln of) Y	β_y(lnY)	0.72069	41.334
lnpl − lnpf [or ln(pl/pf)]	β_1(ln$plpf$)	0.59291	2.898
lnpk − lnpf [or ln(pk/pf)]	β_2(ln$pkpf$)	−0.00738	−0.039
	Adjusted R^2	0.93	

[a]Reestimated using Nerlove (1955) data.

    b. What do you notice about the estimated parameters from Nerlove's log linear Cobb–Douglas specification? Do they seem reasonable?

    c. What do results from the Nerlove specification (log-linear Cobb–Douglas form) indicate in terms of returns to scale?

2. Table 5.3 contains results from Nerlove's original model specification. His data set contained data on total costs, output (in kilowatt hours), and prices of labor (pl), capital (pk), and fuel (pf) for 145 electric utility companies in 1955.

    a. What do you notice about the results?

    b. What do they imply about returns to scale?

FLEXIBLE FUNCTIONAL FORMS

In addition to the previously mentioned properties, a cost function should be flexible enough so as not to restrict the substitution elasticities between inputs. The Cobb–Douglas form restricts these substitution elasticities to equal unity, and the CES function imposes that the elasticities of substitution not vary across observations. While the Leontief form provides the flexibility required, its implication that the marginal productivity of any factor is zero is troubling. As a result, many studies employ the translogarithmic functional form, which is flexible enough to allow the substitution elasticities to vary across observations and conforms easily to meet many (but not all) of the qualifications of a proper cost function.

TRANSLOGARITHMIC COST FUNCTION

The translogarithmic (translog) function is a second-order Taylor's series approximation to any arbitrary cost function. Christensen and Greene (1976) employed this cost specification when they reexamined Nerlove's cost model.

In this case, they employed a single-output, three-input translog cost function, which is given by

$$\ln C = \alpha_0 + \alpha_y \ln y + \sum_i \beta_i \ln p_i + (1/2)\alpha_{yy}(\ln y)^2$$
$$+ (\tfrac{1}{2})\sum_i\sum_j \phi_{ij} \ln p_i \ln p_j + \sum_i \omega_{iy}\ln y \ln p_i, \tag{5.12}$$

where output ($\ln y$) and input prices ($\ln p_i$) enter linearly, as quadratics and as cross-products.

This is one of the specifications estimated later in this chapter, as well as being used in the examples and exercises at the end of the chapter.

COST-SHARE EQUATIONS

Cost-share equations allow for the assumption of cost-minimizing behavior to be imposed on the model. In general, the equation for the ith input price is given by (via Shephard's lemma, which was defined in Chapter 4):

$$s_i = \partial \ln C / \partial \ln p_i, \tag{5.13}$$

that is,

$$s_i = \beta_i + \beta_{ii} \ln p_i + \omega_i \ln Y + (1/2)\sum_j \beta_{ij} \ln p_j \tag{5.14}$$

for $i \neq j$.

Thus, for the three inputs [labor (l), capital (k), and purchased power (p)] employed here, the respective cost-share equations are given by

$$s_P = \beta_p + \beta_{pp} \ln p_p + \omega_P \ln Y + (1/2) \sum_j \beta_{pj} \ln p_j, \tag{5.15}$$

where $j = k, l$;

$$s_l = \beta_l + \beta_{ll} \ln p_l + \omega_l \ln Y + (1/2) \sum_j \beta_{lj} \ln p_j, \tag{5.16}$$

where $j = p, k$; and

$$s_k = \beta_k + \beta_{kk} \ln p_k + \omega_k \ln Y + (1/2) \sum_j \beta_{kj} \ln p_j, \tag{5.17}$$

where $j = l, p$

Again, linear homogeneity in input prices and symmetry are imposed by the following restrictions:

$$\sum_i \beta_i = 1 \text{ and } \sum_i \beta_{ij} = \sum_j \beta_{ji} = \sum_i \omega_{iy} = 0 \tag{5.18}$$

so that the final form to be estimated is given by

$$\ln C = \alpha_0 + \alpha_Y \ln Y + \sum_i \beta_i \ln p_i$$
$$+ 1/2*(\alpha_{YY} \ln Y^2 + \sum_i\sum_j \beta_{ij} \ln p_i \ln p_j) + \sum_i\omega_{iy} \ln Y \ln p_i. \tag{5.19}$$

Once these restrictions are imposed, the system of equations can then be estimated simultaneously by Zellner's method, which was discussed in Chapter 4.

Example 5.2 Translogarithmic Cost Model

Data employed in this example include detailed information on the 711 distribution cooperatives that were RUS borrowers in 1997. However, only 708 were actually used in the estimation procedures due to data irregularities or missing data. In this example, Equation (5.19) was estimated. The variables are defined as:

In C(the natural log of total cost) = the natural log of (the cost of purchased power + distribution expense O&M + customer accounts, service, and information expenses + A&G expense + sales expense + D&A expense + tax expense + interest on long-term debt).

Independent variables (also known as regressors or explanatory variables) include the natural log of factor prices (capital, labor, and purchased power) and the natural log of output (electricity distributed in megawatt hours).

Note: In exercises at the end of the chapter you will add certain cost-shift variables, such as the (natural log of) miles of transmission lines and customer density, which is defined as the number of customers per mile of distribution line.

The first set of regressors corresponds to prices of inputs required in the distribution of electricity—capital, labor, and purchased power—which are defined as

P_k is the price of capital (interest on long-term debt/total long-term debt)
P_l is the price of labor (total payroll expense/total number of hours worked)
P_P is the price of purchased power (cost of power/total MWh purchased)

Dividing through by the price of capital and imposing symmetry yield the cost-share equations to be estimated, which are given by

$$S_p = \beta_p + \beta_{pp} \ln(P_p/P_k) + \omega_{py} \ln Y + \Sigma_j \beta_{pl} \ln (P_l/P_k) \tag{5.20}$$

and

$$s_l = \beta_L + \beta_{ll} \ln(P_l/P_k) + \omega_{ly} \ln Y + \Sigma_j \beta_{lp} \ln(P_p/P_k). \tag{5.21}$$

Thus, a three-equation system characterizes this model: the translogarithmic cost function and the two cost-share equations, which are estimated simultaneously by Zellner's method (see Chapter 4). Estimation results are displayed in Table 5.4.

A PRIORI EXPECTATIONS

Before reviewing the results, it is necessary to discuss the a priori expectations concerning signs of the coefficient estimates, which should accord to economic theory. Recalling from Chapter 4, there are certain characteristics to which a cost model must conform: First, the total effect of a change in output (Y) should cause an increase in the total cost of distributing electricity (i.e., monotonicity in output). Next, it should be nondecreasing in input prices (i.e., share equations should be positive and sum to unity), as an increase in the price of an input should always increase total cost (but at a decreasing rate, which satisfies the concavity in input prices provision). This implies that

1. Own prices must be nonpositive.
2. Cross-price effects are symmetric.

Estimation results of a single-output translog cost model are contained in Table 5.4.

TABLE 5.4 Single-Output Translog Cost Model—1997 Data Update

Variable	Coefficient	Estimated Coefficient	t Statistic
Constant	α_0	2.6192	10.71
Output (Y) (total)	α_y	0.6653	10.23
Output—squared (Y^2)	α_{YY}	0.0175	2.04
Input Prices			
Capital	β_L	0.1641	1.69
Labor	β_K	0.2243	26.30
Purchased power	β_P	0.6117	6.25
Squares and Cross-products			
Capital—squared	β_{LL}	−0.1443	−7.38
Labor—squared	β_{KK}	0.0324	10.24
Purchased power—squared	β_{PP}	−0.0980	−4.95
Labor * capital	β_{LK}	0.0070	1.79
Labor * purchased power	β_{LP}	−0.0394	−12.78
Capital * purchased power	β_{KP}	0.1373	7.01
$Y*P_L$	ω_{ly}	−0.0171	−21.40
$Y*P_K$	ω_{ky}	−0.0462	−2.54
$Y*P_P$	ω_{py}	0.0632	3.48
Adjusted R^2			0.9631

DISCUSSION OF ESTIMATION RESULTS: SINGLE-OUTPUT TRANSLOG COST EQUATION

Estimation results of the single-output translog cost model are, for the most part, as expected in terms of the sign and magnitude of the coefficients. The adjusted R^2 of 0.96 indicates that the model is doing a very good job in explaining the variation in the total cost of distributing electricity. More specifically, the output coefficients are statistically significant and positive in sign.

Because of the inclusion of cross-product terms in this model, these too must be included upon evaluating the total effect of a change in output on cost by evaluating the partial derivative of cost with respect to output (i.e., the degree-of-scale economies, or SCE), that is,

$$SCE = C(Y)/Y*C'(Y)(\text{or } AC/MC). \qquad (5.22)$$

Upon doing this and evaluating the derivative at the variables' sample means, the effect is indeed positive, greater than unity, and is equal to $S = 1.15$.

These results indicate that there were increasing returns to scale in the distribution of electricity for firms in the 1997 sample. In other words, the average cooperative operated in the downward-sloping portion of the average cost curve in 1997, thus implying that marginal cost was less than average cost and that cost minimization was not being achieved.

Next, we turn to evaluation of input price coefficients. Input price coefficients indicate that linear homogeneity has been preserved (since share equations sum to unity).

Note: Because each of the input price coefficients enters both linearly and multiplicatively with other variables in the model, an evaluation of partial derivatives is required to obtain the appropriate interpretations. Appealing to the share Equations (5.15)–(5.17) and evaluating each of the partial derivatives at the variables' sample means, all three are positive in sign (as expected) with that of the price of purchased power the largest in magnitude, which is not surprising given that firms in the data set are distribution cooperatives. To wit:

$$s_P = \beta_p + \beta_{pp} \ln p_p + \omega_P \ln Y + (1/2) \sum_j \beta_{pj} \ln p_j = 0.69$$

$$s_l = \beta_l + \beta_{ll} \ln p_l + \omega_l \ln Y + (1/2) \sum_j \beta_{lj} \ln p_j = 0.09$$

$$s_k = \beta_k + \beta_{kk} \ln p_k + \omega_k \ln Y + (1/2) \sum_j \beta_{kj} \ln p_j = 0.22.$$

Note that share equations sum to unity as required for linear homogeneity in input prices.

SUBSTITUTION ELASTICITIES AMONG INPUTS: HICKS–ALLEN PARTIAL ELASTICITIES OF SUBSTITUTION

As discussed previously in Chapter 4, one of the properties of flexible functional forms is that they place no a priori restrictions on substitution elasticities. Appealing to the Hicks–Allen partial elasticities of substitution between inputs i and j for the translog cost model are equal to

$$\sigma_{ij} = (\beta_{ij} + s_i s_j)/s_i s_j \qquad (5.23)$$

for $i, j = 1, \ldots, n$ but $i \neq j$, and

$$\sigma_{ii} = (\beta_{ii} + s_i^2 - s_i)/s_i^2 \qquad (5.24)$$

for $i = 1, \ldots, n$.

Results from the estimation of Equations (5.23) and (5.24) are displayed in the first column of Table 5.5. As can be seen, σ_{ij}, which represent substitution among the different inputs, are all positive in sign, indicating that these are substitutes while σ_{ii} are complements.

TABLE 5.5 Elasticities of Substitution and Price for Single-Output Translog Model

Inputs	Parameter	Substitution Elasticities	Parameter	Price Elasticities
P_p, P_p	σ_{pp}	−0.66	ε_{pp}	−0.45
P_k, P_k	σ_{kk}	−6.52	ε_{kk}	−1.44
P_l, P_l	σ_{ll}	−6.12	ε_{ll}	−0.55
P_p, P_l	σ_{pl}	0.17	ε_{pl}	0.03
P_p, P_k	σ_{pk}	0.91	ε_{pk}	1.31
P_k, P_l	σ_{kl}	0.07	ε_{kl}	0.12
P_l, P_p	σ_{lp}	0.17	ε_{lp}	0.25
P_k, P_p	σ_{kp}	0.91	ε_{kp}	0.42
P_l, P_k	σ_{lk}	0.07	ε_{lk}	0.30

PRICE ELASTICITIES

Next we turn to price elasticities, which are equal to

$$\varepsilon_{ij} = S_j \sigma_{ij}. \qquad (5.25)$$

As such, this is equivalent to

$$\varepsilon_{ij} = (\beta_{ij} + s_i s_j)/s_i \qquad (5.26)$$

for $i, j = 1, \ldots, n$ but $i \neq j$, and

$$\varepsilon_{ii} = (\beta_{ii} + s_i^2 - s_i)/s_i \qquad (5.27)$$

for $i = 1, \ldots, n$.

The second partial derivatives of a proper cost function represent the own- and cross-price effects of the inputs. More specifically, an increase in the price of an input should decrease the quantity demanded of that input. As displayed in the first three rows of Table 5.5, the own-price elasticities of demand are negative, while cross-price elasticities (displayed in the remaining rows of Table 5.5) are positive (and symmetric) when evaluated at the variables' sample means. Note that the concavity-in-input-prices' criterion is satisfied because the own-price elasticities are negative in sign.

Table 5.5 contains results of Equations (5.23)–(5.27) when evaluated at the sample means of the variables and the estimation results displayed in Table 5.3.

HOMOTHETICITY

It is also informative to look at relationships between output and the input prices, as the input price–output interaction terms allow for the nonhomotheticity of the underlying production function. As described in their 1976 paper, Christensen and Greene write: "A cost function corresponds to a homothetic production function if and only if it can be written as a separable function in output and factor prices."[4]

Homothetic functions are functions of which the marginal technical rate of substitution (slope of the isoquant) is homogeneous of degree zero. Due to this, along rays coming from the origin, the slope of the isoquants will be the same. What this implies for the translogarithmic cost function is that

$$\omega_{iy} = 0 \tag{5.28}$$

for all $i = 1, \ldots, n$.

Coefficients of the input price–output interaction terms measure how a change in an input's price affects its usage and how the change in its usage affects output, which then affects the total cost of distributing electricity. For example, an increase in the wage rate will most likely cause less labor to be employed, thus causing a reduction in output, which, in turn, will affect the total cost of distributing electricity (negatively because the estimated coefficient, $\omega_{ly} = -0.017$, is negative in sign). Likewise, for a positively signed coefficient, for example, ω_{py}, the output–purchased power coefficient, the positive sign indicates that the change in output times purchased power causes a change in total cost in the same direction. That is, in this case,

$$\partial \ln C / \partial \ln YP_p = 0.063 > 0.$$

Again, this may seem contrary to theory, but it is probably due to the requirement that distribution entities procure enough power to serve their native loads. Anyway, because many states have a mechanism in place (known as fuel adjustment clauses), any changes in the cost of power are passed onto rate-payers directly. In effect, this mechanism renders the elasticity of supply extremely inelastic, which could support the positively signed coefficient (or it could be another problem, for example, an incorrect functional form).

END OF SECTION EXERCISES: TRANSLOGARITHMIC COST FUNCTION

1. Using the data set Coops97:
   a. Calculate the summary statistics for the variables in Table 5.3 (you will need this to complete this exercise).

---

[4]Homotheticity was rejected by Christensen and Greene (1976).

b. Estimate Equation (5.19) along with the relevant cost-share equations in Equations (5.20)–(5.21). (Recall from the discussion in Chapter 4 on Zellner's ITSUR method, you must divide $n-1$ of the share equations by the remaining input price variable).

c. Using the results obtained in part b, verify that the cost model accords to theory; that is, evaluate the following at the variables' sample means:

  i. Equation (5.22) Marginal cost part only: $C'(Y)$ (monotonicity in output—marginal cost should be nonnegative)

  ii. Equations (5.15)–(5.17) (share equations—should be nonnegative and sum to unity)

  iii. Equation (5.25) (concave in input prices—should be nonpositive)

  iv. Equations (5.23) and (5.24) (elasticities of substitution and prices)

  v. Verify that the restrictions in Equation (5.18) hold (you must impose them prior to estimating the system of equations)

d. Verify that the average cooperative (in terms of the amount of electricity distributed) operated in the increasing returns-to-scale portion of the average cost curve [i.e., evaluate Equation (5.22)].

2. **Advanced:** Using the same data set,

a. Estimate an equation that includes two cost-shift variables: miles of transmission lines (O_{tr}) and customer density (O_{dm}), which is equal to the number of customers per mile of transmission line. That is, estimate the equation

$$\ln C = \alpha_0 + \alpha_Y \ln Y + \sum_i \beta_i \ln p_i + 1/2^*(\alpha_{YY} \ln Y^2$$
$$+ \sum_i \sum_j \beta_{ij} \ln p_i \ln p_j + \sum_m \sum_n \varphi_{mn} \ln O_m \ln O_n)$$
$$+ \sum_i \omega_{iy} \ln Y \ln p_i + \sum_m \theta_m \ln O_m + \sum_m \delta_m \ln Y \ln O_m$$
$$+ \sum_m \rho_{mi} \ln O_m \ln p_i,$$

(5.19′)

along with the cost-share equations, which are of the general form

$$S_i = \partial \ln C / \partial \ln p_i = \beta_i + \beta_{ii} \ln p_i + \omega_i \ln Y + \sum_j \beta_{ij} \ln p_j$$
$$+ \sum_m \rho_{mi} \ln O_m$$

(5.14′)

for $i \neq j$.

More specifically, for the three inputs employed here, the respective cost-share equations are given by

$$S_P = \beta_p + \beta_{pp} \ln p_p + \omega_P \ln Y + \sum_j \beta_{pj} \ln p_j + \sum_m \rho_{mp} \ln O_m, \quad (5.15′)$$

where $j = K, L$;

$$S_L = \beta_L + \beta_{LL} \ln p_L + \omega_l \ln Y + \sum_j \beta_{ij} \ln p_j + \sum_m \rho_{mL} \ln O_m, \quad (5.16′)$$

where $j = P, K$; and

$$S_K = \beta_K + \beta_{KK} \ln p_k + \omega_K \ln Y + \sum_j \beta_{Kj} \ln p_j + \sum_m \rho_{mK} \ln O_m, \quad (5.17′)$$

where $j = L, P$.

b. Do the signs of the estimated coefficients change as a result of the inclusion of cost-shift variables?

c. Does inclusion of cost-shift variables strengthen or weaken the finding of increasing returns to scale?

Optional:

3. Another measure is returns to density. Using the definition of Caves and co-workers (1984) for the translog functional form, returns to density are given by

$$RTD = 1/\partial \ln C/\partial \ln Y, \tag{5.29}$$

where RTD > 1 indicates that there are increasing returns to density in the distribution of electricity. Evaluating Equation (5.29) at the sample means of data, are returns to density increasing, constant, or decreasing?

Aside: Translogarithmic Cost Model Details: Calculating Average and Marginal Cost

When calculating the marginal and average costs associated with a cost model of the translogarithmic form it is necessary to do the following. Recal that

$$\text{Marginal Cost } (MC) = \partial C/\partial Y. \tag{5.30}$$

However, simply taking the derivative of Equation (5.19), for example, with respect to the (natural log of) output (ln Y), we would obtain

$$\partial \ln C/\partial \ln Y, \tag{5.31}$$

which is equivalent to

$$\partial \ln C/\partial \ln Y = \alpha_Y + \alpha_{YY} \ln Y + \sum_i \omega_{iy} \ln p_i. \tag{5.32}$$

Thus, it is necessary to multiply Equation (5.32) by C/Y [average cost, which is *not* simply Equation (5.19) divided by output] to obtain

$$\partial C/\partial Y = C/Y * \partial \ln C/\partial \ln Y, \tag{5.33}$$

which is the marginal cost.

Proof

By definition,

$$\partial \ln C = \partial C/C \tag{5.34}$$

and

$$\partial \ln Y = \partial Y/Y, \tag{5.35}$$

so that

$$\partial \ln C/\partial \ln Y = \partial C/C * Y/\partial Y. \tag{5.36}$$

This is equivalent to

$$\partial \ln C/\partial \ln Y = \partial C/\partial Y * Y/C \tag{5.37}$$

or

$$\text{Marginal Cost} * (1/\text{Average Cost}). \qquad (5.38)$$

Thus, multiplying Equation (5.37) by C/Y (or Average Cost) yields Marginal Cost, or $\partial C/\partial Y$ QED.

MULTIPRODUCT COST FUNCTIONS

Distributed electricity as a multiproduct cost industry is well established in the literature, some of which was detailed in Chapter 2 and is expanded here. In addition, the concepts of horizontal (i.e., economies of scope) and vertical integration become more relevant. As such, a brief survey of the literature is included here.

LITERATURE REVIEW

Economies of Scale and Integration

Using a sample of 25 Italian electric utilities, Piacenza and Vannoni (2005) use an integrated approach, which simultaneously considers both horizontal and vertical aspects of the technology employed in the generation, transmission, and distribution of electricity. [Note: The methodology is based on the estimation of a *composite* cost function model (Pulley and Braunstein, 1992), which has been proven to be particularly apt for the analysis of cost properties of multiple-output firms.] Their findings were that these utilities, which operate in generation and distribution and serve different categories of users, highlight the presence of both vertical integration gains and scope economies at the downstream stage.

Farsi et al. (2007) explored the economies of scale and scope in electricity, gas, and water utilities. In this paper, an argument is put forth that the potential improvements in efficiency through unbundling should be assessed against the loss of scope economies. Several econometric specifications, including a random coefficient model, were used to estimate a cost function for a sample of utilities distributing electricity, gas, and/or water to the Swiss population. The estimates of scale and scope economies have been compared across different models, and the effect of heterogeneity among companies has been explored. While indicating considerable scope and scale economies overall, results suggest a significant variation in scope economies across companies due to unobserved heterogeneity.

Greer (2008) devised a methodology to estimate the lost vertical economies for rural electric cooperatives, which are not vertically integrated by choice; they are organized as either generation and transmission or distribution-only, and typically with long-term, full-requirements contracts in place. Using 1997 data and her quadratic cost model, she found that, on average, cost savings in excess of 39% could have been realized had the coops adopted a vertically integrated structure.

Fetz and Filippini (2010) employed a quadratic multistage cost function for a sample of electricity companies. In this paper, the authors specify various econometric specifications for panel data, including a random effects and a random coefficients model. Like many other studies on vertical integration, empirical results reflect the presence of considerable economies of vertical integration and economies of scale for most of the companies considered in the analysis. Moreover, results suggest a variation in economies of vertical integration across companies due to unobserved heterogeneity.

Distributed Electricity as a Multiproduct Industry

The motivation for multiproduct cost models is (at least) twofold. First, multiproduct cost specifications have been employed in the literature, particularly for distribution entities. Most studies disaggregate distributed electricity into two categories, or outputs: high voltage and low voltage. Kwoka (1996) specified a quadratic net cost equation for two outputs, generation and distribution, the latter of which is further disaggregated by a voltage requirement. High-voltage customers are those that require little or no voltage reduction and thus entail smaller line losses. Kwoka found that increasing the percentage of electricity distributed to high-voltage customers tended to lower total cost. A 1996 study for the National Rural Electric Cooperative Association by Christensen Associates assumed equal demand elasticities for residential and commercial customers and the fact that both consume relatively small quantities of generation as justification in aggregating them into the small, low-voltage category.

Berry (1994) cited the fact that although industrial customers receive power via the distribution system, some receive power directly from the utility's high-voltage transmission lines and transform the power down to usable levels with their own equipment. As a result, they are less costly to serve than either residential or small commercial users. In addition, industrial users tend to consume a greater percentage of power during off-peak hours (many operate around the clock), the cost of which is substantially lower than electricity consumed during peak hours, as the most expensive generating units come online to satisfy demand during peak hours. In addition, industrial loads tend to provide more stable loads. Residential and commercial customer loads tend to be more volatile, often requiring power during peak hours, especially on hot summer days. The power sold to smaller users must go through the distribution substation and each customer has its own service drop, meter, and billing charges to be calculated. Due to the low density that prevails in many coop territories, costs per unit are rather high.

Karlson (1986), who was quoted earlier, specified a four-output translog cost function (residential, commercial, industrial, and wholesale). He tested for and confirmed that a multiple-output specification was appropriate. Roberts

(1986) found that firms that serve a large proportion of residential and small commercial customers tend to have larger demands for distribution capital, whereas those with large industrial loads have large demands for transmission capital. Along a similar vein, Hayashi et al. (1985) estimated a two-product translog cost function with low-voltage and industrial outputs. They cite that multicollinearity resulted from separate estimation of residential and commercial users. Berry (1994) confirmed this finding, as did Greer (2003).

The second reason for the multiple-output specification is the fact that public utility commissions typically allocate costs in this fashion. It is often the case that large commercial and industrial users are assigned certain charges that neither residential nor small commercial users have to pay, which are known as demand or capacity charges. All customer classes are assigned customer charges and energy charges, both of which vary with the number of customers served. Customer charges are composed of the cost of primary and secondary lines, transformers, services, and metering, and are allocated by the percentage of number of customers in each class. For example, if 70% of the load is residential, then residential users *in theory* would equally share 70% of the total customer cost with each paying the same amount regardless of the cost they impose on the system. Energy charges are based on actual usage, typically measured in cents per kilowatt hour. It is often the case that residential and small commercial users pay higher energy charges than industrial users because they tend to consume power at peak times and have more volatile loads, whereas industrial loads tend to be more stable. The latter tend to use a fairly constant amount of power throughout the course of the day, including off-peak times, and thus are less costly to serve. In addition, industrial customers may choose to be on interruptible contracts, further decreasing the per-kilowatt-hour energy charge. Demand charges, which apply to the maximum amount of electricity consumed in a given period (typically an hour), are predominantly allocated to industrial and large commercial users on a per kilowatt (kW) basis or a per kilovolt ampere (kVA) basis (the latter takes into account the power factor). Because the utility must install enough capacity and other infrastructure to serve these large commercial and industrial loads, the paying of demand charges is one of the ways in which large users subsidize smaller users.

MULTIPRODUCT COST MODELS

You may recall the discussion in Chapter 4 concerning model misspecification, which includes omitted variable bias and incorrect functional form; more specifically, that in the case of distributed electricity, it is more appropriate to estimate a cost function that has (at least) two outputs: electricity distributed to "small" users versus that distributed to "large" users. The distinction here being the voltage level at which the end user receives the electricity (below or above 1000 kVA). Given this, a discussion of such multiproduct specifications ensues.

Multiproduct Translogarithmic Cost Model

Let

$$Y_1 = \text{small users (residential and small commercial customers)}$$

and

$$Y_2 = \text{large users (large commercial or industrial customers)}$$

As such, a multiple-output translog cost function of the form [including input prices (p_i) and cost-shift variables (O_m)] is given by

$$\begin{aligned}
\ln C = \alpha_0 &+ \sum_i \alpha_i \ln Y_i + \tfrac{1}{2}{}^* \alpha_{ij} \ln Y_i\, Y_j + \sum_i \beta_i \ln p_i \\
&+ \tfrac{1}{2}{}^* \sum_i \sum_j \beta_{ij} \ln p_i \ln p_j + \tfrac{1}{2}{}^* \sum_m \sum_n \varphi_{mn} \ln O_m \ln O_n \\
&+ \sum_i \sum_g \omega_{iyg} \ln Y_g \ln p_i + \sum_m \theta_m \ln O_m \\
&+ \sum_g \sum_m \delta_m \ln Y_g \ln O_m + \sum_m \rho_{mi} \ln O_m \ln p_i
\end{aligned} \tag{5.40}$$

for $g, i, j = 1, \ldots, n$, along with cost-share equations, which are of the general form

$$\begin{aligned}
S_i = \partial \ln C / \partial \ln p_i = \beta_i &+ \beta_{ii} \ln p_i + \sum_i \sum_j \omega_{iyg} \ln Y_g \\
&+ \sum_j \beta_{ij} \ln p_j + \sum_m \rho_{mi} \ln O_m
\end{aligned} \tag{5.41}$$

for $g, i, j = 1, \ldots, n$.

Collectively, Equations (5.40) and (5.41) yield a system of equations that could be estimated jointly via Zellner's method, which was described in the appendix to Chapter 4.

More specifically, for the two-output, three-input cost model employed here, the respective cost-share equations are given by

$$S_P = \beta_p + \beta_{pp} \ln p_p + \sum_g \omega_{Pg} \ln Y_g + \sum_j \beta_{pj} \ln p_j + \sum_m \rho_{mp} \ln O_m, \tag{5.42}$$

where $j = L, K$;

$$S_L = \beta_L + \beta_{LL} \ln p_L + \sum_g \omega_{lg} \ln Y_g + \sum_j \beta_{Lj} \ln p_j + \sum_m \rho_{mL} \ln O_m, \tag{5.43}$$

where $j = P, K$; and

$$S_K = \beta_K + \beta_{KK} \ln p_k + \sum_g \omega_{Kg} \ln Y_g + \sum_j \beta_{Kj} \ln p_j + \sum_m \rho_{mK} \ln O_m, \tag{5.44}$$

where $j = P, L$.

(Recall that you must divide through by one of the input prices or the system of equations is singular; as such, it does not have an inverse, which means that the parameters of the models cannot be estimated.[5])

Example 5.3 Estimation Results of Two-Output Translog with Cost-Shift Variables

Again using data in Coops97, estimate the parameters of a multiproduct translog cost function, including the cost-shift variables for miles of transmission lines and customer density. Use Zellner's method to jointly estimate the cost equation along with two of the share equations.

Estimation Results

Table 5.6 displays estimation results of a two-output translogarithmic cost model, including cost-shift variables. As before, a check of a priori expectations of output variables is required, namely the two newly added cost-shift variables. In the case of miles of transmission lines, it is expected that the total effect should be positive in sign, as an increase in transmission lines should increase total cost. However, an increase in customer density, which is defined as the number of customers per mile of distribution line, should cause total cost to decrease. Note: You must calculate the partial derivatives of each of these to assess whether the coefficient estimates are of the appropriate sign.

MULTIPRODUCT COST CONCEPTS (REVISITED)

Ray Average Costs

You may recall from Chapter 2 the discussion on ray average cost (RAC), which requires that Y_1 and Y_2 move in fixed proportions, and that Y, the composite product, is equal to

$$Y = Y_1 + Y_2. \tag{5.45}$$

Appealing to Baumol et al. (1982, p. 48), we can define the average cost of the composite product to be

$$RAC = C(Y)/Y. \tag{5.46}$$

Again, we will use the degree-of-scale economies to assess whether firms in the sample that distributed electricity to both small and large users ($n = 682$) were efficient distributors of electricity according to the translog cost model. The degree-of-scale economies is given by Equation (2.20) in Chapter 2.

Results presented in Tables 5.6A–C are displayed as a histogram in Figure 5.1.

---

[5]Using matrix algebra, B is obtained by

$$B = (X'X)^{-1}/X'Y, \tag{5.44a}$$

where X is an $N \times N$ matrix of independent variables and Y is (a vector of) the dependent variable.

TABLE 5.6A Two-Output Translog Cost Model Estimation Results
(Output-Related Variables)

Coefficient	Variable	Estimate	t Statistic
a_0	Constant	3.2402	12.77
a_1	$\ln Y_1$	0.2681	3.07
a_2	$\ln Y_2$	0.3549	10.51
a_{11}	$\ln Y_1 \text{sq}$	0.1422	9.28
a_{22}	$\ln Y_2 \text{sq}$	0.0748	20.31
a_{21}	$\ln Y_1 Y_2$	−0.0912	−16.21
$w_p y_1$	$\ln P_p Y_1$	0.0487	1.82
$w_k y_1$	$\ln P_k Y_1$	−0.0396	−1.48
$w_l y_1$	$\ln P_l Y_1$	−0.0092	−7.45
$w_p y_2$	$\ln P_p Y_2$	0.0062	0.50
$w_k y_2$	$\ln P_k Y_2$	−0.0005	−0.04
$w_l y_2$	$\ln P_l Y_2$	−0.0057	−11.25

TABLE 5.6B Estimation Results (Price-Related Variables)

Coefficient	Variable	Estimate	t Statistic
b_k	$\ln P_k$	0.3150	2.24
b_l	$\ln P_l$	0.1882	21.16
b_p	$\ln P_p$	0.4968	3.54
b_{pp}	$\ln P_p P_p$	−0.0404	−0.97
b_{kk}	$\ln P_k P_k$	−0.0836	−2.00
b_{ll}	$\ln P_l P_l$	0.0350	9.82
b_{pl}	$\ln P_p P_l$	−0.0391	−12.01
b_{pk}	$\ln P_p P_k$	0.0795	1.91
b_{kl}	$\ln P_l P_k$	0.0040	0.92

TABLE 5.6C Estimation Results (Cost-Shift Variables)

Coefficient	Variable	Estimate	t Statistic
Od	Indens	0.1446	0.93
Ot	lntr	−0.0358	−0.70
oty_1	lntrY$_1$	−0.0038	−0.74
oty_2	lntrY$_2$	−0.0036	−1.77
Ott	lntrsq	0.0101	1.60
Odd	Indenssq	0.0442	1.05
ody_1	IndensY$_1$	−0.0249	−1.31
ody_2	IndensY$_2$	0.0094	1.13
$Otpl$	lntrP$_l$	0.0008	1.87
Otk	lntrP$_k$	0.0273	1.28
Otp	lntrP$_p$	0.0138	1.27
$Odpl$	IndensP$_l$	0.0019	1.04
Odk	IndensP$_k$	−0.0754	−1.06
Odp	IndensP$_p$	−0.0072	−0.17
Otd	lntrdens	−0.0163	−2.07

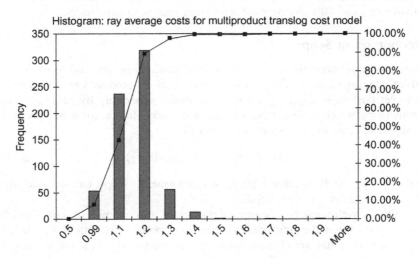

FIGURE 5.1 Histogram of ray average costs for multiproduct translog cost model. As is evident, the majority of firms in the restricted sample data set experienced scale economies greater than unity, indicating that they were operating in the increasing returns-to-scale portion of the average cost curve in 1997.

PRODUCT-SPECIFIC ECONOMIES OF SCALE

In the previous section, it was noted that the concept of ray average cost is relevant only for proportional changes in output. Now, we turn to a measure that allows for variation in an output while holding other quantities of outputs constant. For this, it is necessary to define the concept of *incremental cost* ($IC_i(Y)$) of the output to be varied, that is,

$$IC_i(Y) = C(Y) - C(Y_{N-i}), \tag{5.47}$$

where Y_{N-i} is a vector with a zero component in the place of Y_i.

The average incremental cost $[AIC_i(Y)]$ follows from Equation (5.47):

$$AIC_i(Y) = IC_i(Y)/Y_i. \tag{5.48}$$

As before, it is informative to define the degree-of-scale economies specific to product i. Also known as product-specific returns to scale, this is defined as

$$S_i(Y) = IC_i(Y)/Y_i * C_i \tag{5.49}$$

or

$$S_i(Y) = AIC_i/(\partial C/\partial Y_i). \tag{5.50}$$

Returns to scale of product i at Y are said to be increasing, decreasing, or constant as $S_i(Y)$ is greater than, less than, or equal to unity. Using Equation (5.50) and results displayed in Tables 5.6A–C, 94% of the firms in the sample data set exhibited increasing returns to scale in the distribution of electricity to small users and 70% experienced increasing returns to large users.

Economies of Scope

Economies of scope (or economies of joint production) are said to exist if a given quantity of each of two or more goods can be produced by one firm at a lower cost than if each good were produced separately by two different firms or even two different production processes. That is, for a two-product case, weak economies of scope are given by

$$C(Y_1, Y_2) \le C(Y_1, 0) + C(0, Y_2) \tag{5.51}$$

for all $Y_1, Y_2 > 0$. If not, then there are diseconomies of scope, and separate production of outputs is more efficient.

What is useful here is a measure of the degree of economies of scope, which would allow for, in the presence of economies of scope, capturing of the relative increase in cost that would result from separate production of the two (or more) outputs. Thus, the degree of economies of scope is given by

$$S_c = [C(Y_1, 0) + C(0, Y_2) - C(Y_1, Y_2)]/C(Y_1, Y_2). \tag{5.52}$$

It was previously stated that this particular form is not well suited to modeling cost for multiproduct markets. However, by restricting the sample to those firms that distributed electricity to both types of user (Y_1 and Y_2), estimation results were then used to calculate the degree-of-scope economies.[6]

Evaluating each of these at sample means of the variables, results indicate that separate production is less costly as all but one of the firms in the restricted sample exhibit positive economies of scope.

QUADRATIC COST FUNCTIONS

Another functional form employed in cost estimation for electric utilities is the quadratic cost specification, which, like the translog functional form, imposes no a priori restrictions on elasticities of substitution between inputs. In general, the quadratic cost function is given by

$$C = \alpha_0 + \alpha_y Y + 1/2\, \alpha_{yy} Y^2 + \beta_k p_k + \beta_L p_L + \beta_P p_P. \qquad (5.53)$$

While this is the form often employed in the electric utility cost literature, this equation is not homogeneous of degree one in input prices (you may recall that linear homogeneity in input prices is necessary for a properly defined cost function). Linear homogeneity implies that cost doubles when input prices double, which is not the case for the cost function shown previously.

One way to ensure that a function is linearly homogeneous is to impose the restriction that

$$\sum \beta_i = 1. \qquad (5.54)$$

However, this restriction alone is not sufficient for this model. A proof of this is included in the appendix to this chapter.

Mayo (1984) imposes linear homogeneity by appending to the cost function the product of the input prices times their estimated coefficients, that is,

$$C = (\alpha_0 + \sum_i \alpha_i Y_i + 1/2 \sum_i \sum_j \alpha_{ij} Y_i Y_j) \cdot \prod \beta_i p_i e^\varepsilon. \qquad (5.55)$$

In this form, all outputs enter the equation both as quadratic and as interaction variables.

One concern may be the fact that this model also imposes strict input–output separability, which means that the marginal rate of substitution between any two inputs is independent of the quantities of outputs, and the marginal rate of transformation between any two outputs is independent of the quantities of inputs. However, studies by both Karlson (1986) and Henderson (1985) reject the separability of inputs from outputs in the distribution of electricity.

---

[6]Computation of Equation (5.52) entails the running of three separate regressions: $C = f(Y_1, P, O)$, $C = f(Y_2, P, O)$, and $C = f(Y_1, Y_2, P, O)$, with the latter being Equation (5.52).

A PROPERLY SPECIFIED QUADRATIC COST FUNCTION

Greer (2003) introduced a properly specified quadratic cost model in which input price parameters enter exponentially rather than multiplicatively. Not only does this allow for individual input price parameter estimates to be obtained, but it also preserves the requisite properties to which a proper cost function must conform. This equation is given by

$$C = (\alpha_0 + \sum_i \alpha_i Y_i + 1/2 \sum_i \sum_j \alpha_{ij} Y_i Y_j) \, p_K{}^{\beta k} p_L{}^{\beta l} p_P{}^{\beta p} e^{\varepsilon}. \tag{5.56}$$

This cost function is concave, nondecreasing, and homogeneous of degree one in input prices as well as monotonic in output and, as such, preserves the fundamental properties of a proper cost function. A proof of each is given in the appendix to this chapter.

Estimation of this particular cost function can be somewhat problematic due to the nonlinearity of the specification.[7] One solution is to transform the model so that the parameters enter linearly and the stochastic error term is additive. A logarithmic transformation, which will yield such an error term, is made possible by the creation of a variable, Z, where

$$Z = (\alpha_0 + \sum_i \alpha_i Y_i + 1/2 \sum_i \sum_j \alpha_{ij} Y_i Y_j), \tag{5.57}$$

so that the equation becomes

$$\ln C = \zeta \ln Z + \beta_K \ln p_K + \beta_L \ln p_L + \beta_P \ln p_P + \varepsilon \tag{5.58}$$

or

$$\ln C = \zeta \ln \left(\alpha_0 + \sum_i \alpha_i Y_i + 1/2 \sum_i \sum_j \alpha_{ij} Y_i Y_j \right) + \beta_K \ln p_K$$
$$+ \beta_L \ln p_L + \beta_P \ln p_P + \varepsilon, \tag{5.59}$$

where ζ has been restricted to unity.[8]

Due to the inherent complexity of the nonlinear estimation procedure, it would be informative to expound upon the underlying econometric theory and technique. For this purpose, we will reference Greene (1993) for assistance in this rather complex matter.

Aside: Nonlinear Least-Squares Estimation

For reasons stated earlier, this model must be estimated using a nonlinear estimation procedure, namely nonlinear least squares. In this case, values of

---

[7]This equation is nonlinear in parameters, which obviates the use of the ordinary least-squares estimation technique.

[8]This is the true value of ζ. Equation (5.56) can be written as

$$C = (\alpha_0 + \sum_i \alpha_i Y_i + 1/2 \sum_i \sum_j \alpha_{ij} Y_i Y_j)^{\zeta} * p_K{}^{\beta k} p_L{}^{\beta l} p_P{}^{\beta p} e^{\varepsilon}. \tag{5.59a}$$

the parameters that minimize the sum of squared deviations will be maximum likelihood (as well as nonlinear least-squares estimators). Because the first-order conditions will yield a set of nonlinear equations to which there will not be explicit solutions, an iterative procedure is required,[9] such as the Gauss–Newton method, which is the preferred method.

Probably the greatest concern here is that the estimators produced by this nonlinear least-squares procedure are not necessarily the most efficient (except in the case of normally distributed errors). An excerpt from Greene (1993) illustrates this point nicely:

In the classical regression model, in order to obtain the requisite asymptotic results, it is assumed that the sample moment matrix, (1/n) X'X, converges to a positive definite matrix, Q. By analogy, the same condition is imposed on the regressors in the linearized model when they are computed at the true parameter values. That is, if:

$$plim\ 1/n\ X'X = Q, \tag{5.60}$$

a positive definite matrix, then the coefficient estimates are consistent estimators. In addition, if

$$(1/\sqrt{n})X'\varepsilon \to N[0, \sigma^2 Q], \tag{5.61}$$

then the estimators are asymptotically normal as well. Under nonlinear estimation, this is analogous to:

$$plim(1/n)\underline{X}'\underline{X} = plim(1/n)\sum_i [\partial h(x_i, \beta^0)/\partial\beta^0][\partial h(x_i, \beta^0)/\partial\beta^{0'}] = \underline{Q} \tag{5.62}$$

where **Q** *is a positive definite matrix. In addition, in this case the derivatives in* **X** *play the role of the regressors.*

The nonlinear least-squares criterion function is given by

$$S(b) = \sum_i [y_i - h(x_i, \mathbf{b})]^2 = \sum_i e_i^2, \tag{5.63}$$

where **b**, *which will be the solution value, has been inserted. First-order conditions for a minimum are*

$$g(b) = -2\sum_i [y_i - h(x_i, b)][\partial h(x_i, b)/\partial b] = 0 \tag{5.64}$$

or

$$g(b) = -2\underline{X}'e. \tag{5.65}$$

---

[9]Iterative methods include Gauss–Newton, which is the preferred method. Other methods include quadratic hill-climbing (Goldfeld et al., 1966). Most other methods employ algorithms and grid searches.

This is a standard problem in nonlinear estimation, which can be solved by a number of methods. One of the most often used is that of Gauss–Newton, which, at its last iteration, the estimate of $\mathbf{Q}^{-1}$ *will provide the correct estimate of the asymptotic covariance matrix for the parameter estimates. A consistent estimator of* σ^2 *can be computed using the residuals*

$$\sigma^2 = (1/n)\sum_i [y_i - h(\boldsymbol{x}_i, \boldsymbol{b})]^2. \tag{5.66}$$

In addition, it has been shown that (Amemiya, 1985)

$$\boldsymbol{b} \to N[\beta, \sigma^2/n\,\mathbf{Q}^{-1}], \tag{5.67}$$

where

$$\boldsymbol{Q} = plim(\underline{X}'\underline{X})^{-1}. \tag{5.68}$$

The sample estimate of the asymptotic covariance matrix is

$$Est.\,Asy.\,Var[b] = \underline{\sigma}^2(\underline{X}'\underline{X})^{-1}. \tag{5.69}$$

From these, inference and hypothesis tests can proceed accordingly.

REASONS THAT THE QUADRATIC FORM IS THE "BEST" SUITED FOR MODELING INDUSTRY STRUCTURE

It has been stated that the quadratic is superior to the translog cost specification in estimating cost functions in multiple-output industries. In addition to its allowing for the "unconstrained emergence of economies of scope and subadditivity" (Kwoka, 1996), it is also the case that the Hessian matrix that results from the translogarithmic cost specification varies over input and output levels, while the Hessian matrix for the quadratic form does not.[10] In this sense, restrictions on concavity for the quadratic cost function are global—they do not change with respect to output and input prices. However, the concavity restrictions on the translog are local—fixed at a specific point, because they depend on prices and output levels. As such, in the case of the latter, the various efficiency measures described earlier must be checked at every level of output and input price combination.

Given this, let us now focus on the quadratic form with an example.

Example 5.4 A Basic Single-Output Quadratic Cost Model [Using Equation (5.53)]

Using data contained in the data set named Coops97, estimate the parameters of Equation (5.53). Table 5.7 displays the estimated Equation (5.53).

---

[10]An excerpt from *www.ricardo.ifas.ufl.edu/aeb6184.production/Lecturepercent2020-2005.ppt* (lecture on subadditivity).

TABLE 5.7 Basic Quadratic Cost Model Results

Variable	Coefficient	Estimate	t Statistic
Constant	a_0	−17.081	−7.07
Y	a_y	0.073	55.95
Ysq	a_{yy}	0.000	−17.72
Capital price	b_k	−0.092	−0.28
Purchased power	b_p	0.332	11.49
Labor price	b_l	0.344	3.67
Adjusted R^2		0.917	

Estimation Results

As expected, estimated coefficients of the output variables indicate that cost is increasing at a decreasing rate. (As an exercise you will calculate the degree-of-scale economies.) Despite a rather high adjusted R^2 (0.917), you will note that the results are not as expected. For example, the coefficient on the capital price variable is negative (but it is not statistically different from zero), which violates the nondecreasing-in-input price criteria for a proper cost model. In addition, it is clearly the case that this cost specification is not homogeneous of degree one in prices, as coefficients of the input price variables do not sum to unity as required. Furthermore, one could take issue with the negatively signed constant because its t statistic indicates that it is a significant explanatory variable in modeling the cost of distributing electricity. However, it is likely the case that these unexpected results emanate from some other problem, such as an incorrect functional form or omitted variable bias, both of which were discussed in Chapter 4. (As an exercise you will impose linear homogeneity.)

Example 5.5 Examining Mayo's Specification

Next let's examine Mayo's quadratic cost specification, which is the single-output version of Equation (5.55). Note: To obtain convergence and unbiased estimates from Mayo's cost model it is necessary to

1. Give starting values for the parameter estimates.
2. Impose linear homogeneity on the input price parameters.
3. Use the generalized method-of-moments (or similar) estimation procedure to estimate this equation.[11]

Table 5.8 displays the results of estimating Equation (5.55). (As an exercise you will calculate the degree-of-scale economies.)

---

[11]The generalized method-of-moments estimation procedure is often used in the presence of heteroscedasticity to yield consistent parameter estimates. In the case of Mayo's model, it is likely the case that the multiplicative functional form yields difficulty in estimating the parameters on the input price variables. [It is straightforward to estimate Equation (5.55) in the absence of input prices.]

TABLE 5.8 Mayo Single-Output Quadratic Cost Model

Variable	Coefficient	Estimate	t Statistic
Constant	a_0	0.0210	5.21
Y	a_y	0.0006	19.64
Ysq	a_{yy}	−0.0000001	−4.41
Capital price	b_k	0.357	3.71
Purchased power	b_p	0.332	4.57
Labor price	b_l	0.312	7.08
Adjusted R^2		0.915	

Discussion of Table 5.8 Results

These results are certainly an improvement over the basic quadratic cost model, which is given by Equation (5.53). The adjusted R^2 of 0.915 indicates that the model fits data well, explaining over 90% of the variation in the cost of distributing electricity. However, in terms of the magnitudes of the coefficients, these results do not make sense; for a distribution entity, the cost of purchased power is by and far the largest component of total cost in terms of input prices. As such, the magnitude of the coefficient should reflect this, which is not the case here. This is rectified here using the logarithmic transformation of the quadratic cost function given in Equation (5.59).

Example 5.6 Estimating a Properly Specified Quadratic Cost Model

Table 5.9 displays estimation results for Equation (5.59). (Note: To offer a fair comparison to Mayo's estimation results, the generalized method-of-moments procedure was used to estimate parameters of the cost model.)

TABLE 5.9 Estimation Results—Equation (5.59)

Variable	Coefficient	Estimate	t Statistic
Constant	a_0	0.031229	6.66
Y	a_y	0.002602	12.63
Ysq	a_{yy}	−0.000001	−6.17
Capital price	b_k	0.070645	1.57
Purchased power	b_p	0.698658	17.50
Labor price	b_l	0.230697	6.09
Adjusted R^2		0.959	

Discussion of Table 5.9 Results

The estimation results displayed in Table 5.9 are much improved over the previous two models. Signs of the estimates accord with a priori expectations in that cost increases at a decreasing rate. Furthermore, the input price coefficients are of an appropriate magnitude given the type of firms that are in the data set. This is confirmed by the adjusted R^2, which indicates that 96% of the variation in cost is explained by the variables in the equation.

END OF SECTION EXERCISES

The data set Coops97 contains data on 711 distribution coops that were RUS borrowers in 1997.

1. Obtain summary statistics for the variables in the data set.
2. Estimate the equations used in Examples 5.4–5.6. Confirm results displayed in Tables 5.7–5.9.
   a. Do you find that any of the models fail to converge?
   b. Concerning Mayo's model: Estimate the unrestricted model (the model prior to imposing the three conditions that are specified above). What do you find?
   c. Next, estimate the model using the generalized method-of-moments procedure. Is convergence achieved? Why or why not?
   d. Despite nonconvergence, the previous step should have yielded estimates for the parameters in the model. Using those as starting values, reestimate the equation. Is convergence attained? Why or why not?
   e. Finally, impose linear homogeneity by restricting the estimates of the input prices sum to unity. Does this achieve convergence?

   Note: If not, you will likely need to reset the convergence criterion and/or increase the number of iterations that the estimation procedure requires.

3. Assuming convergence was attained, calculate the degree of SCE using Equation (5.22) and the estimation results from:
   a. The basic cost model [Equation (5.53)].
   b. Mayo's cost model [Equation (5.55)].
   c. Greer's cost model [Equation (5.59)].

   What are your findings? Are they similar?

4. Add the cost-shift variables for miles of transmission lines and density and then reestimate Equations (5.55) and (5.59).
   a. Do the parameter estimates accord with a priori expectations in terms of sign and statistical significance?
   b. Calculate the degree-of-scale economies using the results obtained. Are they different than what was obtained in Exercise 3? If so, how are they different?

CUBIC COST MODELS

Another form is that of the cubic cost equation in which outputs enter linearly, as quadratics, and to the third degree. Not often used in the economics literature, this form of the equation is actually the most correct because it yields appropriately shaped average and marginal cost curves. These are displayed in Figures 5.2 and 5.3.

The reason that this is important is that economic theory dictates that efficiency occurs when price is equal to marginal cost, as this results in optimal allocation of resources, and resources are devoted to their best output (i.e., productive

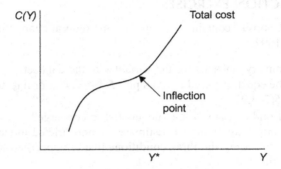

FIGURE 5.2 An appropriately shaped total cost function. A cost function that is cubic in output (Y) yields regions of increasing, constant, and decreasing returns to scale. Y* denotes the inflection point, which is the point at which returns to scale go from being increasing to decreasing.

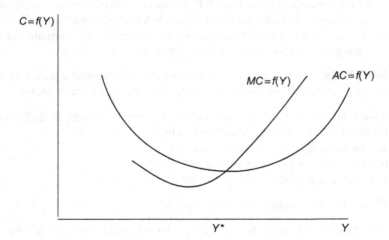

FIGURE 5.3 Average and marginal cost curves. At output levels $Y < Y^*$, increasing returns to scale are indicated by declining average costs [$AC = f(Y)$]. At $Y = Y^*$, returns to scale are constant (and average cost = marginal cost). Beyond this level of output, $Y > Y^*$, diminishing returns to production set in (i.e., marginal cost increases with output) and decreasing returns to scale are experienced.

efficiency). In the case of electricity (and energy services in general), the only way to reduced greenhouse gases (and other emissions) is to price it efficiently and with a rate structure that recovers fixed costs via fixed charges and variables via variable charges. Also known as marginal cost pricing, the concept has been gaining favor in recent years and is the subject of Chapter 8.

In general, a single-output cubic cost function is given by

$$C = \alpha_0 + \alpha_Y Y + \tfrac{1}{2}(\alpha_{YY}Y^2) + \tfrac{1}{3}(\alpha_{YYY}Y^3) + \sum_i \beta_i w_i + \varepsilon, \qquad (5.70)$$

where Y is output (distributed electricity) and w_i are input prices (again capital, labor, and purchased power).

Example 5.7

We could estimate the parameters of this model and test for economies of scale, etc. Using data contained in Coops97, which contains panel data from 1997 to 2001. For the year 1998, the results are obtained and displayed in Table 5.10.

TABLE 5.10 Estimation Results for Cubic Cost Model, 1998 Data

Variable	Estimated Coefficient	t Statistic
Intercept	−15.4932	−4.89
Y	0.059066	9.47
Y^2	4.47E-06	0.61
Y^3	−2.18E-09	−1.74
P_p	0.297248	6.58
P_k	0.426449	1.23
P_l	0.283252	2.84

A Priori Expectations

Before we review the results, let us first review a priori expectations concerning estimated coefficients in the model. In general terms, let the total cost function be given by

$$C(Y) = aY^3 + bY^2 + cY + d. \qquad (5.71)$$

For this to have the appropriate shape (as displayed in Figure 5.2), it is necessary to impose certain restrictions on the parameters a, b, and c.

Equivalently, we know that the marginal cost curve should be positive throughout. In this case, marginal cost is given by

$$\partial C / \partial Y = 3aY^2 + 2bY + c. \qquad (5.72)$$

For this to be positive throughout (as in Figure 5.3), it must be U-shaped with its minimum also greater than zero. In this case, we can ensure this by solving for the level of output that satisfied the first-order condition:

$$\partial^2 C/\partial Y^2 = 6aY + 2b = 0, \tag{5.73}$$

which implies that

$$Y^* = -b/3a. \tag{5.74}$$

(This minimizes marginal cost because

$$\partial^2 MC/\partial Y^2 = 6a \tag{5.75}$$

as long as we restrict $a > 0$.)

What about other parameter restrictions? Substituting $Y*$ into the MC function yields

$$MC \min: \partial C/\partial Y = 3a(-b/3a)^2 + 2b(-b/3a) + c \tag{5.76}$$

or

$$MC \min: \partial C/\partial Y = (3ac - b^2)/3a, \tag{5.77}$$

which implies that

$$3ac > b^2. \tag{5.78}$$

Because we have already restricted that $a > 0$, the aforementioned implies that $c > 0$. (What about b? Is it necessary to impose that $b < 0$? Why or why not?)

Back to the Example at Hand

Given the earlier discussion, it is clearly the case that

$$a_Y > 0, \; a_{YY} < 0, \; \text{and} \; a_{YYY} > 0. \tag{5.79}$$

(Of course, the coefficients, β_i, on the input price variables must be positive to accord with theory.) Do the results obtained accord with Equation (5.79)? Estimation results are displayed in Table 5.10.

Discussion

Despite a rather high adjusted R^2 of 0.93, estimation results do not accord with a priori expectations. The quadratic and cubic terms in output are of the wrong sign. Clearly, this particular equation is not well suited to underlying data, which could be due to the functional form itself or another issue, such as an omitted variable (see Example 4.4 on omitted variable bias in Chapter 4).

Another option is to specify a log-linear model to see if that improves the estimation results. In an exercise you will do this, that is, using the same data as before, to specify a model of the form:

$$\ln C = a_0 + a_1 \ln Y + a_2 \ln Y^2 + a_3 \ln Y^3$$
$$+ b_2 \ln pp + b_1 \ln pk + b_3 \ln pl + o_1 \ln tr + o_2 \ln dens. \tag{5.80}$$

MULTIPLE-OUTPUT MODELS

Recall earlier discussions regarding distributed electricity as a multiple-output product. Given that in the aforementioned specified single-output, could this be the cause of the unexpected results? Let's see.

Example 5.8

Using data in the data set (1998), estimate the parameters of the following model:

$$C = \alpha_0 + \alpha_1 \ln Y_1 + \alpha_2 \ln Y_2 + \alpha_{11}(\ln Y_1)^2 + \alpha_{22}(\ln Y_2)^2 + \alpha_{12}\ln Y_1 \ln Y_2$$
$$+ \alpha_{111}(\ln Y_1)^3 + \alpha_{222}(\ln Y_2)^3 + \alpha_{112}(\ln Y_1)^2 \ln Y_2 \tag{5.81}$$
$$+ \alpha_{221}(\ln Y_2)^2 \ln Y_1 + \Sigma_i \beta_i \ln w_i + \Sigma_m \theta_m \ln O_m,$$

where $\ln Y_1$ is the natural logarithm of Y_1, electricity distributed to small users; $\ln Y_2$ is the natural logarithm of Y_2, electricity distributed to large users; $(\ln Y_1)^2$ is the natural logarithm of Y_1, electricity distributed to small users, squared; $(\ln Y_2)^2$ is the natural logarithm of Y_2, electricity distributed to large users, squared; $\ln Y_1 \ln Y_2$ is the natural logarithm of Y_1 multiplied by natural logarithm Y_2; $(\ln Y_1)^3$ is the natural logarithm of Y_1, electricity distributed to small users, squared; $(\ln Y_2)^3$ is the natural logarithm of Y_2, electricity distributed to large users, squared; $(\ln Y_1)^2 \ln Y_2$ is the natural logarithm of Y_1 squared multiplied by natural logarithm Y_2; $(\ln Y_2)^2 \ln Y_1$ is the natural logarithm of Y_2 squared multiplied by natural logarithm Y_1; $\ln w_i$ is the natural logarithm of the input price, $i = 1, 2, 3$; and $\ln O_m$ is the natural logarithm of the cost-shift variable, $m = 1, 2$.

Table 5.11 displays estimation results (with linear homogeneity imposed).

TABLE 5.11 Multiple-Output Cubic Cost Model Estimation Results

Parameter	Variable	Estimate	t Statistic
a_0	Constant	1.143	1.41
a_1	$\ln Y_1$	1.338	2.84
a_2	$\ln Y_2$	0.226	2.30
a_{11}	$(\ln Y_1)^2$	−0.103	−1.08

(Continued)

TABLE 5.11 Multiple-Output Cubic Cost Model Estimation Results—cont'd

Parameter	Variable	Estimate	t Statistic
a_{12}	$\ln Y_1 \ln Y_2$	−0.098	−1.99
a_{22}	$(\ln Y_2)^2$	0.070	4.30
a_{111}	$(\ln Y_1)^3$	0.008	1.23
a_{222}	$(\ln Y_2)^3$	0.004	5.77
a_{112}	$(\ln Y_1)^2 \ln Y_2$	−0.014	−3.78
a_{221}	$(\ln Y_2)^2 \ln Y_1$	0.009	1.31
b_2	$\ln P_P$	0.687	13.34
b_1	$\ln P_k$	0.086	1.55
b_3	$\ln P_l$	0.227	4.71
o_1	$\ln tr$	0.035	4.53
o_2	$\ln dens$	−0.123	−3.66
Adjusted R^2	0.9611		

Discussion

The adjusted R^2 indicates that the model fits data well. There are, however, several issues that must be addressed. First, the estimated coefficient of $(\ln Y_2)^2$ is not of the expected sign; as such, the implication is that the cost of distributing electricity to large users is increasing at an increasing rate. By Young's theorem, also known as the symmetry of second derivatives, we know that

$$a_{12} = a_{21},$$

and

$$a_{112} = a_{221} = a_{212}.$$

The negatively signed coefficients on the cross-quadratic terms in output indicate that there are economies of scope in the distribution of electricity to both types of end user. (Why?)

In terms of cubic terms in output, do signs of the estimated coefficients accord with theory? Why or why not?

In this model, linear homogeneity in input prices has been imposed and a test of the hypothesis that

$$H_0: \sum_i \beta_i = 1 \tag{5.82}$$

$$H_a: \sum_i \beta_i \neq 1 \tag{5.83}$$

yields a t statistic of 1.22, which indicates that we should not reject the null hypothesis. Were this restriction not imposed on the model, how would the parameter estimates (and t statistics) be different?

MORE COMPLEX MULTIPLE-OUTPUT MODELS

Clearly, there is still an issue with the model that we are trying to specify, as the results obtained thus far do not accord with economic theory. A next step is to specify a model that is nonlinear in parameters to see whether the functional form is the problem. One such model is given by

$$C = [\alpha_0 + \alpha_1 Y_1 + \alpha_2 Y_2 + \tfrac{1}{2}(\alpha_{11} Y_1{}^2 + \alpha_{22} Y_2{}^2 + 2^* \alpha_{12} Y_1 Y_2)$$
$$+ \tfrac{1}{3}(\alpha_{111} Y_1{}^3 + \alpha_{222} Y_2{}^3 + 2^*(\alpha_{112} Y_1{}^2 Y_2 + \alpha_{221} Y_2{}^2 Y_1))] \prod w_i{}^{\beta i} \prod O_m{}^{\theta m} e^\varepsilon.$$
$$(5.84)$$

This model specification uses the same variables as before but the functional form has clearly changed; while the equation in brackets is the same, now input prices and the two cost-shift variables are multiplicatively appended and are nonlinear in parameters. Furthermore, the error term is no longer additive, which obviates the use of ordinary least squares. This is similar to the quadratic model that was introduced earlier and in *Electricity Cost Modeling Calculations* (Greer, 2011).

Certain software has the capability of estimating this model outright. Otherwise, this model can be transformed so that the input prices, cost shifters, and error term enter additively. In other words, a model of the form

$$\ln C = \ln[\alpha_0 + \alpha_1 Y_1 + \alpha_2 Y_2 + \tfrac{1}{2}(\alpha_{11} Y_1{}^2 + \alpha_{22} Y_2{}^2 + 2^* \alpha_{12} Y_1 Y_2)$$
$$+ \tfrac{1}{3}(\alpha_{111} Y_1{}^3 + \alpha_{222} Y_2{}^3 + 2^*(a_{112} Y_1{}^2 Y_2 \qquad (5.85)$$
$$+ \alpha_{221} Y_2{}^2 Y_1))] + \sum_i \beta_i \ln w_i + \sum_m \theta_m \ln O_m + \varepsilon.$$

Estimation results are presented in Table 5.12.

Discussion

All estimates accord with theory in terms of sign although not all are statistically different from zero. The coefficients of the linear terms in output are positive (and different from zero) and those of the quadratics are negative, thus indicating that cost is increasing in output at a decreasing rate, which generates both product-specific economies of scale and economies of scope. While the cubic terms in output are positive in sign as required, only that of output distributed to large customers is significantly different from zero; at lower confidence levels, the other cubed-output coefficients are different from zero.

TABLE 5.12 More Complex Model Results

Parameter	Variable	Estimate	t Statistic
a_0	Constant	19.33128	3.01
a_1	$\ln Y_1$	4.02996	8.54
a_2	$\ln Y_2$	2.89917	5.40
a_{11}	Y_1^2	−0.00191	−2.96
a_{12}	$Y_1 Y_2$	−0.00496	−2.50
a_{22}	Y_2^2	−0.00380	−2.84
a_{111}	Y_1^3	0.000001	1.48
a_{222}	Y_2^3	0.000002	2.44
a_{112}	$Y_1^2 Y_2$	0.000002	0.97
a_{221}	$Y_2^2 Y_1$	0.000003	1.66
b_2	$\ln P_P$	0.19155	3.08
b_1	$\ln P_k$	0.59305	13.24
b_3	$\ln P_l$	0.21539	4.45
o_1	$\ln tr$	0.038878	5.30
o_2	$\ln dens$	−0.09549	−4.10
Adjusted R^2	0.9660		
Restriction	$\sum_i \beta_i = 1$	2.557	0.36

In addition, input price estimates are of the expected sign and magnitude. A test of the null hypothesis of linear homogeneity is affirmed. Cost-shift variables, too, are of the expected sign and are important contributors in explaining the variation in the cost of providing electricity to end users.

OTHER ISSUES

As stated previously, heteroskedasticity tends to be an issue with cross-sectional data. That is also the case here. (You may recall the discussion on this presented in Chapter 4.) Despite using the generalized method of moments (GMM) estimation technique to estimate this model, White's test still indicates that heteroskedasticity is an issue (see appendix on GMM at the end of this chapter). Fortunately, because it does not tend to affect the parameter estimates (only standard errors, and hence t statistics), any conclusions reached concerning economies of scale and scope remain valid.

MEASURES OF EFFICIENCY FOR MULTIPLE-OUTPUT MODELS

You may recall the discussion from Chapter 2 concerning economies of scale. For a multiple-output model, it is no longer the case that scale economies may be obtained by dividing average cost by marginal cost. Now it is not one but several concepts that are relevant for the analysis. A brief review of the relevant concepts is in order.

RAY COST OUTPUT ELASTICITY

The first measure to be computed is that of ray cost output elasticity. Ray average costs describe the behavior of the cost function as output is expanded proportionally along a ray emanating from the origin. Ray average cost declines whenever a small proportional change in output leads to a less-than-proportional change in total cost. Ray average cost is defined as

$$RAC(y) = C(y)/\sum y_i. \tag{5.86}$$

Ray cost output elasticity describes how responsive cost is to a change in output along a ray emanating from the origin. In general, ray cost output elasticities depend not only on the level of output, but also on the output mix between Y_1 and Y_2. Formally, it is defined as

$$\partial C/\partial \sum y_i * Y / C(y). \tag{5.87}$$

When this is less than one, ray economies are present. In the present context, this implies that increasing the amount of electricity to one customer group yields a lower cost to distributing electricity to another customer group.

DEGREE-OF-SCALE ECONOMIES

As the analog to the single-output concept of economies of scale, the degree-of-scale economies, S_N, is equal to the ratio of average cost to marginal cost. In the multiple-output case we have

$$S_N(Y) = C(Y)/Y_i C_i(Y), \tag{5.88}$$

where $C_i(Y)$ is the marginal cost with respect to Y_i for $i = 1, \ldots, n$.

Baumol et al. (1982, p. 51) have shown that

$$S_N = 1/(1 + e), \tag{5.89}$$

where e is the elasticity of RAC (tY) with respect to t at point Y (t is a scalar).

TABLE 5.13 Degree-of-Scale Economies

Year	Percentage $S_1 > 1$	Percentage $S_2 > 1$
1997	98%	72%
1998	98%	73%
1999	98%	74%
2000	97%	75%
2001	95%	75%

Corollary

Returns to scale at the output point y are increasing, decreasing, or locally constant ($S_N > 1$, $S_N < 1$, $S_N = 1$, respectively) as the elasticity of RAC at y is negative, positive, or zero, respectively. Moreover, increasing or decreasing returns at y imply that RAC is decreasing or increasing at Y, respectively.

As such, S_N (the degree-of-scale economies) may be interpreted as a measure of the percentage rate of decline or increase in ray average cost with respect to output (Baumol et al., 1982).

Table 5.13 displays results for the coops in the data set. As can be seen, almost all coops experienced increasing returns in the distribution of electricity to small users and three-fourths experienced increasing returns to large users. These percentages have been consistent over the years included in the data set (1997–2001).

PRODUCT-SPECIFIC ECONOMIES OF SCALE

Because output is not always expanded proportionally for a multiproduct firm, the concept of incremental cost must be examined. This is defined as

$$AIC_i(Y) \equiv IC_i(Y)/Y_i. \tag{5.90}$$

This allows for the identification of returns to scale specific to a particular output, known as product-specific returns to scale, which are expressed as

$$S_i(Y) = IC_i(Y)/Y_i C_i, \tag{5.91}$$

which is equivalent to

$$S_i(Y) = AIC(Y_i)/(\partial C/\partial Y_i). \tag{5.92}$$

According to Baumol et al. (1982), $S_i(Y)$ has a natural interpretation in that revenues collected from the sale of product i ($Y_i^* C_i$) when its price equals marginal cost will exceed, equal, or fall short of the incremental cost (IC_i) incurred by offering that product as $S_i(Y)$ is less than, equal to, or greater than unity, respectively. Thus, returns to scale of product i are said to be increasing, decreasing, or constant as $S_i(Y)$ is greater than, less than, or equal to unity. Because $S_i(Y)$ is the increment in the firm's total cost that results from the addition of an entire product to the firm's set of products, and if the marginal cost is less than the average incremental cost, the latter has a negative derivative and will decline as Y_i increases (Baumol et al., 1982).

Furthermore, the degree of product-specific returns to scale, which measure the economies or diseconomies uniquely associated with the production of a single-output, given that the firm may produce positive amounts of other products for the former, is given by

$$S_i = [\alpha_i Y_i + 1/2(\alpha_{ii} Y_i^2) + \textstyle\sum_{i \neq j} \alpha_{ij} Y_i Y_j] / [\alpha_i Y_i + \alpha_{ii} Y_i^2 + \textstyle\sum_{i \neq j} \alpha_{ij} Y_i Y_j] \qquad (5.93)$$

In general, for the quadratic functional form, the economies or diseconomies associated with production of a single product (given that the firm produces positive amounts of other products) are determined by the sign of the estimated coefficient of the squared term in output; that is,

$S_i > 0$ for α_{jj} (estimated coefficient of squared-output variable) < 0
$S_i < 0$ for α_{jj} (estimated coefficient of output-squared variable) > 0.

The concepts of ray average cost, average incremental cost, and product-specific economies of scale can all be characterized with the behavior of a cost hypersurface in a cross-section of cost and output space.

In addition, a crucial cost concept exists that cannot be characterized in this fashion, as it involves the simultaneous production of several different outputs in a single enterprise. Such economies result from the scope of a firm's operations instead of the scale of its operations. This is the topic of the next section.

ECONOMIES OF SCOPE

Economies of scope (or economies of joint production) are said to exist if a given quantity of each of two or more goods can be produced by one firm at a lower cost than if each good were produced separately by two different firms or even two different production processes. That is, for a two-product case, weak economies of scope are given by

$$C(Y_1, Y_2) \leq C(Y_1, 0) + C(0, Y_2) \qquad (5.94)$$

for all $Y_1, Y_2 > 0$. If not, then there are diseconomies of scope, and separate production of outputs is more efficient.

What is useful here is a measure of the degree of economies of scope, which would allow for, in the presence of economies of scope, capturing of the relative increase in cost that would result from separate production of the two (or more) outputs. Thus, the degree of economies of scope is given by

$$S_c = [C(Y_1, 0) + C(0, Y_2) - C(Y_1, Y_2)]/C(Y_1, Y_2). \qquad (5.95)$$

Due to the regulated nature of this industry, it is important to keep in mind that a divergence exists between *actual* economies experienced and those that could *potentially* result if the industry were not regulated. It is well documented in the literature that regulation reduces the competitiveness or efficiency of firms by lack of entry and via the average cost pricing mechanism imposed upon firms in the industry. Mayo mentions this in his 1984 paper, which, among others, attempts to identify the existence of scope economies in this industry. Using a sample of 200 firms (privately-owned electric and natural gas utilities), he employs a modified (short run, since incorporates fixed costs) multiproduct quadratic cost function to estimate the cost of producing both electricity and gas and confirms the presence of economies of scope for small firms using 1979 data. However, as output increases, he asserts that the absence of competitive pressure leads to cost inefficiencies and eventual diseconomies of scope.

For the multiproduct cubic cost model given in Equation (5.56), the degree of economies of scope, S_c, is given by

$$S_c = (\alpha_0 - \alpha_{12}Y_1Y_2 - \alpha_{112}Y_1^2Y_2 - \alpha_{221}Y_1Y_2^2)/C(Y_1Y_2). \qquad (5.96)$$

For $S_c > 0$, there are economies of scope in the production of both goods.

For virtually every coop in the data set, there are economies of scope in the distribution of electricity to both types of end user. (Not surprising, as signs of estimated coefficients virtually guarantee this.) This can be seen in the case study on cubic cost function estimation.

COST COMPLEMENTARITY

You may also recall from Chapter 2 that cost complementarity, which requires that marginal or incremental costs of any output decline when that output or any other output is increased, is a sufficient condition for economies of scope. For a twice differential multiproduct cost function, cost complementarity exists if

$$\partial^2 C(Y)/\partial Y_i \partial Y_j < 0 \qquad (5.97)$$

for $i \neq j$.

The sign of the estimated coefficient of the interaction term in output determines whether there are cost complementarities (interproduct complementarities) in the distribution of output.

Example 5.4

Using the data set Coops97, estimate the parameters of a two-output, three input-price cost function [given in Equation (5.59)]. Outputs are electricity distributed to small users (Y_1) and electricity distributed to large users (Y_2) with the prices of labor, capital, and purchased power as inputs. Estimation results are displayed in Table 5.14.

TABLE 5.14 Estimation Results—Equation (5.59)

Coefficient	Variable	Estimate	t Statistic
a_0	Constant	0.0188620	7.30
a_1	Y_1	0.0028870	17.05
a_2	Y_2	0.0018150	12.72
a_{11}	Y_1sq	−0.0000007	−3.62
a_{22}	Y_2sq	−0.0000003	−2.22
a_{21}	Y_1Y_2	−0.0000005	−1.17
b_k	$\ln p_k$	0.027516	0.80
b_p	$\ln p_p$	0.650672	23.2
b_l	$\ln p_l$	0.321813	11.68
Adjusted R^2		0.96	

Discussion of Table 5.14 Results

Estimation results accord with a priori expectations in terms of the signs of the estimated coefficients and indicate that cost increases with output at a decreasing rate. Cost complementarity is confirmed by the sign of a_{21}, the estimated coefficient of Y_1Y_2, which is indeed negative as required. The adjusted R^2 square indicates that the model explains over 96% of the variation in the cost of distributing electricity in 1997.

CONCLUSION

This chapter provided a rather extensive overview of several cost models used to estimate efficiency in the distribution of electricity to end users. Exercises at the end of each section provide hands-on experience with data analysis, model estimation, and the interpretation of the results.

END OF SECTION EXERCISES: MULTIPLE-OUTPUT COST MODELS

Again using the data set Coops97, perform the following exercises for the multiproduct cost models described in this chapter.

1. Estimate a multiple-output version of Equation (5.70) in which outputs are electricity distributed to small users (residential and small commercial customers, Y_1) and electricity distributed to large users (large commercial and industrial users, Y_2). Do your results accord with theory? If not, how so?
2. Add the variables "miles of transmission lines" and "customer density." Does this improve your results?
3. Estimate Equation (5.80). Does converting data into logarithms improve your results?
4. Again assuming that convergence was obtained, calculate the degree of economies of scope using Equation (5.84) without cost-shift variables. What do you conclude about the firms in the sample data set?
5. Add the cost-shift variables "miles of transmission line" and "density" and then reestimate Equation (5.85). Do your results accord with those in Table 5.12? If not, what is different?
6. Reestimate Equation (5.81) and impose linear homogeneity in input prices. How does this affect estimation results?

APPENDIX: GENERALIZED METHOD OF MOMENTS (GMM)

The GMM is a very general statistical method for obtaining estimates of parameters of statistical models. It is a generalization, developed by Lars Peter Hansen, of the method of moments.

The term GMM is very popular among econometricians but is hardly used at all outside of economics, where the slightly more general term *estimating equations* is preferred.

Introduction

A typical econometric problem can be formulated in the following terms: Suppose available data consist of a large number of i.i.d. observations

$$\{Y_t\}_{t=1}^{T}, \tag{5.98}$$

where each observation Y_t is an n-dimensional multivariate random variable. Our knowledge of economics dictates a certain econometric model for these data. Such a model is usually defined only up to some parameter, which we will denote by $\theta \in \Theta$. Our main goal is to seek the "true" value of this parameter, θ_0, or at least to find a reasonably close estimate.

To apply GMM there should exist a (possibly vector-valued) function $g(Y, \theta)$ such that

$$m(\theta_0) \equiv E[g(Y_t, \theta_0)] = 0, \tag{5.99}$$

where E denotes expectation and Y_t is just a generic observation, which are all assumed to be i.i.d. Moreover, function $m(\theta)$ must not be equal to zero for $\theta \neq \theta_0$, or otherwise parameter θ will not be identified.

The basic idea behind GMM is to replace theoretical expected value E with its empirical analog—sample average:

$$\hat{m}(\theta) = \hat{E}[g(Y_t, \theta)] \equiv \frac{1}{T} \sum_{t=1}^{T} g(Y_t, \theta). \tag{5.100}$$

By the law of large numbers,

$$\hat{m}(\theta) \approx m(\theta) \tag{5.101}$$

for large values of T, so if we can find a number $\hat{\theta}$ such that

$$\hat{m}(\hat{\theta}) \approx \theta, \tag{5.102}$$

then such number will be a reasonably good estimate for parameter θ_0. So basically all we need to do is to search the parameter space Θ for a number θ that would minimize the distance between $\hat{m}(\theta)$ and zero. For a vector-valued function $\hat{m}$, the notion of distance can be defined in many different ways, and it actually turns out that the obvious Euclidean norm is not the best choice. Instead, a new positive semidefinite "weighting" matrix $\hat{W}_T$ is often used to define the norm as a quadratic form:

$$\|\hat{m}\| = \hat{m}' \hat{W}_T \hat{m}, \tag{5.103}$$

where $'$ denotes matrix transposition. Thus, the GMM estimator can be written as

$$\hat{\theta} = \arg\min_{\theta \in \Theta} \left(\frac{1}{T} \sum_{t=1}^{T} g(Y_t, \theta) \right)' \hat{W}_T \left(\frac{1}{T} \sum_{t=1}^{T} g(Y_t, \theta) \right). \tag{5.104}$$

Under suitable conditions this estimator is consistent, asymptotically normal, and, with right choice of weighting matrix $\hat{W}_T$, asymptotically efficient.

Consistency

Consistency is the most important property of an estimator. It means that having a sufficient number of observations, the estimator will get arbitrarily close to the true value of the parameter:

$$\hat{\theta} \xrightarrow{p} \theta_0 \quad \text{as} \quad T \to \infty \tag{5.105}$$

(see convergence in probability[12]). Necessary and sufficient conditions for a GMM estimator to be consistent are as follows:

1. $\hat{W}_T \xrightarrow{p} W$, where W is a positive semidefinite matrix.
2. $WE[g(Y_t, \theta)] = 0$ only for $\theta = \theta_0$.
3. $\theta_0 \in \Theta$, which is compact.
4. $g(Y, \theta)$ is continuous at each θ with probability one.
5. $E[\sup_{\theta \in \Theta} \|g(Y, \theta)\| < \infty]$.

The second condition here (so-called global identification condition) is often particularly hard to verify. Simpler necessary but not sufficient conditions exist, which may be used to detect a nonidentification problem:

Order condition. The dimension of moment function $m(\theta)$ should be at least as large as the dimension of parameter vector θ.

Local identification. If $g(Y, \theta)$ is continuously differentiable in a neighborhood of θ_0, then matrix $WE[\nabla_\theta g(Y_t, \theta_0)]$ must have full column rank.

---

[12]A sequence $\{X_n\}$ of random variables *converges in probability* toward X if for all $\varepsilon > 0$,

$$\lim_{n \to \infty} \Pr(|X_n - X| \ge \varepsilon) = 0. \qquad (5.105a)$$

Formally, pick any $\varepsilon > 0$ and any $\delta > 0$. Let P_n be the probability that X_n is outside the ball of radius ε centered at X. Then for X_n to converge in probability to X there should exist a number N_δ, such that for all $n \ge N_\delta$, the probability P_n is less than δ. Convergence in probability is denoted by adding the letter p over an arrow, indicating convergence, or using the "plim" probability limit operator:

$$X_n \xrightarrow{p} X, \ X_n \xrightarrow{P} X, \ \plim_{n \to \infty} X_n = X. \qquad (5.105b)$$

For random elements $\{X_n\}$ on a separable metric space (S, d), convergence in probability is defined similarly by

$$\Pr(d(X_n, X) \ge \varepsilon) \to 0, \ \forall \varepsilon > 0. \qquad (5.105c)$$

The properties are:

- Convergence in probability implies convergence in distribution.
- In the opposite direction, convergence in distribution implies convergence in probability only when the limiting random variable X is a constant.
- The *continuous mapping theorem* states that for every continuous function $g(\cdot)$, if

$$X_n \xrightarrow{p} X,$$

then also

$$g(X_n) \xrightarrow{p} g(X).$$

- Convergence in probability defines a topology on the space of random variables over a fixed probability space. This topology is metrizable by the *Ky Fan metric*:

$$d(X, Y) = \inf\{\varepsilon > 0: \Pr(|X - Y| > \varepsilon) \le \varepsilon\} \qquad (5.105d)$$

(Dudley, 2002).

In practice, applied econometricians often simply *assume* that global identification holds, without actually proving it (Newey and McFadden, 1994).

Asymptotic Normality

Asymptotic normality is a useful property, as it allows one to construct confidence bands for the estimator and conduct different tests. Before we can make a statement about the asymptotic distribution of the GMM estimator, we need to define two auxiliary matrices:

$$G = E[\nabla_\theta g(Y_t, \theta_0)], \quad \Omega = E[g(Y_t, \theta_0)g(Y_t, \theta_0)']. \tag{5.106}$$

Then under conditions 1–6 that follow, the GMM estimator will be asymptotically normal with limiting distribution

$$\sqrt{T}(\hat{\theta} - \theta_0) \xrightarrow{d} N[0, (G'WG)^{-1}G'W\Omega WG(G'WG)^{-1}] \tag{5.107}$$

(see convergence in distribution[13]).

---

[13]A sequence $\{X_1, X_2, \ldots\}$ of random variables is said to *converge in distribution*, or *converge weakly*, or *converge in law* to a random variable X if

$$\lim_{n \to \infty} F_n(x) = F(x), \tag{5.107a}$$

for every number $x \in \mathbf{R}$ at which F is continuous. Here F_n and F are the cumulative distribution functions (CDFs) of random variables X_n and X, correspondingly. The requirement that only the continuity points of F should be considered is essential. For example, if X_n are distributed uniformly on intervals $[0, 1/n]$, then this sequence converges in distribution to a degenerate random variable $X = 0$. Indeed, $F_n(x) = 0$ for all n when $x \leq 0$, and $F_n(x) = 1$ for all $n \geq 1/x$ when $x > 0$. However, for this limiting random variable, $F(0) = 1$, even though $F_n(0) = 0$ for all n. Thus, the convergence of CDFs fails at the point $x = 0$ where F is discontinuous. Convergence in distribution may be denoted as

$$X_n \xrightarrow{d} X, \quad X_n \xrightarrow{D} X, \quad X_n \xrightarrow{\mathscr{L}} X, \quad X_n \xrightarrow{d} \mathscr{L}_x, X_n \rightsquigarrow X, \quad X_n \Rightarrow X, \mathscr{L}(X_n) \to \mathscr{L}(X), \tag{5.107b}$$

where $\mathscr{L}_x$ is the law (probability distribution) of X. For example, if X is standard normal we can write

$$X_n \xrightarrow{d} N(0, 1). \tag{5.107c}$$

For random vectors $\{X_1, X_2, \ldots\} \subset \mathbf{R}^k$ the convergence in distribution is defined similarly. We say that this sequence *converges in distribution* to a random k-vector X if

$$\lim_{n \to \infty} \Pr(X_n \in A) = \Pr(X \in A) \tag{5.107d}$$

for every $A \subset \mathbf{R}^k$, which is a continuity set of X. The definition of convergence in distribution may be extended from random vectors to more complex random elements in arbitrary metric spaces, and even to the "random variables" that are not measurable—a situation that occurs, for example, in the study of empirical processes. This is the "weak convergence of laws without laws being defined"—except asymptotically (Bickel et al., 1998). In this case the term *weak convergence* is preferable, and we say that a sequence of random elements $\{X_n\}$ converges weakly to X (denoted as $X_n \Rightarrow X$) if

$$E * h(X_n) \to Eh(X) \tag{5.107e}$$

for all continuous bounded functions $h(\cdot)$ (van der Vaart and Wellner, 1996). Here E^* denotes the *outer expectation*; that is, the expectation of a "smallest measurable function g that dominates $h(X_n)$."

Conditions

1. θ is consistent (see previous section).
2. θ_0 lies in the interior of set Θ.
3. $g(Y, \theta)$ is continuously differentiable in some neighborhood N of θ_0 with probability one.
4. $E[\|g(Y_t, \theta)\|^2] < \infty$.
5. $E[\sup_{\theta \in N} \|\nabla_\theta g(Y_t, \theta)\|] < \infty$.
6. Matrix $G'WG$ is nonsingular.

Efficiency

So far we have said nothing about the choice of matrix W, except that it must be positive semidefinite. In fact, any such matrix will produce a consistent and asymptotically normal GMM estimator; the only difference will be in the asymptotic variance of that estimator. It can be shown that taking

$$W \infty \Omega^{-1} \tag{5.108}$$

will result in the most efficient estimator in the class of all asymptotically normal estimators. Efficiency in this case means that such an estimator will have

---

The properties are:

- Since $F(a) = \Pr(X \leq a)$, the convergence in distribution means that the probability for X_n to be in a given range is approximately equal to the probability that the value of X is in that range, provided n is sufficiently large.
- In general, convergence in distribution does not imply that the sequence of corresponding probability density functions will also converge. As an example, one may consider random variables with densities $f_n(x) = (1 - \cos(2\pi n x)) 1_{\{x \in (0,1)\}}$. These random variables converge in distribution to a uniform $U(0, 1)$, whereas their densities do not converge at all (Romano and Siegel, 1985).
- Portmanteau lemma provides several equivalent definitions of convergence in distribution. Although these definitions are less intuitive, they are used to prove a number of statistical theorems. The lemma states that $\{X_n\}$ converges in distribution to X if and only if any of the following statements are true:
  - $Ef(X_n) \rightarrow Ef(X)$ for all bounded, continuous functions f.
  - $Ef(X_n) \rightarrow Ef(X)$ for all bounded, Lipschitz functions f.
  - $\limsup\{Ef(X_n)\} \leq Ef(X)$ for every upper semicontinuous function f bounded from above.
  - $\liminf\{Ef(X_n)\} \geq Ef(X)$ for every lower semicontinuous function f bounded from below.
  - $\limsup\{\Pr(X_n \in C)\} \leq \Pr(X \in C)$ for all closed sets C.
  - $\liminf\{\Pr(X_n \in U)\} \geq \Pr(X \in U)$ for all open sets U.
  - $\lim\{\Pr(X_n \in A)\} = \Pr(X \in A)$ for all continuity sets A of random variable X.
- Continuous mapping theorem states that for a continuous function $g(\cdot)$, if the sequence $\{X_n\}$ converges in distribution to X, then so does $\{g(X_n)\}$ converge in distribution to $g(X)$.
- Lévy's continuity theorem states that the sequence $\{X_n\}$ converges in distribution to X if and only if the sequence of corresponding characteristic functions $\{\varphi_n\}$ converges pointwise to the characteristic function φ of X, and $\varphi(t)$ is continuous at $t = 0$.
- Convergence in distribution is metrizable by the Lévy–Prokhorov metric.
- A natural link to convergence in distribution is the Skorokhod's representation theorem.

the smallest possible variance (we say that matrix A is smaller than matrix B if $B - A$ is positive semidefinite).

In this case, the formula for the asymptotic distribution of the GMM estimator simplifies to

$$\sqrt{T}(\hat{\theta} - \theta_0) \xrightarrow{d} N[0, (G'\Omega^{-1}G)^{-1}]. \tag{5.109}$$

(The proof that such a choice of weighting matrix is indeed optimal is quite elegant and is often adopted with slight modifications when establishing efficiency of other estimators. As a rule of thumb, a weighting matrix is optimal whenever it makes the "sandwich formula" for variance collapse into a simpler expression.)

Implementation

One difficulty with implementing the outlined method is that we cannot take $W = \Omega^{-1}$ because, by the definition of matrix Ω, we need to know the value of θ_0 to compute this matrix, and θ_0 is precisely the quantity we don't know and are trying to estimate in the first place.

Several approaches exist to deal with this issue, the first one being the most popular.

1. Two-step feasible GMM:
   Step 1. Take $W_T = I$ (the identity matrix) and compute preliminary GMM estimate $\theta_{(1)}$. This estimator is consistent for θ_0, although probably not efficient.
   Step 2. Take

   $$\hat{W}_T = \left(\frac{1}{T} \sum_{t=1}^{T} g(Y_t, \hat{\theta}_{(1)}) g(Y_t, \hat{\theta}_{(1)})' \right)^{-1}, \tag{5.110}$$

   where we have plugged in our first-step preliminary estimate $\hat{\theta}_{(1)}$. This matrix converges in probability to Ω^{-1}; therefore, if we compute $\hat{\theta}$ with this weighting matrix, the estimator will be asymptotically efficient.
2. Iterative GMM. Essentially the same procedure as the two-step GMM except that the matrix W_T is recalculated several times; that is, the estimate obtained in step 2 is used to calculate the weighting matrix for step 3 and so on. Asymptotically, no improvement can be achieved through such iterations, although certain Monte Carlo experiments suggest that finite sample properties of this estimator are slightly better.
3. Continuously updating GMM (CUE). Estimates θ simultaneously with estimating the weighting matrix W:

$$\hat{\theta} = \arg \min_{\theta \in \Theta} \left(\frac{1}{T} \sum_{t=1}^{T} g(Y_t, \theta) \right)' \left(\frac{1}{T} \sum_{t=1}^{T} g(Y_t, \theta) g(Y_t, \theta)' \right)^{-1} \left(\frac{1}{T} \sum_{t=1}^{T} g(Y_t, \theta) \right). \tag{5.111}$$

Another important issue regarding the implementation of minimization procedure is that the function is supposed to search through (possibly high-dimensional) parameter space Θ and find the value of θ, which minimizes the objective function. No generic recommendation for such a procedure exists, as it is a subject of its own field, numerical optimization.

J Test

When the number of moment conditions is greater than the dimension of the parameter vector θ, the model is said to be *overidentified*. Overidentification allows us to check whether the model's moment conditions match data well or not. (This is discussed further in Chapter 9.)

Conceptually, we can check whether $\hat{m}(\theta)$ is sufficiently close to zero to suggest that the model fits data well. The GMM method has then replaced the problem of solving the equation

$$\hat{m}(\theta) = 0, \tag{5.112}$$

which chooses θ to match the restrictions exactly by a minimization calculation. Minimization can always be conducted even when no θ_0 exists, such that $m(\theta_0) = 0$. This is what the J test does. The J test is also called a *test for overidentifying restrictions*.

Formally, we consider two hypotheses:

$$H_0: m(\theta_0) = 0 \tag{5.113}$$

(the null hypothesis that the model is "valid"), and

$$H_1: m(\theta) \neq 0, \forall \theta \in \Theta \tag{5.114}$$

(the alternative hypothesis that the model is "invalid"; data do not come close to meeting the restrictions).

Under hypothesis H_0, the following so-called J statistic is asymptotically χ^2 with $k - l$ degrees of freedom. Define J to be

$$J \equiv T \cdot \left(\frac{1}{T} \sum_{t=1}^{T} g(Y_t, \hat{\theta})\right)' \hat{W}_T \left(\frac{1}{T} \sum_{t=1}^{T} g(Y_t, \hat{\theta})\right) \xrightarrow{d} \chi^2_{\kappa - \ell} \tag{5.115}$$

under H_0, where $\hat{\theta}$ is the GMM estimator of the parameter θ_0, k is the number of moment conditions (dimension of vector g), and l is the number of estimated parameters (dimension of vector θ). Matrix $\hat{W}_T$ must converge in probability to Ω^{-1}, the efficient weighting matrix (note that previously we only required that W be proportional to Ω^{-1} for the estimator to be efficient; however, to conduct the J test, W must be exactly equal to Ω^{-1}, not simply proportional).

Under the alternative hypothesis H_1, the J statistic is asymptotically unbounded:

$$J \xrightarrow{p} \infty \tag{5.116}$$

under H_1.

To conduct the test, we compute the value of J from the data. It is a nonnegative number. We compare it with (say) the 0.95 quantile of the $\chi^2_{\kappa-\ell}$ distribution:

- H_0 is *rejected* at 95% confidence level if $J > q^{\chi^2_{\kappa-\ell}}_{0.95}$.
- H_0 cannot be rejected at 95% confidence level if $J < q^{\chi^2_{\kappa-\ell}}_{0.95}$.

APPENDIX: PROOFS

I. The quadratic cost equation of the following form is not linearly homogeneous in input prices:

$$C = \alpha_0 + \alpha_y Y + 1/2\alpha_{yy}Y^2 + \beta_k p_k + \beta_L p_L + \beta_P p_P. \qquad (5.117)$$

Proof

As an example, let $Y = 1$, $p_K = \$5$, $p_L = \$10$, and $p_P = \$5$. Furthermore, assume that the following estimates are obtained for the input price coefficients: $\beta_K = \beta_L = \beta_P = 0.333$. Therefore, the predicted cost is given by

$$C = Z + 0.33(\$5) + 0.33(\$10) + 0.33(\$5) \qquad (5.118)$$

or

$$C = Z + \$6.6,$$

where

$$Z = \alpha_0 + \alpha_y + (1/2)\alpha_{yy}. \qquad (5.119)$$

Now let prices double. It follows that

$$C = Z + 0.33(\$10) + 0.33(\$20) + 0.33(\$10)$$

or

$$C = Z + \$12.12.$$

Clearly, total cost has not doubled (it has only increased by $5.52).

II. The following equation is a proper cost function, as defined in Chapter 4 and in this chapter:

$$C = (\alpha_0 + \alpha_Y Y + 1/2{}^*\alpha_{YY}Y^2)p_K{}^{\beta k}p_L{}^{\beta L}p_P{}^{\beta P}e^\varepsilon. \qquad (5.120)$$

Proofs

1. Increasing factor prices: Let $x_i(p, Y)$ be the firm's conditional factor demand for input i at price p_i. By Shephard's lemma,

$$\partial C(p, Y)/\partial p_i = x_i(p, y) \geq 0. \qquad (5.121)$$

In this case, the conditional factor demands for capital, labor, and purchased power are given by

$$\partial C(\boldsymbol{p}, Y)/\partial p_k = x_k(\boldsymbol{p}, Y) = (\beta_k)p_K{}^{(\beta k-1)}(\alpha_0 + \alpha_Y Y \\ + 1/2\alpha_{YY}Y^2)p_L{}^{\beta L}p_P{}^{\beta P} \geq 0 \tag{5.122}$$

$$\partial C(\boldsymbol{p}, Y)/\partial p_L = x_L(\boldsymbol{p}, Y) = (\beta_L)p_L{}^{(\beta L-1)}(\alpha_0 + \alpha_Y Y \\ + 1/2\,\alpha_{YY}Y^2)p_K{}^{\beta K}p_P{}^{\beta P} \geq 0 \tag{5.123}$$

$$\partial C(\boldsymbol{p}, Y)/\partial p_P = x_P(\boldsymbol{p}, Y) = (\beta_P)p_P{}^{(\beta P-1)}(\alpha_0 + \alpha_Y Y \\ + 1/2\,\alpha_{YY}Y^2)p_L{}^{\beta L}p_K{}^{\beta K} \geq 0. \tag{5.124}$$

Because nonnegative amounts of inputs (x_i) at nonnegative prices (p_i) are required to produce nonnegative quantities of output (Y) at nonnegative marginal costs (which implies that the quadratic term in parentheses is nonnegative) **AND**, by linear homogeneity in input prices; that is,

$$\beta_K + \beta_L + \beta_P = 1, \tag{5.125}$$

where $\beta_K, \beta_L, \beta_P \geq 0$, then this cost function is increasing in factor prices.
2. Proof of concavity in input prices: This requires that the matrix of second derivatives of the cost function with respect to input prices must be a symmetric, negative semidefinite matrix. This has several implications:
   a. Cross-price effects are symmetric (by Young's theorem).
   b. Own-price effects are negative, that is,

$$\partial^2 C(\boldsymbol{p}, Y)/\partial p_i^2 \leq 0. \tag{5.126}$$

In the case of the modified quadratic equation, own-price effects are as follows.
Capital:

$$\partial^2 C(\boldsymbol{p},Y)/\partial p_K{}^2 = (\beta_k - 1)(\beta_k)p_K{}^{(\beta k-2)}(\alpha_0 + \alpha_Y Y + 1/2\,\alpha_{YY}Y^2)p_L{}^{\beta L}p_P{}^{\beta P}. \tag{5.127}$$

Labor:

$$\partial^2 C(\boldsymbol{p},Y)/\partial p_L{}^2 = (\beta_L - 1)(\beta_L)p_L{}^{(\beta L-2)}(\alpha_0 + \alpha_Y Y + 1/2\,\alpha_{YY}Y^2)p_K{}^{\beta k}p_P{}^{\beta P}. \tag{5.128}$$

Purchased power:

$$\partial^2 C(\boldsymbol{p}, Y)/\partial p_P{}^2 = (\beta_P - 1)(\beta_P)p_P{}^{(\beta P-2)}(\alpha_0 + \alpha_Y Y + 1/2\,\alpha_{YY}Y^2)p_K{}^{\beta k}p_L{}^{\beta L}. \tag{5.129}$$

Again, because $C(p, Y)$ is linear homogeneous in input prices, then the first term on the right side is negative. The remaining terms are positive:

i. $\beta_K, \beta_L, \beta_P \geq 0$.

ii. By Shephard's lemma:

$$\partial C(p, Y)/\partial p_i = x_i(p, y) \geq 0. \tag{5.130}$$

Positive amounts of inputs (p_i) are required to produce positive quantities of output, so $p_i \geq 0$.

iii. Monotonicity in output (as well as positive marginal cost) ensures that the quadratic term in the parentheses is positive.

Thus, own price effects are negative as required.

3. Monotonic in output (Y): An increase (decrease) in output will always increase (decrease) total cost; that is, marginal cost must be nonnegative for the range of outputs in the sample. In this case, marginal cost is given by

$$\partial C(p, Y)/\partial Y = (\alpha_Y + \alpha_{YY} Y) p_K{}^{\beta k} p_L{}^{\beta L} p_P{}^{\beta P} \geq 0, \tag{5.131}$$

which holds as long as $Y \geq -\alpha_Y/\alpha_{YY}$. (This must be checked for every observation and level of output.)

4. Linear homogeneous in input prices: Let

$$Z = (\alpha_0 + \alpha_y Y + (1/2)\alpha_{yy} Y^2) \tag{5.132}$$

and impose that

$$\sum \beta_i = 1. \tag{5.133}$$

In addition, let $Y = 1$, $p_K = \$5$, $p_L = \$10$, and $p_P = \$5$. Furthermore, suppose that the following estimates are obtained for the input price coefficients: $\beta_K = \beta_L = \beta_P = 0.333$. Therefore, the predicted cost is given by

$$C = (Z)(5^{0.33} 10^{0.33} 5^{0.33})$$

or

$$C = Z*(5.288).$$

Next, let input prices double, which implies that predicted cost is given by

$$C = Z*(10^{0.33} 20^{0.33} 10^{0.33})$$

or

$$C = Z*12.567.$$

Clearly, total cost has doubled as a result of input prices that have doubled and linear homogeneity is preserved by this modification.

Aside

While it is true that taking logs preserves all of the properties of the original function, the author offers a proof of linear homogeneity of the log-linearized version of (1) simply as a check. However, in this case, the author let the estimated coefficients vary in magnitude so that $\beta_K = 0.15$, $\beta_L = 0.25$, and $\beta_P = 0.60$.

Proof

In logs, the equation to be estimated is given by

$$\ln C = \ln Z + \beta_K \ln p_K + \beta_L \ln p_L + \beta_P \ln p_P + \varepsilon. \tag{5.134}$$

Evaluating at sample means (ln$Z$ is constant), the predicted cost is given by

$$\ln C = \text{Constant} + 0.15(1.6055) + 0.25(2.895) + 0.60(1.38) \tag{5.135}$$

or

$$\ln C = \text{Constant} + 0.2408 + 0.72375 + 0.828 = 1.79.$$

Next, doubling each input price (and taking logs) and then evaluating at sample means yields

$$\ln C = \text{Constant} + 0.15(2.2986) + 0.25(3.583) + 0.60(2.073)$$

or

$$\ln C = \text{Constant} + 0.3448 + 0.896 + 1.244 = 2.4846.$$

Doubling input prices increases ln$C$ by 0.6946, the exponential of which is equal to 2.003, indicating that total cost has doubled from a doubling of input prices. Thus, linear homogeneity in input prices is preserved.

Chapter 6

Case Study[1]

A Multiproduct Cubic Cost Model to Estimate the Marginal Cost of Distributing Electricity in the United States: An Application to Rate Making and the Reduction of Greenhouse Gas Emissions

INTRODUCTION

The issue of global climate change and its consequences has become one of growing concern in recent years. As a result, there has been an increased focus on energy efficiency and on the development of alternative sources of energy, particularly renewables,[2] but also nuclear and clean-coal technologies. "Going green" has become the buzzword of the early 21st century. As a result, much of the work being performed at utilities has been focused on the potential impacts of conservation and energy efficiency on load forecasts, resource planning that includes noncarbon supply resources, and shareholder value. At state utility commissions, regulators are asking tough questions about ratepayer impacts of utility investment in demand-side management, cost recovery, and efficiency of investment. In addition, many states are crafting their own energy policies, including legislating mandates on renewable portfolios and energy-efficiency resource standards.

There is little doubt that any meaningful limit or reduction of carbon dioxide emissions will have a significant impact on the electric supply industry. For example, in the United States the electric power sector accounted for over 40% of the total carbon dioxide emissions in 2008. Also, these emissions from power plants have increased by over 30% from 1990 to 2008 as the demand of electricity has increased (U.S. Energy Information Administration, 2009).[3] This is displayed in Figure 6.1.

---

[1]This chapter was the basis for a presentation at the National Association of Regulatory Utility Commissions conference on climate change in December 2009. It is presented here as a case study.

[2]Renewable technologies, such as wind turbines, have the added benefit of not being subject to the price volatility of fossil fuels, but may have drawbacks that include intermittent availability and high initial capital costs.

[3]Based on data from the EIA report "Emissions of Greenhouse Gases in the United States," Table 6, 2009.

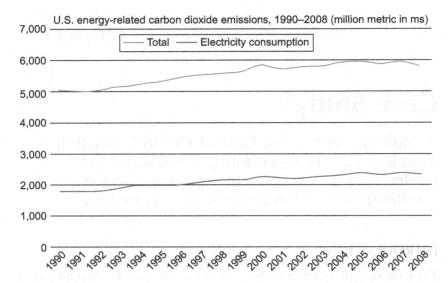

FIGURE 6.1 Emissions of carbon dioxide in the United States. Emissions from generating electricity account for about 40% of total emissions.

Note: Total emissions decline in 2008 were predominantly due to industrial and transportation sectors, both of which were hit hard by the 2007 recession.

Addressing the concerns raised by electric suppliers[4] without causing harm to ratepayers will require compromises among regulators, utilities, and consumers. This is a global issue that transcends regulatory structure—whether one is an electric customer in a deregulated European market or in the United States where the price of electricity is subject to a mix of regulation and market mechanisms, the end result will be the same: Decarbonizing electricity will entail extra costs that will be reflected in rates, which should be set to incentivize consumers to make appropriate choices and to promote the efficient allocation of resources. This is to say that prices should be set to reflect the true (i.e., marginal) cost of providing service to the end user, which must be estimated in the absence of real-time prices. Only a properly specified cost model will yield appropriately shaped average and marginal cost curves, the

---

[4]Under the current regulatory structure in the United States, rates are typically set based on the average cost of service; for investor-owned firms, the allowed rate of return is earned by selling electricity—the more sold, the higher the return. For investor-owned utilities (IOUs), which have a fiduciary responsibility to shareholders, any action geared toward reducing sales tends to be avoided.

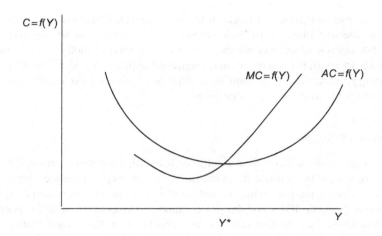

FIGURE 6.2 Appropriately shaped average and marginal cost curves.

latter of which can be used to price electricity. These are displayed in Figure 6.2. Such a cost model is introduced and estimated in this case study.

At output levels $Y < Y^*$, increasing returns to scale are indicated by declining average costs $(AC = f(Y))$. At $Y = Y^*$, returns to scale are constant (and average cost = marginal cost). Beyond this level of output, $Y > Y^*$, diminishing returns to production set in (i.e., marginal cost increases with output) and decreasing returns to scale are realized.

THEORY OF EFFICIENT PRICING

The concept of using marginal cost to price electricity is not a new one. Indeed, it was more than a century ago when an engineer (not an economist) made the case for time-of-use rates in terms that "came quite close to holding that prices should equal marginal costs" (United States Association of Energy Economists Dialogue, August 2007). To wit: The motivation—and contribution—of this study is the estimation of an appropriate cost model; this not only means that the model itself is theoretically correct but also that it yields appropriately shaped marginal and average cost curves. More specifically, this chapter makes an important contribution to the literature in that it employs a cubic cost function to estimate the marginal cost of distributing electricity. Such a model, which has not been fully developed in the literature before, conforms to economic theory and provides a methodology for the determination of the optimal pricing of electricity. This is crucial given the current regulatory climate and the desire to reduce greenhouse gas (GHG) emissions to below 1990 levels. Clearly, price is the mechanism that incentivizes consumers to make optimal choices in terms of not only the quantity of electricity to consume, but also in the investment decisions made in terms of household appliances, vehicles, and the thermal integrity of their homes. Model results can be employed to calculate optimal electricity prices,[5] which can be used to motivate consumers to make appropriate choices in energy-efficient appliances and in the wise use of energy, which are important first steps in combating the environmental problems with which we are now faced.

STUDY DESIGN

In this case study, data on rural electric distribution cooperatives from 1997 to 1999 were used to estimate the parameters of a properly specified cubic cost model. Another unique feature is that most studies on the reduction of GHG emissions focus on IOUs and the preservation of shareholder value [a seeming paradox given the nature of rate-of-return regulation in the United States (i.e., the more sold, the higher the profit)]. In this case, cooperatives provide an interesting case study in that there is not (presumably) a profit motive, although their customers will be impacted by legislative mandates nonetheless. The next section provides more background on rural electric cooperatives in the United States.

REASONS THAT COOPERATIVELY-OWNED UTILITIES ARE DIFFERENT

Unlike IOUs, not all states regulate electric cooperatives. In fact, fewer than 20 states have jurisdiction over coops, and the degree of regulation is not consistent among these states.[6] This being said, coops will still be subject to the same environmental legislative mandates as their investor-owned counterparts, which could create a special challenge for federal and state policy makers in the United States (Greer, 2003, 2008). However, coops have been quite proactive in their pursuit of energy efficiency, providing financial incentives for consumer investment in energy efficiency, upgrading power lines and transformers, and advanced meter infrastructure. According to a recent National Rural Electric Cooperative Association article, the Federal Energy Regulatory Commission has reported that "market penetration of advanced meter infrastructure [13%] is highest among rural electric cooperatives" (*Rural Electric Magazine*). This infrastructure, known as AMI technology, or smart meters, enables the meter in one's home or business to signal changes in price so that the end user can respond by changing the load pattern. For example, on very hot days one

---

[5]It is important to note that optimal means that fixed costs are recovered by fixed charges while marginal costs by variable charges.

[6]For example, in some states, public regulatory commissions do not regulate rates charged by coops, but rather the terms of their service obligations.

might defer laundry to the evening or early morning hours when price is typically the lowest (see Table 6.1, which displays an example of a midwestern utility that charges time-of-use rates to a subset of its residential customers). This requires (at the very least) prices that change throughout the day (or even hourly), which is known as time-of-use or real-time pricing. Note that in the last row is the critical peak price, which is three and a half times that during the afternoons on weekdays. Such prices reflect the changing marginal cost of supply under various conditions, as displayed in Figure 6.3.

Note: This is discussed in more detail in the case study on real-time pricing.

More specifically, Figure 6.3 shows a supply curve for a utility with several types of generation, which are brought online according to the marginal cost of generating electricity (at least in theory).

Generating units are dispatched according to marginal cost, which is based predominantly on the cost of fuel. This is represented by **S**, the supply

TABLE 6.1 Typical Time-of-Use Electric Tariff (Summer Months, in Cents/kWh)

	12 a.m.–9 a.m.	10 a.m.–1 p.m.	1 p.m.–6 p.m.	6 p.m.–12 a.m.
Weekday	5.0	5.5	10.0	5.5
Weekend	5.0	5.5	5.5	5.0
Critical peak price (3 p.m.–6 p.m., weekdays only)	35.0			

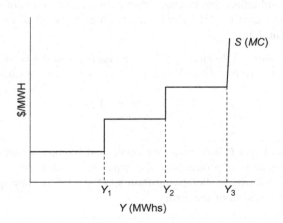

FIGURE 6.3 Supply curves for an electric utility with various generation sources.

function. In the region of the supply curve where Y (electric load) $\leq Y_1$, base load capacity is the first to be dispatched, which is typically pulverized coal, hydro (if in the northwestern part of the United States or if one of the federally-owned power administrations is the supplier), or nuclear generation. When $Y > Y_1$, the next generating units to come online will likely be natural gas fired, such as combined-cycle gas turbine, which is typically used for peaking capacity. Finally, given the cost, renewable resources such as wind or solar may be the last to be dispatched *if there is no renewable portfolio standard*, which is discussed in some detail in Chapter 3. Again, it depends on the marginal cost of the fuel at the time, which affects the stacking order of the generating units. Of course, it may be the case that it is less costly to purchase from the market than it is to run the utility's own generating units, especially should some type of national energy policy that deals with the emissions of carbon and/or other hazardous air pollutants be legislated.

LITERATURE REVIEW

Few studies employ the cubic functional form to model costs, especially in multiproduct industries. Prior to this study, Martins and colleagues (2006) employed such a specification to examine the cost structure of the Portuguese water industry. Like similar studies (i.e., cost estimation) for public utilities, their objective was to estimate economies of scale and scope and, ultimately, to test for whether the electric industry is a natural monopoly. Their model is given by

$$C(Y) = a + \sum_i b_i y_i + 1/2 \left(\sum_i c_i y_i \right)^2 + 1/3 \left(\sum_i d_i y_i \right)^3. \qquad (6.1)$$

However, and unlike this chapter, Martins et al. (2006) do not include input prices into their cost model, which violates a basic tenet of cost functions, which, by definition:

A cost function measures the minimum cost of producing a given level of output (Y) for some fixed factor price (w). As such, and at the very least, a cost model is given by

$$C(w, Y) \equiv wx(w, Y) \qquad (6.2)$$

(Varian, 1992).

Laband and Lentz (2005) employ a cubic cost function to estimate higher-education costs and the production of extension, which is often important in agriculture-related studies. Following Baumol et al. (1982), they specify a flexible fixed cost equation of the form:

$$C_i = \alpha + \sum_i \alpha_i F_i + \sum_i \beta_i Y_i + 1/2 * \sum_i \sum_j c_{ij} Y_i Y_j + \varepsilon_i, \qquad (6.3)$$

where C_i are total expenditures by the higher-education institution i in 1996 and $Y_{i,j}$ are outputs: undergraduate education, graduate education, and research and extension.[7]

Creedy et al. (2003) also studied the cost of higher-education costs in Australia. They specify the form

$$C_k = \alpha + \sum_i \beta_{1i} L_{ik} + \sum_i \beta_{2i} L^2{}_{ik} + \sum_i \beta_{3i} L^3{}_{ik} + \sum_j \delta_j D_{jk} + \varepsilon_k, \qquad (6.4)$$

where L is the number of students and D is the student share of a discipline group. Again, no input prices are included.

The only paper the author has found that uses a cubic function in the electric utility industrial is that of Yoo (1988). Estimating a two-output (electricity and gas) cost model of the form

$$C = \alpha_0 + \alpha_1 y_1 + \alpha_2 y_2 + \beta_1 y_1{}^2 + \beta_2 y_1 y_2 + \beta_3 y_2{}^2 + \tau y_1{}^3 + \tau y_2{}^3 + \varepsilon, \qquad (6.5)$$

Yoo found that in 1983 product-specific economies of scale for electricity were prevalent up to 10 GWh (the coefficient on the cubed output term is not statistically different from zero). Nonetheless, appropriately shaped average and marginal cost curves result. However, again no input prices are included, as his is a long-run model, which, as stated previously, does not accord with economic theory.

This chapter is unique in its contribution to the literature, offering a theoretically sound and properly specified multiproduct cost function in which outputs enter cubically and, as such, generate the appropriately shaped average and marginal cost curves, as displayed in Figure 6.2. This is a necessary condition for the efficient pricing of services (i.e., prices are equal to the marginal cost of providing service), which in turn is necessary to properly signal and incentivize consumers to make appropriate investments in energy efficiency and conservation in general in the consumption of electricity.

ESTIMATING COST MODELS

Modeling the cost of distributing electricity as a multiple-output process is well documented in the economic literature [e.g., see studies by Greer (2003, 2008), Kwoka (1996), Hollas et al. (1994), Berry (1994), Karlson (1986), and Roberts (1986), to name a few].

The motivation for this is several-fold: First, large industrial customers tend to have more stable load patterns than their residential or small commercial counterparts, the latter of whom tend to increase demand during peak times when more expensive generating units are brought online to serve load.

---

[7]The author does not see this as being a cubic form!

Another reason for modeling the distribution of electricity as a multiproduct market is that public utility commissions tend to set rates with this distinction; that is, it is typically large commercial and industrial customers that pay demand charges, not residential or small commercial rate classes.

Finally, it is often the case that large users can accept electricity at higher voltages than their smaller counterparts; as such, they tend to be less costly to serve and they experience lower line losses. Given all of this, a two-output approach is taken here.[8]

DATA

The National Rural Electric Cooperative Association provided panel data used in this study. Contained herein is all pertinent information on the distribution cooperatives that distributed electricity in the United States from 1997 to 2001. Given the nature of data, cost models (rather than production functions) are appropriate.[9]

The distribution of electricity requires a large infrastructure investment in poles and conductors, in voltage transformers, and in the transmission and distribution lines themselves. Because of the low customer density in the territories that are typically served by distribution cooperatives, the average cost of providing such service is quite high. To estimate the cost of distributing electricity, numerous variables are included and estimated, with the final specification given by Equation (6.7). Summary statistics are displayed in Table 6.2.

Note in particular the large variation in some of the variables among the distribution cooperatives, namely the electricity distributed to large users (Y_2).

Data provided by the Rural Utilities Service (RUS) contain data on all rural electric cooperatives that were RUS borrowers during the study period (around 700 distribution coops and 50 generation and transmission coops [G&Ts]). Customer classes are distinguished by two main subgroups: customers served at voltage levels below 1000 kVA (residential and small commercial) and those served at above 1000 kVA (large commercial and industrial). In addition, data on fuel prices, labor, capital, and purchased power costs are included. Other relevant variables included are miles of transmission lines and customer density.

Coops serve mostly residential loads (90%) and small commercial loads. In fact, 23% of coops do not serve any large industrial customers. Table 6.2 contains summary statistics for the coops in the data set for 1997–1999.

---

[8]In several studies (e.g., Karlson, 1986; Roberts, 1986), the distribution of electricity is modeled as a three- or even four-output product model. When modeled as a three-output product, the author found that the separate estimation of residential and commercial users was highly collinear, which is not surprising given that data separated customer classes between those supplied at below or above 1000 kVA (residential and small commercial falling into the former).

[9]See Chapter 4 for more detail.

TABLE 6.2 Summary Statistics for RUS Borrowers, 1997–1999

Variable	Variable Name	Mean	Standard Deviation	Minimum	Maximum
C	Total cost	19,812.12	23,281.71	428.85	209,840.32
Y_1	Electricity distributed to small users (MWh)	232.24	285.42	1.61	2,552.76
Y_2	Electricity distributed to large users (MWh)	57.27	162.63	0.00	2,878.36
P_p	Price of purchased power ($/MWh)	39.28	8.74	0.00	90.40
P_k	Price of capital (%)	4.91	0.79	0.00	7.66
P_l	Price of labor ($/MWh)	14.16	2.73	6.85	28.61
T_R	Miles of transmission lines	42.22	77.34	0.00	585.00
Density	Customer density	5.42	3.54	0.67	25.73

Note: The data set actually contains data through 2001. The author chose not to include data for the last two years here because it has not been as well vetted and likely contains errors, which resulted in the odd results obtained when these data were included.

COST MODELS

In specifying a proper cost model, several properties must be ensured. First, it must be nonnegative, nondecreasing, concave, and linear homogeneous in input prices. Second, it should be capable of estimation with zero values of some outputs, which means that it should allow for economies of scale, scope, and subadditivity. While no single form satisfies all of these conditions, the two most commonly employed in cost estimation are the translogarithmic and the quadratic, both of which are flexible enough to avoid a priori restrictions on the elasticities of substitution among the input variables (Greer, 2003, 2011).

It has been stated that the quadratic is superior to the translog cost specification in estimating cost functions in multiple-output industries. In addition to its allowing for the "unconstrained emergence of economies of scope and subadditivity" (Kwoka, 1996), it is also the case that the Hessian matrix that results from the translogarithmic cost specification varies over input and output levels,

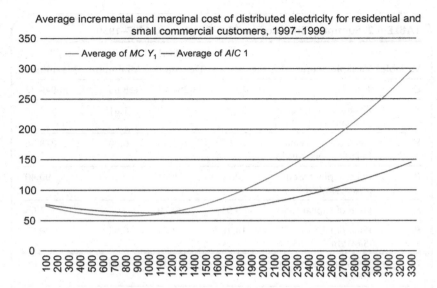

FIGURE 6.4   Appropriately shaped average and marginal cost curves for rural electric cooperatives generated by cubic cost function estimated with 1997–1999 data.

while the Hessian matrix for the quadratic form does not.[10] In this sense, the restrictions on concavity for the quadratic cost function are global—they do not change with respect to output and input prices. However, the concavity restrictions on the translog are local—fixed at a specific point because they depend on prices and output levels. As such, in the case of the latter, the various efficiency measures described earlier must be checked at every level of output and input price combination.

As stated previously, this chapter elevates the economics literature one step further: It estimates a cubic cost function, which not only conforms to economic theory but also yields appropriately shaped average and marginal cost curves. Figure 6.4 displays estimated average (incremental, as it is two-output) and marginal cost curves for output distributed to small users (Y_1) from 1997 to 1999.

A Properly Specified Cost Model

Previous work has employed a properly specified and theoretically sound multiproduct quadratic cost model of the form

$$C = (\alpha_0 + \alpha_i Y_i + \alpha_j Y_j + 1/2 * \sum_i \sum_j \alpha_{ij} Y_i Y_j) \Pi w_m^{\beta m} \Pi O_n^{\theta n} e^{\varepsilon} \quad \text{for } i, j = 1, 2, \quad (6.6)$$

---

[10]An excerpt from *www.ricardo.ifas.ufl.edu/aeb6184.production/Lecturepercent2020-2005.ppt* (lecture on subadditivity).

where C is total cost: operating deductions plus interest on long-term debt; Y_1 is electricity distributed to residential and small commercial users in GWh; Y_2 is electricity distributed to large commercial and industrial users in GWh; W_m is input prices: capital (k), labor (l), and purchased power (p); and O_n is other (cost-shift) variables: miles of transmission line and customer density.

This cost function is concave, nondecreasing, and homogeneous of degree one in input prices[11] as well as monotonic in output and, as such, preserves the fundamental properties of a proper cost function (Greer, 2003, 2011). Taking this one step further yields the multiproduct cubic model, which is given by

$$C = \{\alpha_0 + \alpha_1 Y_1 + \alpha_2 Y_2 + 1/2(\alpha_{11} Y_1{}^2 + \alpha_{22} Y_2{}^2 + 2^*\alpha_{12} Y_1 Y_2) + 1/3[\alpha_{111} Y_1{}^3 + \\ 2^*(\alpha_{112} Y_1{}^2 Y_2 + \alpha_{221} Y_2{}^2 Y_1) + \alpha_{222} Y_2{}^3]\} \prod w_i{}^{\beta i} \prod O_m{}^{\theta m} e^\varepsilon \tag{6.7}$$

Estimation of this particular equation is somewhat problematic. Because it is highly nonlinear, the stochastic error term does not enter additively as required. Hence, the model must be transformed so as to allow it to be additive. A logarithmic transformation, which will yield such an error term, is made possible by the creation of a variable, Z, where

$$Z = \{\alpha_0 + \alpha_1 Y_1 + \alpha_2 Y_2 + 1/2(\alpha_{11} Y_1{}^2 + \alpha_{22} Y_2{}^2 + 2^*\alpha_{12} Y_1 Y_2) + 1/3[\alpha_{111} Y_1{}^3 + \\ 2^*(\alpha_{112} Y_1{}^2 Y_2 + \alpha_{221} Y_2{}^2 Y_1) + \alpha_{222} Y_2{}^3]\} \tag{6.8}$$

so that (6.7) becomes

$$\ln C = \zeta \ln Z + \beta_K \ln w_K + \beta_L \ln w_L + \beta_P \ln w_p + \theta_{tm} \ln O_{tm} + \theta_d \ln O_d + \varepsilon \tag{6.9}$$

Discussion of Cost Models

To estimate the parameters of the model displayed in Equation (6.9), a nonlinear estimation procedure is required. A further complication is the panel structure of the data set, which presents a further complication (see Appendix A at the end of this chapter for details). More specifically, not only is it the case that the econometric issues are complex, but also the estimation software used is not capable of estimating the parameters of the equation specified here and impose linear homogeneity in input prices. However, the author was able to transpose data in such a way as to simultaneously estimate different cost equations (one for each year) and impose cross-equation restrictions on the parameters of the equations. The estimation results are discussed next.

---

[11]Linear homogeneity is ensured by imposing that $\beta_K + \beta_L + \beta_P = 1$ (where β_i are input price coefficients).

ESTIMATION RESULTS

Table 6.3 contains results for estimation of the multiple-output cubic cost model, the expanded equation of which is given by Equation (6.9):

$$\ln C_t = \zeta \ln Z_t + \beta_K \ln w_{Kt} + \beta_L \ln w_{Lt} + \beta_P \ln w_{Pt} + \theta_{tm} \ln O_{tmt} + \theta_d \ln O_{dt} + \varepsilon_t,$$

(6.10)

where t is the time frame for the (annual) data, 1997–1999.

TABLE 6.3 Estimation Results—Cubic Cost Model

Variable	Coefficient	Estimate	t Statistic
Constant	α_0	14.6907	7.27
Outputs (MWh)			
Y_1—electricity distributed to small users	α_1	3.8240	19.02
Y_2—electricity distributed to large users	α_2	2.7855	12.38
Y_1^2—squared output (small users)	α_{11}	−0.0022	−4.62
Y_2^2—squared output (large users)	α_{22}	−0.0059	−4.71
$Y_1 Y_2$ ($Y_2 Y_1$)—output cross-product	$\alpha_{21}(\alpha_{12})$	−0.0025	−2.49
Y_1^3—cubed output (small users)	α_{111}	0.000001	2.04
$Y_1^{2*}Y_2$—cross-product	α_{112}	0.000002	1.15
$Y_2^{2*}Y_1$—cross-product	α_{221}	0.000002	1.52
Y_2^3—cubed output (large users)	α_{222}	0.000003	4.62
Input Prices			
Labor, w_L	β_L	0.2543	11.61
Capital price, w_K	β_K	0.1335	5.38
Purchased power, w_P	β_P	0.6122	29.45
Cost Shifters			
Transmission lines (miles)	θ_{TM}	0.0379	7.81
Density	θ_D	−0.0996	−8.81
Other Statistics			
AR(1) on 1997 cost model	ρ_1	0.1735	2.58
Test: Linear homogeneity		−1.4074	−0.08
Adjusted R^2: 0.967, 0.969, 0.971			

Should an autoregressive disturbance term (AR) be required, it would be of the form:

$$\varepsilon_t = \rho_1 \varepsilon_{t-1} + \mu_t \qquad (6.11)$$

[this is an AR(1) term].

Discussion of Estimation Results

Estimation results accord with a priori expectations in terms of the signs and statistical significance of the estimates. First and foremost, total cost increases initially with output at a decreasing rate (as indicated by the negatively signed coefficients on the squared output terms). However, as diminishing returns set in, total cost increases with output at an increasing rate, which is consistent with economic theory.

The quadratic terms are negative in sign, indicating that this cost model is concave in output and generates ray economies and product-specific economies of scale. In addition, the negatively signed (and statistically significant) coefficient on the interaction term (α_{ij}) between outputs indicates cost complementarity, which is a sufficient condition for economies of scope.

Also as prescribed by economic theory, coefficients of the cubic terms are positive in sign and, with the exception of α_{112}, coefficient estimates are statistically different from zero at the 90% level. Finally, coefficients of the input price and cost-shift variables are of the appropriate sign and statistical significance. Overall, the models fit data well, as is indicated by the adjusted R^2 of each equation, which varies from 0.966 to 0.975.

Other issues, however, need to be resolved. One is the clear evidence of heteroskedasticity among the error terms, which often occurs in cross-sectional data. The White test statistic indicates that the null hypothesis of homoscedastic errors should be rejected. Also, because of the methodology employed to estimate this panel data set (Proc Panel in SAS™ will not estimate nonlinear models so the author employed the Proc Model and simultaneously estimated three separate cost models, one for each year), the generalized method of moments (see the appendix to Chapter 4 for details) estimation procedure was not able to be used (you may recall that this can help correct for heteroskedasticity). Another SAS procedure was developed to estimate the heteroskedasticity-consistent covariance matrix (HCCME); unfortunately, at the time of this writing, the procedure was not working properly so it could not be invoked. Nonetheless, details have been provided in Appendix B at the end of this chapter.

Another issue is that of serially correlated errors, which is evidenced by results of a Godfrey Lagrange multiplier test in which the null hypothesis is that residuals are white noise. Despite inclusion of an autoregressive correction [AR(1)] in Equation (6.10), there is still some evidence that serial correlation may be an issue.

EFFICIENCY MEASURES

This notwithstanding, it is informative to view the estimated cost curves, which are displayed in Figure 6.4. As expected, the cubic cost equation generates U-shaped marginal and average cost curves, which could be used to price the variable charge of electricity. Also keep in mind that the average incremental cost is relevant here; because there are two outputs, this measure replaces the concept of average cost in the single-output case. Appealing to Baumol et al. (1982), the average incremental cost is given by

$$AIC(y_i) = [C(Y_N) - C(Y_{N-i})]/Y_i, \tag{6.12}$$

where $C(Y_{N-i})$ is the cost of producing all N of the multiproduct firm outputs except product i.

Note: These are explained in much more detail in Chapter 2.

This specification allows the identification of returns to scale that are specific to a particular output. Hence, product-specific returns to scale, which are analogs of scale economies in the single-output case, are given by

$$S_i(y) = AIC(y_i)/(\partial C/\partial y_i), \tag{6.13}$$

where $\partial C/\partial y_i$ is the marginal cost with respect to product i.

Returns to scale of product i at y are said to be increasing, decreasing, or constant as $S_i(y)$ is greater than, less than, or equal to unity, respectively.

Evaluating the equations given previously, there are product-specific returns to scale for over 95% of the coops in the sample in the distribution of electricity to small users and over 65% in the distribution to large users. Not surprising, beyond a certain level of output, decreasing returns to scale set in, which occurs more frequently with larger (industrial) users.

Yet another measure is that of economies of scope, which are said to exist if a given quantity of each of two or more goods can be produced by one firm at a lower cost than if each good were produced separately by two different firms or even two different production processes. For a two-product case, weak economies of scope are given by

$$C(Y_1, Y_2) \le [C(Y_1, 0) + C(0, Y_2)] \tag{6.14}$$

for all $Y_1, Y_2 > 0$. And the degree of economies of scope is given by

$$S_c = [C(Y_1, 0) + C(0, Y_2) - C(Y_1, Y_2)]/C(Y_1, Y_2). \tag{6.15}$$

For the firms in the sample, over 98% experienced economies of scope over the 1997–1999 timeframe.

DISCUSSION OF FIGURE 6.4—AVERAGE INCREMENTAL COST AND MARGINAL COST

Figure 6.4 displays the average incremental and marginal cost curves generated by the total cost model in Equation (6.10). To display these results, it was necessary to compute a composite output, v, where $v = Y_2/Y_1$. In the case of Figure 6.4, $v = 0.2$, which covers most of the coops in the sample. Average cost is falling until distributed electricity is equal to about 1000 MWhs (1 GWh). After this, the rising marginal cost causes average cost to rise, which accords with economic theory. The relevant issue here is that the price paid by the end user should reflect this rising marginal cost and that the rates paid by customers in the United States are typically not set in this fashion (see Chapter 8 for more details).

GENERAL IMPLICATIONS OF ESTIMATION RESULTS

Estimation results indicate that this cost model could be employed to estimate the marginal cost of providing electricity to various end users, the application of which has been missing both in practice and in the literature thus far, despite the fact that for over a century marginal cost pricing has been demonstrated to result in economic efficiency. As aptly stated by John Kelly of the American Public Power Administration in a recent edition of Rural *Electric Magazine*,

It (marginal cost pricing) recognizes the implications of such pricing for economic efficiency of electricity production and lowering costs, specifically by designing rate structures to improve the utilization of electricity plants and to lower average costs. It seems that engineers, like (Alfred) Gibbings and (W. S.) Barstow, were the first to recognize the important connection between time-differentiated rates, capacity utilization, and costs.

Given that one of the best ways to reduce greenhouse gas (GHG) emissions is via energy efficiency (i.e., megawatts), price is certainly a key factor in consumers' decisions, be it investment in more efficient appliances, thermal integrity of homes (e.g., insulation, new windows, etc.), or just conservation in general (e.g., thermostat setting, turning off lights, etc.). The bottom line is that price must reflect the true marginal cost of the electricity consumed; otherwise, consumers will not make appropriate decisions and an inefficient use of resources will continue to occur with little to no reduction in GHG emissions. This effort will require a partnership among consumers, producers, regulators, and policy makers.

CONCLUSION

A fundamental change must occur in the existing paradigm of utility rate making, as current rate structures do not provide incentives to consumers or producers to promote energy conservation and environmental stewardship. Customer-driven economic efficiency dictates that price must be set in accordance with marginal costs and that variable costs should be recovered by variable charges and fixed costs by fixed charges. This chapter deployed a unique cubic cost model methodology to estimate the marginal cost of distributing electricity to distinct end users.[12] This is an important first step in promoting the efficient market-based allocation of resources, which in turn will help mitigate GHG emissions.

APPENDIX A: PANEL DATA

Time-series cross-sectional data used to estimate the parameters of the cost model introduced here require the use of special econometric techniques. Also known as panel data, a data set in this form contains multiple dimensions of phenomena observed over multiple time periods for the same firms or individuals. An excerpt from Greene, Lecture on Panel Data,[13] provides a nice overview. It is summarized here.

In matrix algebra form, the classic linear regressions model is given by

$$Y = X'\beta + \varepsilon, \tag{6.16}$$

where there are N observations and matrix X has K columns, including a column of ones.

In addition, standard assumptions are that

$$
\begin{array}{ll}
1. & E(\varepsilon/X) = 0. \\
2. & E(\varepsilon) = 0. \qquad\qquad (6.17) \\
3. & \mathrm{cov}(\varepsilon, X) = 0.
\end{array}
$$

A Caveat

There could be many reasons that the aforementioned do not hold. The one to focus on here is that of the endogeneity problem, which is discussed in some detail in Chapter 9 and implies that one or more independent variables are correlated with the error term so that

$$E(\varepsilon/X) \neq 0. \tag{6.18}$$

This may result from a number of issues: omitted variables, unobserved heterogeneity, measurement error, or simultaneity, the latter of which is discussed in more detail in Chapter 9.

---

[12]It is straightforward to estimate a three-output model, which may be required for rate-making purposes.

[13]See *http://pages.stern.nyu.edu/~wgreene/Econometrics/PanelDataNotes.htm.*

While it is surely the case that panel data observations are correlated across time, the benefits of using panel data may outweigh the costs. These benefits include:

1. Time and individual variation in behavior unobservable in cross-sections or aggregate time series.
2. Observable and unobservable individual heterogeneity.
3. Rich hierarchical structures (nested).
4. More complicated models (e.g., nonlinear, discrete choice, and censored data).
5. Features that cannot be modeled with only cross-section or time-series data alone.
6. Dynamics in behavior.

It is the unobservable individual heterogeneity that causes the issue here, as it is the same as omitted variable bias. That is, the true regression equation to be estimated is given by

$$Y_{it} = X_{it}'\beta + C_i + \varepsilon_{it}. \tag{6.19}$$

Furthermore, there are fixed effects, in which C_i is correlated with the included variables (X_{it}), and random effects, in which C_i it is not correlated with the included variables so that $\text{cov}(c_i, x_{it}) = 0$. The former, which is also known as a dummy variable model, can be written as

$$Y_{it} = \alpha_i + X_{it}'\beta + \varepsilon_{it}, \tag{6.20}$$

whereas the latter, which is also known as an error components model, can be written as

$$Y_{it} = X_{it}'\beta + \varepsilon_{it} + \mu_i. \tag{6.21}$$

The potential bias or inconsistency of ordinary least-squares (OLS) estimation on β depends on which type of effect. This, in turn, affects the choice of estimation procedure to be used. Because the former is merely a regression with a dummy (or binary) variable included, least-squares is the best linear, unbiased estimator. However, a random effects model requires a two-stage approach. In the first stage, the variance components are calculated using methods described by Fuller and Battese (1974; among others), and in the second stage, variance components are used to standardize data, and OLS regression is performed.

APPENDIX B: HETEROSCEDASTICITY-CONSISTENT COVARIANCE MATRIX ESTIMATION

Homoscedasticity is required for ordinary least-squares regression estimates to be efficient. A nonconstant error variance, heteroscedasticity, causes the OLS estimates to be inefficient, and the usual OLS covariance matrix, $\sum$, is generally invalid:

$$\sum = \sigma^2 (X'X)^{-1}. \tag{6.22}$$

When the variance of the errors of a classic linear model

$$Y = X\beta + \varepsilon \tag{6.23}$$

is not constant across observations (heteroscedastic), so that

$$\sigma_i^2 \neq \sigma_j^2 \tag{6.24}$$

for some, $j > 1$.

In this case, the ordinary least-squares estimator

$$\beta_{OLS} = (X'X)^{-1}X'Y \tag{6.25}$$

is unbiased but is inefficient. Models that take into account the changing variance can make more efficient use of data. When variances, σ_i^2, are known, then generalized least-squares (GLS) can be used and the estimator

$$\beta_{GLS} = (X'\Omega X)^{-1}X'\Omega^{-1}Y, \tag{6.26}$$

where the sigma matrix is given by

$$\Omega = \begin{vmatrix} \sigma_i^2 & 0 & 0 \\ 0 & \sigma_i^2 & 0 \\ 0 & 0 & \sigma_i^2 \end{vmatrix}, \tag{6.27}$$

is unbiased and efficient. However, GLS is unavailable when variances are unknown.

To solve this problem, White (1980) proposed an HCCME,

$$\Sigma = (X'X)^{-1}X'\Omega X(X'X)^{-1}, \tag{6.28}$$

that is consistent as well as unbiased, where the sigma matrix is given by

$$\Omega_0 = \begin{vmatrix} \varepsilon_1^2 & 0 & 0 \\ 0 & \varepsilon_2^2 & 0 \\ 0 & 0 & \sigma_t^2 \end{vmatrix} \tag{6.29}$$

and

$$\varepsilon_t = Y_t - X_t\beta_{OLS} \tag{6.30}$$

(refer to SAS Help for more detail).

Chapter 7

Case Study: Cost Models to Illustrate KLEM Data

COBB–DOUGLAS COST MODEL

Much of the following is taken from Berndt (1991). It is cited here with permission of the author.

Recalling from Chapter 5 on cost models, the Cobb–Douglas production function with constant returns to scale is given by

$$\ln Q = \ln A + \beta_K \ln K + \beta_L \ln L, \tag{7.1}$$

where K is capital and L is labor.

In addition, the assumption of constant returns to scale (or homogeneity of degree one in input quantities) implies that

$$\beta_K + \beta_L = 1. \tag{7.2}$$

Substituting this restriction into Equation (7.1) yields

$$\ln Q/L = \ln A + \beta_K \ln (K/L). \tag{7.3}$$

This relates the average productivity of labor to the ratio of capital to labor. Denote the prices of K, L, and Q as P_K, P_L, and P, and take the exponential of Equation (7.1), which yields

$$Q = A\,K^{\beta}L^{1-\beta}. \tag{7.4}$$

Next, take the partial derivatives of Q with respect to K and L, which yield an expression for the marginal products:

$$\partial Q/\partial K = \beta A\,K^{\beta-1}L^{1-\beta} \tag{7.5}$$

$$\partial Q/\partial L = (1-\beta)A(K/L)^{\beta}. \tag{7.6}$$

Under the assumption that firms maximize profits, marginal products are equal to real input prices, P_K and P_L. It follows that

$$\partial Q/\partial K = P_K/P \tag{7.7}$$

$$\partial Q/\partial L = P_L/P, \tag{7.8}$$

where P is the price of output.

Rearranging, we have

$$\beta_K = (\beta_K K / PQ) \tag{7.9}$$

$$\beta_L = (\beta_L L / PQ). \tag{7.10}$$

One important implication of the log–log specification is that parameters β_K and β_L must equal value shares of inputs in the value of output. In their study on U.S. manufacturing over the 1899–1922 time period, Cobb and Douglas (1928) argued that if markets were competitive, if firms chose inputs so that marginal products equaled real prices, and if production technology in U.S. manufacturing over this timeframe followed the constant returns to scale log–log specification described earlier, then the ordinary least-squares (OLS) estimators of β_K and β_L should be approximately equal to 0.25 and 0.75, respectively.

In fact, these estimates were consistent with actual shares of total product published by the National Bureau of Economic Research, which were 0.259 and 0.741 for K and L, respectively.

ELASTICITIES OF SUBSTITUTION FOR COBB–DOUGLAS

You may also recall from Chapter 5 the discussion that the substitution elasticities of the Cobb–Douglas model always equal unity. Again, the elasticity of substitution between inputs, which measures the degree of substitutability between inputs (x_i, x_j), is given by

$$\sigma_{ij} = \partial \ln (x_i/x_j) / \partial \ln (F_i/F_j), \tag{7.11}$$

where F_i, F_j are the marginal products of x_i, x_j.[1]

To get around this limitation, an important paper by Arrow et al. (1961) asked: "For what functional form would σ be constant but not constrained equal to unity?" The solution involved integrating Equation (7.11), which resulted in

$$\ln (K/L) = \text{constant} + \sigma \ln (F_K/F_L), \tag{7.12}$$

where F_K/F_L is the marginal rate of technical substitution between K and L.

Next, they integrated the marginal rate of substitution (F_K/F_L) to obtain the implied production function. This generated the constant elasticity of substitution (CES) production function, which, with constant returns to scale imposes, is given by

$$Y = A^*[\delta K^{-\rho} + (1 - \delta)L^{-\rho}]^{-1/\rho}, \tag{7.13}$$

where $\sigma = 1/(1 + \rho)$.

---

[1] The marginal product of an input (P_i) is equal to the partial derivative of y with respect to that input (x_i):

$$\partial y / \partial x_i = P_i. \tag{7.11a}$$

Cobb–Douglas is a special case of the CES function in the limiting case in which $\rho \to 0$ and $\sigma \to 1$.

It is interesting to note that the CES specification had been derived almost a quarter of a century in the literature on consumer demand analysis (Bergson, 1936). More specifically, Equation (7.13) is an example of the mean value function:

$$Y^{-\rho} = A\left(\sum_i \delta_i X_i^{-\rho}\right). \tag{7.14}$$

Let us now move on to some more popular cost models that allow for substitution elasticities to vary with input prices.

TRANSLOGARITHMIC COST FUNCTION

Years later, the translogarithmic cost model (translog) emerged as a choice functional form for many important papers in the economics literature; after all, it offered the desirable property of flexibility (in that it does not place restrictions on substitution elasticities like the Cobb–Douglas or the constant elasticity of substitution forms). As this chapter is a case study, it is appropriate to examine this form here.

The translog cost function is a second-order Taylor series approximation of logarithms to an arbitrary cost function. This is a nonhomothetic function, which means that the ratios of cost-minimizing input demands are allowed to depend on the level of output.[2] This is given by

$$\ln C = \ln \alpha_0 + \alpha_y \ln y + \sum_i \beta_i \ln p_i + (1/2)\alpha_{yy}(\ln y)^2$$
$$+ (1/2)\sum_i \sum_j \gamma_{ij} \ln p_i \ln p_j + \sum_i \omega_{iy} \ln y \ln p_i, \tag{7.15}$$

where

1. Output ($\ln y$) and input prices ($\ln p_i$) enter linearly, as quadratics and as cross-products.
2. $\gamma_{ij} = \gamma_{ji}$.

For a cost function to be well behaved, it must be homogeneous of degree one in prices, given output. This implies that the following restrictions apply:

$$\sum_i \beta_i = 1 \tag{7.16}$$

---

[2]In contrast, with homothetic functions, relative input demands are independent of the level of output. In addition, further restrictions need to be imposed on the translog cost function. To be homothetic, it is necessary and sufficient to impose that

$$\omega_{iy} = 0 \tag{7.14a}$$

for all $i = 1, \ldots, n$.

and

$$\sum_i \gamma_{ij} = \sum_j \gamma_{ji} = \sum_i \omega_{iy}, \tag{7.17}$$

where $i, j = 1, \ldots, n$.

Gains in efficiency are realized by estimating the optimal, cost-minimizing input demand equations, which are given by the cost-share equations, along with the cost function. These are given by

$$\partial \ln C / \partial \ln P_i = P_i / C * \partial C / \partial P_i = P_i X_i / C = \beta_i + \sum_i \gamma_{ij} P_j + \omega_{iy} \ln y, \tag{7.18}$$

where

$$C = \sum_i P_i X_i \tag{7.19}$$

for $i = 1, \ldots, n$.

Next, defining cost shares as

$$S_i = P_i X_i / C, \tag{7.20}$$

it follows that

$$\sum_i S_i = 1, \tag{7.21}$$

which has important implications for econometric estimation.

In their seminal paper, Berndt and Wood (1975) used a four-factor model and annual data from U.S. manufacturing to estimate the parameters of a translogarithmic cost model and share equations to calculate substitution elasticities among inputs capital (K), labor (L), energy (E), and materials (M). For this, share equations associated with the translog cost model are given by

$$S_k = \beta_k + \gamma_{kk} \ln p_k + \gamma_{kl} \ln p_l + \gamma_{ke} \ln p_e + \gamma_{km} \ln p_m + \omega_{ky} \ln Y \tag{7.22}$$

$$S_l = \beta_l + \gamma_{kl} \ln p_k + \gamma_{ll} \ln p_l + \gamma_{le} \ln p_e + \gamma_{lm} \ln p_m + \omega_{ly} \ln Y \tag{7.23}$$

$$S_e = \beta_e + \gamma_{ek} \ln p_k + \gamma_{el} \ln p_l + \gamma_{ee} \ln p_e + \gamma_{em} \ln p_m + \omega_{ey} \ln Y \tag{7.24}$$

$$S_m = \beta_m + \gamma_{mk} \ln p_k + \gamma_{ml} \ln p_l + \gamma_{me} \ln p_e + \gamma_{mm} \ln p_m + \omega_{my} \ln Y \tag{7.25}$$

In the absence of symmetry conditions, there are 24 parameters to be estimated, 6 in each of the share equations. Imposing symmetry conditions, this drops to 18. The symmetry conditions are

$$\gamma_{kl} = \gamma_{lk} \tag{7.26}$$

$$\gamma_{ke} = \gamma_{ek} \tag{7.27}$$

$$\gamma_{km} = \gamma_{mk} \tag{7.28}$$

$$\gamma_{le} = \gamma_{el} \tag{7.29}$$

$$\gamma_{lm} = \gamma_{ml} \tag{7.30}$$

$$\gamma_{em} = \gamma_{me}. \tag{7.31}$$

In addition, the restriction that the cost function be homogeneous of degree one in input prices requires further restrictions on the parameters. These restrictions reduce the number of parameters to be estimated to 12 (from 18); again, efficiency gains in estimation are realized by these. More specifically, this implies that

$$\beta_k + \beta_l + \beta_e + \beta_m = 1 \tag{7.32}$$

$$\gamma_{kk} + \gamma_{kl} + \gamma_{ke} + \gamma_{km} = 0 \tag{7.33}$$

$$\gamma_{kl} + \gamma_{ll} + \gamma_{le} + \gamma_{lm} = 0 \tag{7.34}$$

$$\gamma_{ek} + \gamma_{el} + \gamma_{ee} + \gamma_{em} = 0 \tag{7.35}$$

$$\gamma_{mk} + \gamma_{ml} + \gamma_{me} + \gamma_{mm} = 0 \tag{7.36}$$

$$\omega_{ky} + \omega_{ly} + \omega_{ey} + \omega_{my} = 0. \tag{7.37}$$

To estimate this system of share equations empirically, it is necessary to specify a stochastic framework, which can be done by appending an additive random disturbance term to each share equation. The resulting disturbance vector $e \equiv \{e_k, e_l, e_e, e_m\}$ is assumed to be multivariate normally distributed with mean zero and constant covariance matrix Ω. In addition to being empirically necessary, the rationale for the stochastic specification consists of the simple fact that firms make random errors in choosing their cost-minimizing bundles. Having worked in the "real world," the author can attest to the fact that many firms have no idea how to choose such bundles; in fact, calculus rarely enters into any type of analysis that is performed. Given this, such a specification makes perfect sense.

As discussed previously in Chapter 4, the translog share equation system possesses a special property in that the sum of dependent variables sums to unity for each observation. This means that only $n - 1$ of them are linearly independent, which has several important implications. Paraphrasing Berndt (1991):

1. Because the shares sum to unity and only $n - 1$ are linearly independent, the sum of the disturbances across equations must always equal zero for each observation. This implies that the disturbance covariance matrix Ω is singular and nondiagonal.
2. Because the shares sum to unity at each observation, when the symmetry restrictions are not imposed, equation-by-equation OLS yields parameter estimates that obey the following column adding up conditions:

$$\beta_k + \beta_l + \beta_e + \beta_m = 1, \tag{7.38}$$

TABLE 7.1 Translog Share Equations in U.S. Manufacturing

Parameter	OLS	IZEF	Parameter	OLS	IZEF
B_k	0.0279	0.057	β_e	0.205	0.044
Γ_{kk}	0.045	0.030	γ_{ek}	(0.004)	(0.010)
Γ_{kl}	0.031	(0.000)	γ_{el}	0.029	(0.004)
Γ_{ke}	0.000	(0.010)	γ_{ee}	0.011	0.019
Γ_{km}	(0.015)	(0.019)	γ_{em}	(0.013)	(0.004)
γ_{kY}	(0.043)	0.000	γ_{eY}	(0.031)	0.000
B_l	0.398	0.253	β_m	0.119	0.645
Γ_{lk}	0.021	(0.000)	γ_{mk}	(0.062)	(0.019)
γ_{ll}	0.101	0.075	γ_{ml}	(0.161)	(0.071)
γ_{le}	0.041	(0.004)	γ_{me}	(0.053)	(0.004)
γ_{lm}	(0.123)	(0.071)	γ_{mm}	0.150	0.094
γ_{lY}	(0.028)	0.000	γ_{mY}	0.102	0.000

$$\gamma_{kk} + \gamma_{lk} + \gamma_{ek} + \gamma_{mk} = \gamma_{kl} + \gamma_{ll} + \gamma_{el} + \gamma_{ml}$$
$$= \gamma_{ke} + \gamma_{le} + \gamma_{ee} + \gamma_{me}$$
$$= \gamma_{km} + \gamma_{lm} + \gamma_{em} + \gamma_{mm} \tag{7.39}$$
$$= \omega_{ky} + \omega_{ly} + \omega_{ey} + \omega_{my} = 0.$$

These relationships imply that the OLS residuals e_i across equations will sum to zero at each observation; that is,

$$e_k + e_l + e_e + e_m = 0. \tag{7.40}$$

Estimation results are presented in Table 7.1.

3. Because the disturbance covariance and the residual cross-products matrices will both be singular, maximum likelihood estimation[3] is not feasible because the determinant will be zero for any set of parameters that satisfy Equation (7.39). The most common procedure is to drop one of the equations. For example, one could impose the homogeneity restrictions, delete

---

[3]Maximum likelihood, also called the maximum likelihood method, is the procedure of finding the value of one or more parameters for a given statistic that makes the *known* likelihood distribution a maximum.

(i.e., divide through by) the materials share equation, and then estimate the following:

$$S_k = \beta_k + \gamma_{kk} \ln(p_k/p_m) + \gamma_{kl} \ln(p_l/p_m) + \gamma_{ke} \ln(p_e/p_m) + \omega_{ky} \ln Y \quad (7.41)$$

$$S_l = \beta_l + \gamma_{kl} \ln(p_k/p_m) + \gamma_{ll} \ln(p_l/p_m) + \gamma_{le} \ln(p_e/p_m) + \omega_{ly} \ln Y \quad (7.42)$$

$$S_e = \beta_e + \gamma_{ke} \ln(p_k/p_m) + \gamma_{le} \ln(p_l/p_m) + \gamma_{ee} \ln(p_e/p_m) + \omega_{ey} \ln Y. \quad (7.43)$$

Algebraically, one could then recover the estimates of the six other parameters related to the materials equation (β_m, γ_{km}, γ_{lm}, γ_{em}, γ_{my}, and γ_{mm}). Although dropping one share equation appears to be reasonable, it begs the question: Does it matter which equation is dropped (in terms of the parameter estimates)? The answer to this question is that as long as one employs a maximum likelihood estimation procedure, then the parameter estimates, log-likelihood values, and standard errors do not vary by the choice of which equation is deleted.[4]

Estimation results for Equation (7.15) are displayed in Table 7.1.

SUBSTITUTION ELASTICITIES FOR THE TRANSLOG FORM: HICKS–ALLEN PARTIAL ELASTICITIES OF SUBSTITUTION

As discussed previously in Chapter 5, one of the properties of flexible functional forms is that they place no a priori restrictions on substitution elasticities. Appealing to the Allen partial elasticities of substitution between inputs i and j for the translog cost model are given by

$$\sigma_{ij} = (\gamma_{ij} + S_i S_j)/S_i S_j \quad (7.44)$$

for $i, j = 1, \ldots, n$ but $i \neq j$, and

$$\sigma_{ii} = (\gamma_{ii} + S_i^2 - S_i)/S_i^2 \quad (7.45)$$

for $i = 1, \ldots, n$.

Results from the estimation of Equations (7.38)–(7.39) are displayed in Table 7.1. As can be seen, estimates of γ_{ij}, which represent substitution among the different inputs, vary in sign; those that are positive imply that they are substitutes, whereas those that are negative indicate they are complements in

---

[4]It is important to note that this only holds for the one-step (as opposed to the iterated) Zellner-efficient estimator (ZEF), with the caveat that the first-round estimate of Ω^* (the disturbance covariance matrix with the nth row and column deleted) is based on equation-by-equation least-squares estimation without the symmetry conditions imposed. However, if one-step ZEF is the methodology and the estimate of Ω^* is based on OLS estimation of a stacked $n-1$ equation system with symmetry restrictions imposed, then the parameter estimates of the ZEF estimator will vary depending on which equation is deleted.

production. From these and the fitted values of the share equations, substitution elasticities can be calculated. You will have the opportunity to do this in the exercises at the end of the chapter.

PRICE ELASTICITIES

Next we turn to price elasticities, which are equal to

$$\varepsilon_{ij} = (S_j S_i + \gamma_{ij})/S_i \tag{7.46}$$

for $i, j = 1, \ldots, n$ but $i \neq j$, and

$$\varepsilon_{ii} = (\gamma_{ii} + S_i^2 - S_i)/S_i \tag{7.47}$$

for $i = 1, \ldots, n$.

Discussion of Results

Because parameter estimates and fitted values replace the γ_{ij}'s, γ_{ii}'s, S_i's, and S_j's when computing the estimates of σ_{ij} and ε_{ij}, the estimated elasticities will vary across observations (unlike those resulting from the Cobb–Douglas and constant elasticity of substitution functional forms). In addition, while parameter estimates and fitted shares have variances and covariances, estimated substitution elasticities also have stochastic distributions. According to Berndt (1991), these elasticities are highly nonlinear functions of the γ_{ij}'s, γ_{ii}'s, S_i's, and S_j's, which implies that it is difficult to obtain estimates of the variances of these elasticities. Finally, the estimated translog cost function should be checked at every observation to ensure that it is monotonically increasing and strictly quasiconcave in input prices, as required by economic theory (see the appendix to this chapter for details).

Now let's move onto another flexible form that illustrates the duality between cost and production functions. Known as the generalized Leontief functional form, first introduced by Erwin Diewert in his dissertation, this form was the first in a series of developments in the theory of dual cost and production [for more detail, see Berndt (1991)].

GENERALIZED LEONTIEF COST FUNCTION

Another flexible functional form that places no a priori restrictions on substitution elasticities is the generalized Leontief. This section explores this model in more detail and works some examples/exercises. With constant returns to scale imposed, this cost function can be written as

$$C = Q\left(\sum_i \sum_j \beta_{ij} P_i^{0.5} P_j^{0.5}\right), \tag{7.48}$$

where Q is output, P_i, P_j are input prices, and β_{ij} are parameters to be estimated.

From Shepherd's lemma, the optimal, cost-minimizing demand for input i can be obtained by differentiating the cost function with respect to P_i; that is,

$$X_i = \partial C / \partial P_i \qquad (7.49)$$

or

$$X_i = Q\sum_j \beta_{ij}(P_j/P_i)^{0.5}. \qquad (7.50)$$

According to Berndt (1991), this can be estimated by dividing through by Q, which yields the optimal input–output demand equations, denoted by a_i:

$$a_i = X_i/Q = \sum_j \beta_{ij}(P_j/P_i)^{0.5}. \qquad (7.51)$$

Note that when $i = j$, $(P_j/P_i)^{0.5} = 1$, so β_{ij} is a constant term in the ith input–output equation.

EMPIRICAL ESTIMATION

To estimate the GL equation, it is necessary to append an additive disturbance term to each of the input–output equations.[5] One may assume that the resulting disturbance vector is independently and identically normally distributed (i.i.d.) with mean vector zero and constant, nonsingular covariance matrix Ω.[6] As Berndt aptly explains:

These disturbances could simply reflect optimization errors on the part of firms. Alternatively, firms could be envisaged as differing from each other according to

---

[5]An additive disturbance term is attractive so that one might use OLS to estimate the parameters of the model.

[6]Covariance matrix: If the entries in the column vector

$$X = \begin{bmatrix} X_1 \\ \vdots \\ X_n \end{bmatrix} \qquad (7.51a)$$

are random variables, each with finite variance, then the covariance matrix Σ (as opposed to Ω) is the matrix of which the (i, j) entry is the covariance

$$\Sigma_{ij} = \mathrm{cov}(X_i, X_j) = E[(X_i - \mu_i)(X_j - \mu_j)], \qquad (7.51b)$$

where

$$\mu_i = E(X_i) \qquad (7.51c)$$

is the expected value of the ith entry in vector X. In other words, we have (7.51d). The inverse of this matrix, Σ^{-1}, is the inverse covariance matrix. The elements of this matrix have an interpretation in terms of partial correlations and partial variances. This is given by

$$\Sigma^{-1} = \begin{bmatrix} E[(X_1 - \mu_1)(X_1 - \mu_1)] & E[(X_1 - \mu_1)(X_2 - \mu_2)] & \cdots & E[(X_1 - \mu_1)(X_n - \mu_n)] \\ E[(X_2 - \mu_2)(X_1 - \mu_1)] & E[(X_2 - \mu_2)(X_2 - \mu_2)] & \cdots & E[(X_2 - \mu_2)(X_n - \mu_n)] \\ \vdots & \vdots & \ddots & \vdots \\ E[(X_n - \mu_n)(X_1 - \mu_1)] & E[(X_n - \mu_n)(X_2 - \mu_2)] & \cdots & E[(X_n - \mu_n)(X_n - \mu_n)] \end{bmatrix}^{-1}. \qquad (7.51d)$$

parameters that are known by the firms' managers but not by the econometrician examining the aggregated data. To the econometrician, such firm effects can manifest themselves as random parameters in the GL cost function and as additive disturbances in the input demand functions.

As an example, Berndt (1991) presents an example in which there are four inputs: capital (K), labor (L), energy (E), and nonenergy immediate materials (M). In this case, there would be four input–output equations, which, given the additive disturbance term, one might attempt to use OLS to estimate the parameters of this system of equations (they are linear in parameters). However, given that there are cross-equation symmetry constraints that need to be imposed (to conform to a properly specified cost model), this essentially obviates the use of OLS to estimate the parameters of the model, which need to be estimated as a system of equations rather than a set of independent equations. The example is quoted with permission here.

In this example, energy (E) and materials (M) are intermediate inputs, which means that the appropriate measure of output quantity Q is gross output (sales quantity plus net changes in output inventory quantities), not value added. In the four-input case, the GL cost-minimizing input–output equations are given by

$$a_i = X_i/Q = \sum_j \beta_{ij}(P_j/P_i)^{0.5} \tag{7.52}$$

or

$$a_K = X_K/Q = \beta_{KK} + \beta_{KL}(P_L/P_K)^{0.5} + \beta_{KE}(P_E/P_K)^{0.5} + \beta_{KM}(P_M/P_K)^{0.5} \tag{7.53}$$

$$a_L = X_L/Q = \beta_{LL} + \beta_{KL}(P_K/P_L)^{0.5} + \beta_{LE}(P_E/P_L)^{0.5} + \beta_{LM}(P_M/P_L)^{0.5} \tag{7.54}$$

$$a_E = X_E/Q = \beta_{EE} + \beta_{KE}(P_K/P_E)^{0.5} + \beta_{LE}(P_L/P_E)^{0.5} + \beta_{EM}(P_M/P_E)^{0.5} \tag{7.55}$$

$$a_M = X_M/Q = \beta_{MM} + \beta_{KM}(P_K/P_M)^{0.5} + \beta_{LM}(P_L/P_M)^{0.5} \\ + \beta_{EM}(P_E/P_M)^{0.5}. \tag{7.56}$$

These are subject to the cross-equation constraints:

$$\beta_{KL} = \beta_{KL}$$

$$\beta_{KE} = \beta_{KE}$$

$$\beta_{KM} = \beta_{KM}$$

$$\beta_{LM} = \beta_{LM} \tag{7.57}$$

$$\beta_{LE} = \beta_{LE}$$

$$\beta_{EM} = \beta_{EM}.$$

Note that while each of the four equations has four parameters, the six cross-equation symmetry constraints [in Equation (7.57)] reduce the number of free parameters to be estimated from 16 to 10.

Although equation-by-equation OLS estimation might appear attractive [Equations (7.53)–(7.56) are linear in parameters], cross-equation symmetry constraints require use of a system's estimator, such as Zellner's seemingly unrelated regression estimator, also known as ZEF. This is described in more detail here.

Even if one were to ignore the cross-symmetry constraints, one would still expect the ZEF system's estimator to yield different parameter estimates than those from equation-by-equation OLS. The reason for this is twofold:

1. It is to be expected that disturbances across input–output equations would be contemporaneously correlated, which implies that the disturbance covariance matrix would be nondiagonal.
2. As seen in Equations (7.53)–(7.56), each input–output equation contains different regressors.

For both of these reasons, in large samples the ZEF estimator provides more efficient parameter estimates than OLS. This is described in more detail in the next section.

ITERATED ZELLNER-EFFICIENT ESTIMATOR

In the cost function estimation literature, one popular estimation technique is that of Zellner's iterated seemingly unrelated regression. One nice feature of the general Leontief specification is that, via Shephard's lemma, optimal input demand equations can be derived and then estimated simultaneously, which yields estimates that are more efficient than equation-by-equation ordinary least squares. This method requires an estimate of the cross-equation covariance matrix, which increases the sampling variability of the estimator and yields estimates that are numerically equivalent to the maximum likelihood estimator (Berndt, 1991, p. 463). But before estimation can proceed, several precautions must be made.

1. Because shares always sum to unity and only $n - 1$ of the share equations are linearly independent, for each observation the sum of the disturbances across equations must always equal zero. This implies that the disturbance covariance matrix is singular and nondiagonal. Thus, one of the equations must be deleted and its parameters inferred from the homogeneity condition. This raises the question of whether the parameter estimates are invariant to the choice of the equation to be dropped. However, as long as either maximum likelihood or Zellner's method (one step or iterated seemingly unrelated regressions) estimation is performed, then the estimates are invariant to the choice of the equations to be estimated.

TABLE 7.2 Equation-by-Equation OLS and IZEF/ML Estimates of Parameters in Generalized Leontief Input–Output Demand Equations [Equations (7.53)–(7.56)]*

B_{ij}	Capital Equation (a_K)		Labor Equation (a_L)		Energy Equation (a_E)		Materials Equation (a_M)	
	OLS	IZEF	OLS	IZEF	OLS	IZEF	OLS	IZEF
$j = K$	0.0232	0.0263	0.0485	0.0517	−0.0139	−0.0111	−0.0550	−0.0542
	(0.0157)	(0.0143)	(0.0269)	(0.0245)	(0.0097)	(0.0088)	(0.0459)	(0.0420)
$j = L$	0.0048	0.0036	−0.0692	−0.0719	−0.0041	−0.0048	−0.1372	−0.1374
	(0.0097)	(0.0088)	(0.0166)	(0.0151)	(0.0058)	(0.0053)	(0.0281)	(0.0258)
$j = E$	0.0605	0.0649	0.2183	0.2200	0.0373	0.0403	0.0385	0.0399
	(0.0331)	(0.0301)	(0.0523)	(0.0476)	(0.0201)	(0.0183)	(0.0933)	(0.0855)
$j = M$	−0.0381	−0.0443	0.0281	0.0264	0.0199	0.0150	0.7420	0.7401
	(0.0469)	(0.0426)	(0.0743)	(0.0676)	(0.0285)	(0.0259)	(0.1325)	(0.1214)

*Standard errors are in parentheses.

2. To preserve the linear homogeneity of the system, both cost-share equations must be normalized by dividing each input price by the input price that corresponds to the deleted cost-share equation. An excerpt from Berndt (1991) is provided:

In effect, the ZEF estimator uses equation-by-equation OLS to obtain an estimate of the disturbance covariance matrix Ω and then does generalized least squares, given this initial estimate of Ω, on an appropriately "stacked" set of equations. Furthermore, one can update the estimates of Ω and iterate the Zellner procedure until changes from one iteration to the next in the estimated parameters and the estimated Ω become arbitrarily small. This iterated Zellner-efficient estimator is typically termed IZEF, and in this case it yields parameter estimates that are numerically equivalent to those of the maximum likelihood (ML) estimator.

Table 7.2 provides a comparison of the equation-by-equation OLS and the IZEF estimates without symmetry imposed. While the parameter estimates are not dissimilar, the IZEF standard errors are generally smaller. Needless to say, imposing cross-equation symmetry constraints would yield even more efficient parameter estimates.

GENERALIZED LEONTIEF COST FUNCTION—ELASTICITIES

As stated previously, one attractive feature of the generalized Leontief cost function is that there are no a priori restrictions on substitution elasticities.

TABLE 7.3 Hicks–Allen Elasticities for Generalized Leontief Cost Model

Parameter	Cross-Price Elasticities, σ_{ij}	Parameter	Own-Price Elasticities, σ_{ii}
σ_{KL}	6.84	σ_{KK}	−5.50
σ_{KE}	−1.98	σ_{LL}	−1.27
σ_{KM}	−0.87	σ_{EE}	−13.72
σ_{LM}	1.04	σ_{MM}	−0.09
σ_{EM}	0.15		
σ_{LE}	11.14		

In this case, Hicks–Allen partial elasticities of substitution for a general dual-cost function with n inputs are computed as

$$\sigma_{ij} = C * C_{ij} / C_i * C_j, \tag{7.58}$$

where i, j are the first and second partial derivatives of the cost function with respect to input prices (p_i, p_j) and $i, j = 1, \ldots, n$.

More specifically, Hicks–Allen substitution elasticities are given by

$$\sigma_{ij} = 1/2 * C\beta_{ij}(P_jP_i)^{-0.5}/Q \, a_i a_j \tag{7.59}$$

for all $i, j = 1, \ldots, n$ but $i \neq j$, and the own Hicks–Allen elasticities by

$$\sigma_{ii} = -1/2 * C\sum_j \beta_{ij} P_j^{0.5} P_i^{-1.5}/Q \, a_i^2 \tag{7.60}$$

for $i = 1, \ldots, n$.

Hicks–Allen elasticities can be computed once estimates of the generalized Leontief model parameters are available (Table 7.3).

Discussion of Results

These estimates suggest that capital and labor in U.S manufacturing in 1971were substitutes but capital was complementary with energy and materials, whereas labor, energy, and materials were substitutes.

If instead one wanted to compute price elasticities, they are given by

$$\varepsilon_{ii} = \partial X_i/\partial P_i * P_i/X_i. \tag{7.61}$$

In the case of the generalized Leontief cost function, we have:

1. Cross-price:

$$\varepsilon_{ij} = 1/2 * \beta_{ij}(P_i/P_i)^{-0.5}/a_i \tag{7.62}$$

for $i, j = 1, \ldots, n$.

2. Own price:

$$\varepsilon_{ii} = -1/2 * \sum \beta_{ij}(P_i/P_i)^{-0.5}/a_i \qquad (7.63)$$

for all $i = 1, \ldots, n, j = 1, \ldots, n$ but $i \neq j$.

Aside: Some Things to Note on These Elasticities

1. Because input prices and a_i vary across observations, in general the estimates of price and substitution elasticity will also vary over different observations.

2. Although by construction

$$\sigma_{ji} = \sigma_{ij},$$

it is not necessarily the case that

$$\varepsilon_{ji} = \varepsilon_{ij}$$

(i.e., price elasticities are not symmetric, unlike Hicks–Allen elasticities).

3. An examination of Equations $(7.59)(\sigma_{ij})$ and $(7.62)(\varepsilon_{ij})$ reveals that inputs i and j are substitutes, independent, or complements depending on whether the estimated β_{ij} is positive, zero, or negative, respectively. In addition, for the own-price elasticity to be negative, it is necessary that the estimated coefficients (the β_{ij}'s) in Equation $(7.63)(\varepsilon_{ii})$ be positive.

4. Because C and a_i, a_j are in the elasticity equations, estimated elasticities are based on estimated parameters and predicted values of C, a_i, and a_j, not on their observed values. Given that it is easy to make computational errors using these equations, a check involves the following summation of elasticities, which must always hold:

$$\sum \varepsilon_{ij} = 0 \qquad (7.64)$$

for $i = 1, \ldots, n$.

5. As required by theory, which states that a proper cost function should be monotonically increasing and strictly concave in input prices, one must verify that fitted values for all of the input–output equations are positive and that the $n \times n$ matrix of substitution elasticities is negative semidefinite at every observation.

6. Because computed elasticities depend on estimated parameters, estimated elasticities also have variances and covariances.[7]

---

[7]According to Berndt, substantial nonlinearities inherent in elasticity calculations have forced researchers to employ approximation techniques in calculating these variances. Furthermore, because distribution properties of such elasticity estimates have not been derived, the basis for employing statistical inference on them does not yet exist (at least at the time of this writing).

Aside: Statistical Inference and Measure of Fit in Equation Systems

There are several ways by which statistical inference on the validity of parameter restrictions in systems of equations can be undertaken. Three common test statistics are the:

1. Wald test statistic
2. Likelihood ratio (LR)
3. Lagrange multiplier (LM)

Many econometric software programs will compute these statistics, but for more detail, see Berndt (1991). In terms of preference by econometricians, in practice, it appears to be the case that the LR test procedure appears to be used most frequently.

There is another issue that merits discussion: the goodness of fit, or R^2 statistic. In the single-equation context, in which the regression model is given by

$$Y = X\beta + e, \tag{7.65}$$

most computer software programs compute R^2 as

$$R^2 = 1 - e'e/(Y - Y^*)'(Y - Y^*), \tag{7.66}$$

where Y^* is the sample mean of Y.

Note that in the single-equation context,

$$e'e = (e - e^*)'(e - e^*). \tag{7.67}$$

This is the case because least-squares estimation ensures that the sum of residuals equals zero and therefore their mean also equals zero.

There are two reasons that this measure of R^2 is not appropriate in equation systems.

1. It is possible that the R^2 from a particular equation could be negative, as it is not necessarily the case with system estimation that the sum of residuals is zero within each equation. As such, it is possible that the numerator of Equation (7.66) is larger than the denominator, which results in a negative R^2.
2. While single-equation least squares minimize $e'e$ and therefore maximize R^2, in general, system estimation methods do not minimize $e'e$. As an exercise, you will verify this using data for generalized Leontief cost shares (i.e., obtain negative R^2).

Given this, a different goodness-of-fit measure should be employed. The generalized variance of the matrix Y is defined as

$$Y = \text{the determinant of } y'y, \tag{7.68}$$

where $y \equiv (Y - Y^*)$ and

$$Y = \text{a } T \times n \text{ matrix of observations on the } Y_i \tag{7.69}$$

for $i = 1, \dots, n$.

Then the generalized measure R^2, denoted as R^{2*}, indicates the proportion of the generalized variance of Y that is explained by the variation in the right variables in the system of equations and is computed as

$$R^{2*} = 1 - (|E'E| / |y'y|), \tag{7.70}$$

where $|E'E|$ is the determinant of the residual cross-products matrix, which the maximum likelihood estimator minimizes; as such, it also maximizes R^{2*}.

QUADRATIC COST MODEL

In the author's previous book (*Electricity Cost Modeling Calculations*), much time was spent discussing a properly specified quadratic cost model that the author developed in the writing of her dissertation. Let us revisit this form and use KLEM data to further illustrate the concepts described in this chapter.

NONLINEAR QUADRATIC COST MODEL

Chapter 5 presented a nonlinear, single-output quadratic cost model, which is given by

$$C = (\alpha_0 + \alpha_1 Q + 1/2\,\alpha_2 Q^2)\prod P_i^{\beta i}, \tag{7.71}$$

where Q is output, P_i is input price, and α_i and β_i are parameters to be estimated.

For the KLEM model, the cost-minimizing, optimal level of inputs $x_i{}^*$ are given by Shephard's lemma:

$$x_k^* = \partial C/\partial P_k = \beta_k P_K^{(\beta k - 1)}(\alpha_0 + \alpha_1 Q + 1/2\,\alpha_2 Q^2)P_L^{\beta l}P_M^{\beta m}P_E^{\beta e} \tag{7.72}$$

$$x_l^* = \partial C/\partial P_l = \beta_l P_l^{(\beta l - 1)}(\alpha_0 + \alpha_1 Q + 1/2\,\alpha_2 Q^2)P_K^{\beta k}P_M^{\beta m}P_E^{\beta e} \tag{7.73}$$

$$x_e^* = \partial C/\partial P_e = \beta_e P_e^{(\beta e - 1)}(\alpha_0 + \alpha_1 Q + 1/2\,\alpha_2 Q^2)P_K^{\beta k}P_M^{\beta m}P_L^{\beta l} \tag{7.74}$$

$$x_m^* = \partial C/\partial P_m = \beta_m P_m^{(\beta m - 1)}(\alpha_0 + \alpha_1 Q + 1/2\,\alpha_2 Q^2)P_K^{\beta k}P_E^{\beta e}P_L^{\beta l}. \tag{7.75}$$

Again, cross-equation symmetry constraints can be imposed and estimated jointly along with the cost function itself.

Estimation

Earlier models in this chapter had constant returns to scale imposed. This model does not. First, let us estimate the parameters of the cost function alone, which are displayed in Table 7.4.

Discussion of Results

An adjusted R^2 of 0.99 indicates that the model fits data well, explaining over 99% of the variation in cost. In addition, parameter estimates are of the expected

TABLE 7.4 KLEM Data Estimated via a Nonlinear Quadratic Cost Model

Parameter	Variable	Estimate	t Statistic
a_0	Constant	33.34795	2.32
a_1	Output	0.769603	9.69
a_{11}	Output squared	−0.00036	−1.74
b_k	Price of capital	0.000417	0.02
b_m	Price of materials	0.585341	6.07
b_l	Price of labor	0.333219	4.56
b_e	Price of energy	0.081023	1.06

sign but neither energy nor capital appears to be an important explanatory variable. A test of the homogeneity-in-input-prices restriction, which is given by

$$\sum \beta_i = 1, \qquad (7.76)$$

indicates that the null hypothesis (linear homogeneity) should not be rejected.

ELASTICITIES

It is informative to examine the price and substitution elasticities that result from the quadratic form.

Price Elasticities

For the purpose here it is necessary to distinguish two types of price elasticity: own- and cross-price.

Own Price

$$\varepsilon_{ii} = \partial X_i / \partial P_i * P_i / X_i. \qquad (7.77)$$

This yields

$$\varepsilon_{ii} = \beta_i - 1. \qquad (7.78)$$

For the KLEM model, own-price elasticities are

$$\varepsilon_{mm} = -0.3721$$

$$\varepsilon_{ll} = -0.7266$$

$$\varepsilon_{kk} = -0.9468$$

$$\varepsilon_{ee} = -0.9546.$$

Note that own-price elasticities are negative as required by economic theory.

Cross-Price

Cross-price elasticities, which measure the effect of a change in the price of one input on the demand for another input, are given by

$$\varepsilon_{ji} = \partial X_i / \partial P_j * P_j / X_i. \tag{7.79}$$

For the quadratic cost model given in Equation (7.71), this becomes

$$\varepsilon_{ij} = \beta_j. \tag{7.80}$$

These are displayed in Table 7.4.

Substitution Elasticities

It was stated previously that the Allen–Uzawa elasticities of substitution for inputs j and i are given by ([Equation (7.58)]:

$$\sigma_{ij} = C * C_{ij} / C_i * C_j,$$

where i, j are the first and second partial derivatives of the cost function with respect to input prices (p_i, p_j) and $i, j = 1, \ldots, n$.

This is equivalent to

$$\sigma_{ij} = \varepsilon_{ij} / S_j, \tag{7.81}$$

where

$$S_j = P_j * \partial C / \partial P_j / C. \tag{7.82}$$

As an exercise, you will calculate the Allen–Uzawa elasticities of substitution.

MORISHIMA ELASTICITIES OF SUBSTITUTION

Instead of using Allen–Uzawa elasticities, let us look at the Morishima elasticities of substitution, which are nonsymmetric, as the value depends on the direction of a change in price and the impact on the ratio of prices P_j / P_i. According to Mundra and Russell (2004, p. 8):

The conceptual foundations of Allen-Uzawa and Morishima taxonomies of substitutes and complements are, of course, quite different. The Allen-Uzawa taxonomy classifies a pair of inputs as direct substitutes (complements) if an increase in the price of one causes an increase (decrease) in the quantity demanded of the other, whereas the Morishima concept classifies a pair of inputs as direct substitutes (complements) if an increase in the price of one causes the quantity of the other to increase (decrease) relative to the quantity of the input whose price has changed. For this reason, the Morishima taxonomy leans more toward substitutability. Furthermore, if two inputs are direct substitutes according to the Allen-Uzawa criterion, theoretically they must be direct substitutes according to the Morishima criterion, but if two inputs are direct complements according to the Allen-Uzawa criterion, they can be either direct complements or direct substitutes according to the Morishima criterion. ... This relationship can be seen algebraically from 7.81 and 7.84. If i and j are direct Allen-Uzawa substitutes, in which case

$$\varepsilon_{ij} > 0, \tag{7.83}$$

then concavity of the cost function (and hence negative semi-definiteness of the corresponding Hessian) implies that

$$\varepsilon_{ij} - \varepsilon_{jj} > 0, \tag{7.84}$$

so that j is a direct Morishima substitute for i. Similar algebra establishes that two inputs can be direct Morishima substitutes when they are direct Allen-Uzawa complements.

CONCLUSION

This concludes our discussions on the various cost models used to estimate price and substitution elasticities. Now you may move onto the end-of-chapter exercises, which give you the opportunity to work through the concepts introduced in this chapter. In the case study on time-of-use pricing in Chapter 10, more sophisticated models will be introduced; the author says that they are more sophisticated in that they attempt to estimate intraweek and intraday elasticities of substitution (i.e., the use of real-time or time-of-use prices to estimate the degree of which consumers shift electricity usage to off-peak hours—either from weekdays to weekends or from midafternoon to late evening on a hot summer day). In other words, these are used to estimate substitution elasticities for *demand* (i.e., the consumer side) as opposed to supply (or cost, which is the supplier's side of the equation).

APPENDIX: QUASICONCAVITY IN INPUT PRICES

Strict quasiconcavity in input prices, which means that the σ_{ij} matrix is negative semidefinite, requires that the following conditions hold[8]:

1. All four own σ_{ii} are negative at each observation.
2. The six 2×2 matrices of which the elements consist of

$$\begin{vmatrix} \sigma_{ii} & \sigma_{ij} \\ \sigma_{ji} & \sigma_{jj} \end{vmatrix}$$

each has a positive **determinant** at every observation.

---

[8]For an in-depth discussion of this concept, see Chiang (1984).

3. The four possible 3×3 matrices whose elements consist of

$$\begin{vmatrix} \sigma_{ii} & \sigma_{ij} & \sigma_{ik} \\ \sigma_{ji} & \sigma_{jj} & \sigma_{jk} \\ \sigma_{ik} & \sigma_{jk} & \sigma_{kk} \end{vmatrix}.$$

Each has a negative **determinant** at every observation.

The 4×4 matrix consisting of all of the σ_{ij} ($i, j = K, L, E$, and M in the KLEM model) has a determinant whose value is zero at each annual observation.

EXERCISES

To get an understanding of the cost models presented in this chapter, please locate KLEM.xls data, which contain annual data on prices (indexed to 1947), inputs, total production cost, and output. As always, you should plot data and check for reasonableness.

1. Using data in KLEM.xls, form the input–output coefficients and square root transformations required to estimate Equations (7.53)–(7.56), which are based on the generalized Leontief cost equation. Estimate the parameters of the equations by using:
   a. Equation-by-equation ordinary least squares (verify that your results replicate those presented in Table 7.2).
   b. Zellner's iterated method without imposing cross-symmetry constraints (again, verify that your results accord with those presented in Table 7.2).
   c. Are the estimates generated by each estimation technique the same? If not, why?
   d. Optional: Generate the cross-products matrix of the residuals (a 4×4 matrix). When divided by the sample size, this yields an estimate of Ω, the covariance matrix. Does this give an indication as to the reason that the estimates differ? (Hint: Standard errors of the IZEF estimates should be smaller than those generated via equation-by-equation OLS.)

2. Again, one important take-away from this section is the computation of substitution elasticities. Using data in KLEM:
   a. Estimate the parameters of the input–output equations from the generalized Leontief model using Zellner's iterated method. Impose cross-symmetry constraints and compute fitted values for a_i, the predicted input–output coefficients. Verify that these are all positive in sign (for monotonicity).
   b. Using Equations (7.62) and (7.63), compute own- and cross-price elasticity estimates for each year in the sample. In 1971, which inputs appear to be substitutes and which are complements?
   i. Are the own-price elasticities negative as required by economic theory?
   ii. Do the cross-price elasticities sum to zero [Equation (7.64)]?

   c. On the basis of the parameter estimates obtained in 2a, compute the Allen partial elasticities of substitution σ_{ij} for each year in the sample using Equations (7.59) and (7.60).

Note: You will have to calculate the predicted value for average cost (C/Q). Because σ_{ij} are symmetric, it is only necessary to compute 10 substitution elasticities.

   d. Advanced: Verify that the cost function is strictly quasiconcave in input prices, which means that the σ_{ij} matrix is negative semidefinite. This requires:

     i. All four own σ_{ii} are negative at each observation.

     ii. The six 2×2 matrices of which the elements consist of

$$\begin{vmatrix} \sigma_{ii} & \sigma_{ij} \\ \sigma_{ji} & \sigma_{jj} \end{vmatrix}$$

   for $i, j = K, L, E, M$, but $i \neq j$.

Each has a positive determinant at every observation.

     iii. The four possible 3×3 matrices of which the elements consist of

$$\begin{vmatrix} \sigma_{ii} & \sigma_{ij} & \sigma_{ik} \\ \sigma_{ji} & \sigma_{jj} & \sigma_{jk} \\ \sigma_{ik} & \sigma_{jk} & \sigma_{kk} \end{vmatrix}$$

   for $i, j, k = K, L, E, M$, but $i \neq j, i \neq k, j \neq k$.

Each has a negative determinant at every observation.

     iv. The 4×4 matrix, consisting of all of the σ_{ij}, $i, j = K, L, E,$ and M, has a determinant of which the value is zero at each annual observation.

3. Repeat Exercise 2 using the translogarithmic cost function and the relevant share equations.

4. This exercise requires estimation of Equation (7.71) (nonlinear quadratic cost function) and the respective cost shares, along with own-price, cross-price, and substitution elasticities.

   a. Estimate the nonlinear quadratic cost function displayed in Equation (7.71). Verify the parameter estimates displayed in Table 7.4.

   b. Now estimate this model, along with the relevant share equations, as a system of equations. What can you tell about the explanatory power of the model itself? (Hint: Do the goodness-of-fit values make sense? Why or why not?)

   c. Calculate the relevant elasticities:

     i. Own price, as given in Equation (7.78).

     ii. Cross-price, as given by Equation (7.80).

     iii. Substitution, as given by Equation (7.81).

What can you tell about the relationship among the input variables in terms of whether they are substitutes or complements? Does this make sense? Why or why not?

5. Prove that, for the quadratic cost model given in Equation (7.71), Equation (7.83) always holds.

6. Prove that Equation (7.77) yields Equation (7.78).

7. Prove that Equation (7.79) yields Equation (7.80).

8. Using estimation results, calculate returns to scale for firms in the KLEM data set. Are returns to scale increasing, constant, or decreasing for these firms?

Chapter 8

Efficient Pricing of Electricity

INTRODUCTION

This chapter examines the various methodologies by which rates are set under regulation and the reasons that these rarely lead to a pareto-efficient outcome, which occurs when no one is made better off if someone else is made worse off. In fact, it is only by pricing at marginal cost, which is efficient both allocatively and productively (and maximizes welfare), that such an outcome results; this is the theme of this chapter. In addition, a new program being developed at the Electric Power Research Institute is examined. This program is all about understanding the electric utility customer and is composed of two parts, one of which is an understanding of how various customer groups respond to different price signals.

THEORY OF EFFICIENT PRICES

The introductory chapter (and the common theme of this book) has purported that pricing at marginal cost is efficient both productively and allocatively. But what, exactly, does this mean? Let us start with a basic definition and build upon it. First and foremost, what is economic efficiency? How is it attained? Simply put, economic efficiency means maximizing the level of output while minimizing the amount (and cost) of input; the latter is also known as factors of production (e.g., labor and capital). But is this all? Is it really this simple or is there some deeper, more complex underlying issue here? Actually, the answers to these questions are yes and no.

Figure 8.1 displays a market in equilibrium, which yields price P^* and output level Y^*. In this situation, both producer (PS) and consumer (CS) surplus are maximized; as such, a pareto-efficient outcome emerges.

But is it really this simple? In the case of regulated utility industries, the answer is clearly no; such entities are regulated because a substantial portion of the capital investment renders them "natural monopolies"; that is, requiring investments that are highly sunk so that duplication would be wasteful and inefficient and thus competition infeasible. As such, both fixed and variable costs must be incurred (and recovered) to provide service to end users, which

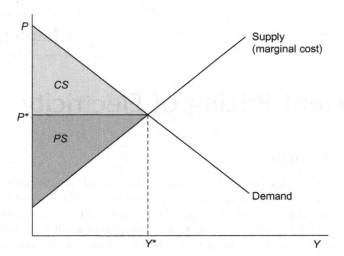

FIGURE 8.1   Consumer and producer surplus in a competitive market paradigm. Marginal cost pricing [the supply curve above average variable cost (not shown)] maximizes total welfare, which is equal to the consumer surplus (*CS*) plus producer surplus (*PS*). Thus, a pareto-efficient outcome is attained.

is the reason that regulated firms often charge both fixed and variable charges to end-use customers,[1] the latter of which should be based on marginal cost.

DEBATE ON THE OPTIMAL PRICING OF ELECTRICITY: A BRIEF HISTORY

The idea of marginal cost pricing is not new; for centuries, economists have espoused that pricing goods and services at marginal cost is efficient both allocatively and productively. In the case of electricity, it was actually two engineers who, in the late 19th century, argued for marginal cost pricing, also known as *time-of-use* or *real-time* pricing. The following excerpt contains a nice synopsis of the history of the debate on the optimal pricing of electricity, which requires an accurate estimation of the marginal cost of providing service to various types of end users (residential, commercial, industrial, etc.). In the introduction to his treatise on demand response and efficient pricing that appeared in the *USAEE Dialogue* in August 2007, John Kelly (president of the American Public Power Association) wrote:

Renewed Interest in Demand Response, but "Whither the Economic Rationale for Efficient Pricing?"

---

[1]This is not the quarrel here but rather the fact that it is often the case that variable charges do not necessarily reflect the true marginal cost of supplying electricity.

The time for implementing practical marginal cost pricing programs is long overdue, and almost anything that encourages marginal cost pricing is beneficial, whether under the guise of demand response or otherwise. However, it is appropriate to ask whether current analyses and discussions of demand response—which have become a cottage industry—are producing more heat than light on the subject because they stray from basic notions about economic costs.

This is an excellent point; in the next two sections he provides a brief history of the concept of marginal cost pricing applied to electricity, which is reprinted here with Mr. Kelly's consent.

II. Demand Response Proposal More Than a Century Old

Professors John Neufeld and William Hausman tell us that in 1894 engineer Alfred Gibbings made the case for time-of-use rates in terms that "came quite close to holding that prices should equal marginal costs." He criticized rates based on demand charges on grounds that are essentially the same as those modern economists use to criticize such charges. Neufeld and Hausman go on to note that W. S. Barstow, an engineer like Gibbings, was another early advocate of time-of-use rates and argued for the principle of marginal cost pricing. In 1895 at a meeting of the Association of Edison Illuminating Companies (AEIC), he argued that a utility should charge "customers a low rate during the light load" periods. Particularly interesting is Barstow's rationale for adopting time-of-day rate structures: The two-rate system seems to produce two desired results:

1. Broadening of the maximum peak.
2. An equally important result, the increasing of minimum peaks; that is, it encourages the forming of peaks during the minimum period of the load curve.

This justification is especially important because it states a central rationale for the principle of marginal cost pricing. It recognizes the implications of such pricing for economic efficiency of electricity production and lowering costs, specifically by designing rate structures to improve the utilization of electricity plants and to lower average costs. It seems that engineers, like Gibbings and Barstow, were the first to recognize the important connection between time-differentiated rates, capacity utilization, and costs. Hausman and Neufeld "found no evidence that professional economists had any input into the electric power industry's discussion about rate structures" during the very early years of the industry. But when economists eventually enter discussions about rate structures "they immediately embraced time-of-day rates on the basis of marginal cost considerations, even though they did not use the term 'marginal cost.'" For example, in a 1911 paper titled "Rates for Public Utilities," John Maurice Clark advocated prices based on marginal cost: If consumers can make extra demands on the utility without paying as much as the extra expense they are causing, they are likely to make wastefully large demands on it. ... But any consumers who cannot make extra use of the utility without paying many times more than the extra expense they would be causing, will skimp on their use, and the tendency will be to keep the plant in wasteful idleness."

About 10 years later, economist George Watkins wrote one of the first books, *Electric Rates*, devoted solely to the pricing of electricity. Hausman and Neufeld found that "the justification for differential rates was clear" to Watkins; it "was to improve the efficiency of resource allocation. Differential rates existed solely to improve the utility load factor and Watkins emphasized rates reflecting marginal costs as the way to achieve this greater efficiency and enhance social welfare." The main reason customers should be charged more during peak periods than nonpeak periods, Watkins said, was to encourage consumption during nonpeak hours, thereby making better use of utility plants and lowering the average cost of electricity.

The emphasis on the importance of marginal cost as a guide to efficient pricing and the efficient use of existing resources was continued in the 1930s with Harold Hotelling's classic paper "The General Welfare in Relation to Taxation and of Railway and Utility Rates." James Bonbright at the time considered Hotelling's paper "one of the most distinguished contributions to rate-making theory in the entire literature of economics."

From the late 1940s to the end of the 20th century, Professor William Vickrey was among the leading proponents of efficient pricing of utility services. He urged that electric rate structures should "be developed by careful weighing of the relevant factors with a view of guiding consumers to make efficient use of facilities that are available." He argued that electricity should be priced based on short-run marginal cost, and although the principle need "not in practice to be followed absolutely, it must play a major and even dominant role in the elaboration of any scheme of rates or prices that seriously pretends to have a major motive of the efficient utilization of available resources and facilities." More broadly, marginal cost principles are recognized as the starting point for the proper pricing of goods and services.

Thomas Nagle and Ronald Holden tell managers in other industries that "not all costs are relevant for every pricing decision." Relevant costs are "costs that are incremental (not average), avoidable (not sunk)." They go on to note that relevant costs are "those that actually determine the profit impact of the pricing decision."

Unfortunately, most discussions of demand response obscure the compelling economic logic that prices should reflect the time-varying cost—the marginal cost—of electricity service so that existing facilities are used more efficiently and rates are lower than under existing ratemaking practices based on fully allocated cost accounting practices.

III. A Renewed Interest in Demand Response, but "Whither the Economic Rationale for Efficient Pricing?"

The last time there was such high interest in demand response as there is today was in the late 1970s after the Public Utility Regulatory Policies Act was enacted. A large part of the recent interest is due to the disconnect between the time-varying prices of electricity in spot markets and the essentially nonvarying prices charged in retail markets.

In 2004, the Government Accountability Office issued a report that concluded that increased use of demand response would improve efficiency in the electric utility industry and recommended that state utility commissions do more to promote demand response programs. The Energy Policy Act of 2005 directed the Secretary of Energy to provide "Congress with a report that identifies and quantifies the national benefits of demand response and make a recommendation on achieving specific levels of such benefits." The U.S. Department of Energy (DOE) early the following year issued a report titled "Benefits of Demand Response in Electricity Markets and Recommendations for Achieving Them." The Act also instructed the Federal Energy Regulatory Commission to assess the use of demand response programs and related metering technologies in the nation, and in August 2006 the FERC released the results of an industry-wide survey. In addition, states have showed renewed interest in demand response. For example, the state of California commissioned a study to evaluate the benefits of so-called critical peak pricing that would allow sharply higher prices during critical peak times.

But what does the term *demand response* mean? The term is variously defined, but DOE's definition is representative: Changes in electricity usage by end-use customers from their normal consumption patterns in response to changes in the price of electricity over time, or to incentive payments to induce lower electricity use at times of high wholesale market prices or when system reliability is jeopardized.

The DOE report says that states should consider aggressive implementation of price-based demand response a "high priority." They should do this because "flat, average cost retail rates that do not reflect the actual cost to supply power lead to inefficient capital investment in new generation, transmission, and distribution infrastructure and higher electric bills for consumers."

More specifically, it states that:

[The] disconnect between short-term marginal electricity production costs and retail rates paid by consumers leads to an inefficient use of resources. Because customers don't see the underlying short-term cost of supplying electricity, they have little or no incentive to adjust their demand or supply-side conditions. Thus, flat electricity prices encourage customers to over-consume relative to an optimally efficient system in hours when electricity prices are higher than average rates, and under-consume in hours when the cost of producing electricity is lower than average rates. As a result electricity costs may be higher than they would otherwise be because high-cost generators must sometimes run to meet the non-price responsive demands of consumers.

These are cogent words, but how do we put them into practice? And that is where the topic of rate design comes into play.

RATE DESIGN

In Chapter 1, the utility's *revenue requirement* was introduced; that is, the amount of dollars that must be collected from ratepayers to recover the utility's expenses (and required return, in the case of investor-owned utilities) for the period during which such rates would be in effect. However, until now, there has been no discussion of how the revenue requirement would be allocated among different customer classes nor has a discussion ensued about the various methodologies by which said revenue requirement would be recovered. This is one objective of this chapter: to present the various methodologies employed in the process of rate recovery and some of the consequences that could emerge as the result of nonoptimal rate-making mechanisms. In essence, what we are once again talking about is how to design rates to motivate both producers and consumers to make appropriate choices and behave in an optimal manner in terms of consumption, investment, and conservation.

Formally, the *rate design* process is that which determines how the revenue requirement will be allocated among the various customer classes. At best (and at the highest level) it is presumably determined according to the cost that each class imposes upon the system, which is also known as cost-of-service regulation. However, politics often enter into the mix, and certain customer classes do not necessarily pay the full cost that servicing them imposes upon the system (i.e., their cost of service). As such, the issue of *cross-subsidy* is pervasive and an important consideration. Residential customers are voters and are well represented at utility rate proceedings (by the states' attorney general); as such, they are often subsidized by the larger classes, which is discussed later in this chapter.

MORE ABOUT RATE DESIGN: IN THEORY

As said, the rate design process is that which determines the portion of the revenue requirement that will be recovered by each customer class and the methodology by which it will be recovered; that is, fixed versus variable charges.

In most cases, it begins with a cost-of-service study performed by the utility. It attempts to classify the costs of generation, transmission, and distribution among the components of such costs (i.e., customer, energy, and demand charges) by the various customer classes (residential, commercial, industrial, and other). In some cases, these costs are further delineated by season (winter or summer) and then even by time of use (peak vs. off peak, or more frequent).

More formally, the rate design process consists of the following steps:

- Determination of total costs and revenue requirements.
- Functionalization of costs.
- Classification of costs.

- Identification of rate classes.
- Design of end-user rates.

Each is discussed in turn.

OVERVIEW OF RATE DESIGN PROCESS

Total Revenue Requirements

Total revenue requirements are total cost incurred by the utility in the provision of service. It is the amount to be recovered from ratepayers as authorized by the state's public regulatory commission. In determination of this amount, costs are grouped into capital, operations and maintenance, administrative, and taxes. The revenue requirement (or total cost of service) is the sum of the return on undepreciated capital investment and all other expenses. The standard equations for revenue requirements are[2]:

$$RR = (RB)*r + E + D + T + O \qquad (8.1)$$

and

$$RB = (PV - CD), \qquad (8.2)$$

where RR is revenue requirements, r is allowed rate of return, RB is rate base, E is operating expenses, D is annual depreciation, T is taxes, O is other expenses, PV is plant value (investment in plant), and CD is cumulative depreciation.

Functionalization and Classification of Costs

Utilities are required to keep a detailed accounting of its costs, which can be grouped by major category, such as utility plant, operating expenditures, and taxes. Under each of these a number of subaccounts exist; for example, utility plant may include land and right of way, plant equipment, and other structures and improvements. For the purpose of rate design, costs from the different categories are grouped by operating function: generation, transmission, and distribution. This is the process of functionalization.

Once functionalized, these costs are broken down further by their consumption or cost causation characteristics, which include demand (or capacity), energy related (the cost of fuel), customer related (metering and billing), and revenue related (tax receipts and some overhead costs) (Harunuzzaman and Koundinya, 2000).

Identification of Rate Classes

The next step is to separate customers into rate classes so that the costs of servicing each can be determined. Rate classes are defined by certain characteristics

---

[2]See "Cost Allocation and Rate Design for Unbundled Gas Services," NRRI 00-08, for more details.

common among members, such as size (or usage level), load factor,[3] and customer type (i.e., residential, commercial, industrial).

Once this is done, costs are then allocated to each rate class. In some cases, the causation is clear cut: installing a meter in a residence is a customer-related cost to the residential rate class. However, it is often the case that the delineation is not so evident. Joint or common costs characterize public utilities; in fact, these attributes give rise to their being natural monopolies. In the case of electricity, transmission lines provide service to all customer rate classes. Clearly, the allocation of transmission and related costs is a difficult undertaking.

Joint (or Common) Cost Allocation

The most common method of allocating joint costs is the fully distributed cost (FDC), which assigns costs on the basis of the relative demand of each rate class. Based on embedded costs, this method uses various techniques to allocate costs to each classification of service (Harunuzzaman and Koundinya, 2000). The classifications of embedded costs are:

- Demand or capacity costs—including coincident and noncoincident peak, and average and excess (again see Harunuzzaman and Koundinya, 2000).
- Commodity or energy costs—typically based on the share of total energy consumed by each customer class.
- Customer costs—generally tied to the number of customers in a given class.

Allocation of Fixed Costs

While seemingly simple, the allocation of fixed costs among customer classes can be difficult. In the case of electricity, it is often the case that fixed costs (i.e., customer charges or entry fees) paid by residential customers are different from those paid by commercial and industrial consumers. In addition, the latter tend to pay demand charges based on the maximum demand (monthly) that serving them imposes on the system.

Later it will be demonstrated that two-part tariffs could be beneficial and increase total surplus (although not necessarily increase consumer and producer surplus equally, which is a different matter; rates are not designed for equality but for fairness—the distinction is important). Seemingly, it was straightforward to calculate an appropriate entry fee (and in the exercises at the end of the chapter

---

[3]Load factor is an index of a customer's consumption pattern and is defined as the ratio of average consumption to peak consumption. Low load factor customers, such as residential and small commercial customers, tend to have a spiked consumption pattern, characterized by high peak consumption relative to their average consumption. High load factor customers, however, tend to have a flatter consumption pattern, with their peak consumption closer to their average consumption. Load factor is an important determinant of cost allocation. It generally costs more to deliver a unit of energy to a low load factor customer than to a high load factor customer, as the former imposes a relatively high capacity cost on the system, which needs to be recovered from fewer units of energy (Harunuzzaman and Koundinya, 2000).

you will). In the real world, demand functions are not necessarily known (nor are supply or cost functions) so other methodologies must be employed.

As stated, one methodology of allocating common costs, such as costs of generation, transmission, and distribution, is known as fully distributed cost pricing. Under this method, the regulator (1) allocates the costs to serve a particular customer to that customer and (2) divides common costs among customers.

For many years utilities have been using cost, output, and revenue data from the most recent 12 months (the test period) to be used in the allocation of costs by function and by customer class. In some cases, price elasticities have been used but these are often difficult to ascertain, especially under the conditions that prevailed in the 1980s and early 1990s; declining (or flat) energy costs yield little in the measurement of customer response to price changes. According to Brown and Sibley (1986, p. 49), "price elasticities of demand have no place in setting FDC rates, except perhaps in forecasting revenue, so FDC prices will generally be much different from Ramsey prices."

Although the most widely used, FDC pricing methodology provides no incentive to increase efficiency because it is an average cost rather than setting prices based on marginal cost. And then there is the issue of cross-subsidization, which is discussed in an upcoming section.

Design of End-User Rates

Rates (or tariffs) are typically composed of fixed charges (customer access/entry fee) and variable charges (those that apply on a per-unit consumed basis), which are generally called energy charges. However, it is often the case that energy charges include some demand-related costs. This is especially true in residential and smaller commercial rate classes.

The energy charge can be constructed in a variety of ways:

- Block rates (uniform, or linear, increasing, or declining), which may vary by season, day of the week, or time of day.
- Marginal cost pricing.
- Average cost pricing.

As said, these often vary by class of customer, but which of these yield prices (i.e., rates) that result in the most efficient allocation of resources and send the appropriate signal to end users to incent them to use electricity wisely? In other words, which represents an optimal rate design?

EFFICIENT PUBLIC UTILITY PRICING

In the adoption of efficient pricing for a regulated firm, three concepts are worth noting:

1. Efficient prices are those that maximize total welfare.
2. Changes in prices can create "winners" and "losers." However, it is possible that "winners" can compensate "losers" in some fashion so as to render them better off than before the change.

3. Because the firm typically must break even out of its own sales revenues (i.e., no governmental subsidy or taxes), then it is likely the case that total welfare will be reduced.

Regarding the first point, the absence of competition creates the ability for producers to gain at the expense of consumers. Figure 8.2 illustrates this point nicely. (Note: Marginal cost is constant for simplicity.)

In the absence of regulation, the monopolist maximizes profit by producing a level of output, Y^M, which equates marginal revenue (*MR*) and marginal cost (*MC*). Because the demand curve (*D*) slopes downward, the price charged in the market is P^M, which is well above the price that would result in a competitive market (P^*). In addition, market output is below that which would occur were competition present (Y^*).

What has occurred is that there has been a transfer of surplus from the consumer to the producer, which is equal to the area of the rectangle above P^* below the demand curve, that is,

$$\Delta PS = (P^M - P^*)*Y^M. \tag{8.3}$$

In addition, the consumer (and society) has lost the surplus associated with the area of the triangle, which is given by

$$\Delta CS = 1/2*(Y^* - Y^M)*(P^M - P^*). \tag{8.4}$$

The latter is known as *dead-weight loss* and represents the lost output that has value to society. To the point expressed in the second concept, even if it

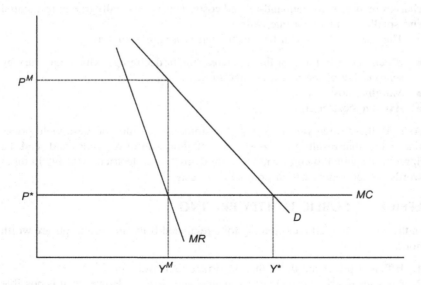

FIGURE 8.2 Monopoly supplier in absence of regulation.

were feasible for some type of tax or subsidy to be created to compensate consumers the initial loss in surplus [given by Equation (8.3) or that lost to the producer], dead-weight loss is not able to be compensated and is truly a loss to society.

Example 8.1

A monopolist has the following cost structure:

$$TC = 1000 + 40*Q. \tag{8.5}$$

Market demand for its product is given by

$$P = 200 - 2Q. \tag{8.6}$$

What are the profit-maximizing output and price that will prevail in the market? Setting $MR = MC$ and solving for $Q*$ and $P*$ yield:

$$Q* = 40$$

and

$$P* = 120,$$

which yields

$$\text{Profit} = \text{total revenue} - \text{total cost, or}$$

$$\text{Profit, or producer surplus} = \$2200$$

In this case, consumer surplus has been reduced and transferred to the producer in the form of profit (i.e., producer surplus). In addition, there is a dead-weight loss, which is absorbed entirely by the consumer.

RAMSEY PRICES: A SECOND-BEST OPTION[4]

The discussion thus far illustrates the reason(s) that price regulation is necessary in the case of public utilities, which also motivates a discussion on the third concept introduced earlier. The first-best option of marginal cost pricing will not work; the presence of fixed costs means that a regulatory structure that sets price equal to marginal costs implies that all costs will not be recovered in rates. One solution (in the case of other countries) is that the government imposes a tax (or a subsidy) so that fixed costs are recovered. However, in the United States, the absence of taxes or government subsidies implies that utilities must at least break even from their sales revenues. In addition, in the case of investor-owned firms, earn a "fair return" to shareholders (this is typically determined by the public utility regulatory commission in each state).

This is where reality also sets in: marginal costs are not constant and, over the relevant range of output, it is likely the case that average costs (embedded,

---

[4]Formally, Ramsey pricing is a linear pricing scheme designed for the multiproduct natural monopolist (see Ramsey, 1927).

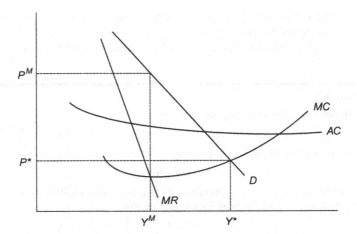

FIGURE 8.3 Profit-maximizing price and output for a natural monopoly. Average costs exceed marginal costs over the relevant range of output so that marginal cost pricing does not allow the firm to break even. One option is the two-part tariff, which charges a fixed charge that approximates the differential between average and marginal cost. In reality, this is far more complex an undertaking.

or fixed costs) are above marginal costs so that marginal cost pricing will not allow the firm to break even. Furthermore, neither marginal nor average cost is constant (or even linear, for that matter), which adds an additional element of complexity. This situation is illustrated in Figure 8.3.

A breakeven constraint implies the presence of fixed costs that must be recovered in some fashion. The fact that such costs are common to all classes of customers necessitates the ability to allocate these costs among these classes (e.g., residential, commercial, industrial). For now, let us focus on one solution to this problem.

An example from Brown and Sibley (1986) illustrates this point nicely.

Example 8.2: A Two-Period Electricity Pricing Example

Suppose that there are two periods—off peak and on peak—and that quantity demanded (Q_i) in each period is given by

$$Q_1 = 720 - 4000 * P_1 \tag{8.7}$$

and

$$Q_2 = 180 - 1000 * P_2. \tag{8.8}$$

Note: The fact that there are two equations implies that there are two different prices (and levels of output).

Fixed costs are $2.00 and marginal costs (c_i) are equal to

$$c_1 = \$0.09$$

and

$$c_2 = \$0.02.$$

Also note that the marginal cost of electricity is much less expensive in the off-peak period, which is not surprising. Why?

The firm's objective then is to earn total revenue (TR) so that total costs (TC) are covered, which implies that

$$TR = TC.$$

More specifically,

$$TR = P_1(720 - 4000*P_1) + P_2(180 - 1000*P_2) \qquad (8.9)$$

and

$$TC = \$0.09*(720 - 4000*P_1) + \$0.02*(180 - 1000*P_2) + 2.00 \qquad (8.10)$$

In this example, several pairs of prices satisfy the firm's breakeven constraint and the objective then becomes to find that pair that yields the lowest dead-weight loss, which, according to Brown and Sibley (1986), is that $P_1 = \$0.09$ (equivalent to marginal cost) and $P_2 = \$0.034$ (above marginal cost); in essence, the off-peak users absorb the entire fixed cost. Does this appear to be an efficient outcome? Why or why not? (As an exercise, you will show that this minimizes dead-weight loss and, as such, is the most efficient outcome.)

RAMSEY PRICING—THE SECOND-BEST OPTION

Because the first-best option (i.e., marginal cost pricing) is not necessarily feasible in the presence of fixed costs, a "second-best" option is now presented. In the most basic form the most efficient uniform[5] second-best prices are those that

- Maximize total surplus with respect to price (P_1, P_2, ..., P_n).
- Are subject to $PS = F$.
- F is fixed costs of the firm.

What this entails is, in essence, finding the markup over marginal cost in each market (or customer class) that reduces total surplus by the least amount. And what this amounts to is increasing price more in markets (i.e., customer classes) that are less sensitive to changes in price, which is equivalent to a lower price elasticity of demand,[6] that is,

$$\text{Markup} = (P_i - c_i)/P_i = \lambda/\varepsilon_i, \qquad (8.12)$$

---

[5]Uniform (or linear) prices are those that do not vary with output.

[6]Formally, price elasticity of demand is given by

$$\varepsilon = \partial Q/\partial P * P/Q. \qquad (8.11)$$

This measures the percentage change in quantity demanded that results from a change in price. When ε is less than unity (in absolute value), quantity demanded is inelastic (insensitive) to changes in price. When greater than unity (in absolute value), quantity demanded is said to be elastic.

where P_i is price in market I, c_i is marginal cost, λ is a proportionality constant, and ε_i is price elasticity of demand in market i (or customer class i).

Also known as the inverse elasticity rule, this pricing rule is a well-known result in the literature on efficient public utility pricing. Formally, this rule states that the price that maximizes social welfare (TS) subject to a profit constraint will exceed marginal cost by an amount that is inversely proportional to elasticity of demand.

Another way of expressing Equation (8.10) is that (for a two-output market)

$$\lambda = [(P_i - c_i)/P_i]^*\varepsilon_i = [(P_j - c_j)/P_j]^*\varepsilon_j \qquad (8.13)$$

for $j \neq i$.

What this implies is that for any pair of markets (or customer classes), the percentage increase over marginal cost, weighted by the price elasticities of demand, should be equal to λ, which is known as the Ramsey number (Brown and Sibley, 1986).

Example 8.1 (Continued): Marginal Cost Pricing in the Presence of Fixed Costs

An electric utility has the following cost structure:

$$TC = 1000 + 40^*Q. \qquad (8.14)$$

Market demand for its electricity is

$$P = 200 - 2Q. \qquad (8.15)$$

If price is set at marginal cost, what is the electric utility's profit?

Solution: Setting the price equal to marginal cost implies that $P = 40$ and $Q = 80$. As such,

Profit = Total Revenue − Total Cost or
Profit = $(200 − 2Q)^*Q − 1000 − 40^*Q$.

At $Q = 80$, then

Profit = −$1000.

Not surprising (due to the presence of fixed costs), the first-best optimal pricing (i.e., price equal to marginal cost) strategy does not allow the firm to recover all of its costs and a loss of $1000 occurs. This situation is displayed in Figure 8.4.

ANOTHER OPTION: AVERAGE COST PRICING

At first blush, one might think that a feasible solution is to set price at average cost; after all, this seemingly addresses the utility's need to recover all costs associated with providing service. Also, given the situation depicted in Figure 8.1, the price charged (and the quantity delivered) does not seem to diverge much

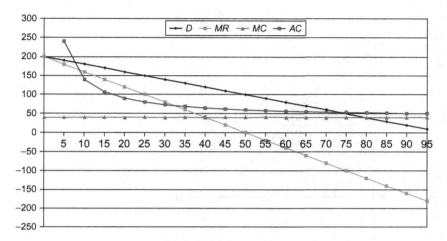

FIGURE 8.4 Demand (*D*), marginal revenue (*MR*), marginal cost (*MC*), and average cost (*AC*) curves for the regulated firm depicted in Example 8.1.

from the optimum, which occurs when price equals marginal cost. Appearances can be deceiving, however, as shown in continuation of this example. (Nonetheless, average cost pricing is one of the most popular rate-making mechanisms employed by utilities and accepted by regulators in the United States.)

Example 8.1 (Continued)

Next, if price is set at average cost, what are the equilibrium price and output? Setting price equal to average cost yields

$$200 - 2Q = 40 + 1000/Q. \tag{8.16}$$

Solving for the equilibrium output, Q^*, requires use of the quadratic formula, which implies that

$$-2Q^2 + 160*Q - 1000 = 0 \tag{8.17}$$

Solving for Q^* yields two solutions but only one is feasible. Why? This yields a price that is significantly higher than the first-best solution (which sets price equal to marginal cost).

What is the dead-weight loss associated with this pricing scheme? In this example, the differential between the average cost pricing scheme (or mechanism, which is often employed in the United States) and that which represents the first-best solution (or even marginal cost pricing with fixed charges to recover fixed costs) is substantial as you will see; not only does a significant dead-weight loss occur but also the price paid by consumers is significantly higher than it would have been under an optimally designed pricing scheme.

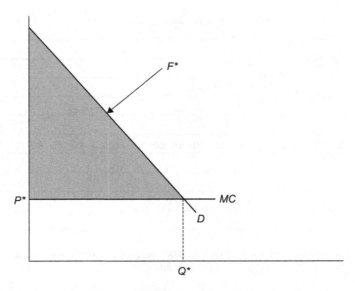

FIGURE 8.5 An "optimal" two-part tariff: Perfect price discrimination allows the producer to extract the entire consumer surplus as a fixed (or entry) fee. Clearly, it is optimal for the producer but not for the consumer.

TWO-PART TARIFFS

Example 8.1 showed that despite the fact that average cost pricing allows the firm to recover all of its costs, allocating fixed costs to a variable charge results in a dead-weight loss to society. Another option, and one that is used throughout the industry, is that of the *two-part tariff*, which allows that fixed costs be recovered via fixed charges while variable costs are recovered by marginal cost pricing. Originally suggested by R. H. Coase (1946), the structure of this tariff is that the usage charge is set equal to marginal cost and the entry charge (or fixed component) is set equal to the regulated firm's total fixed costs, which are divided by the number of users so that each customer pays the firm's average fixed cost. As such, the firm's total costs are covered, and because the price paid for each unit (the usage charge) is equal to marginal cost, dead-weight loss is eliminated. Total surplus is unaffected (there is, however, a transfer of surplus from the consumer to the producer). This is often called an "optimal" two-part tariff. (Note: It is optimal for the producer who extracts the entire consumer surplus.) This is displayed in Figure 8.5.

Example 8.1 (Continued): A Two-Part Tariff

The firm depicted earlier requests that the state public regulatory commission allow it to recover its costs via a two-part tariff, which, they argue, is more efficient than average cost pricing. Why?

Given this, what fixed charge (or entry fee[7]) will appear on the customer's bill? Recalling that the consumer surplus is the shaded area in Figure 8.5, we have

$$CS = 1/2*(100-20)*80.$$

This yields

$$CS = \$3200.$$

Under this pricing mechanism, the consumer is sent the appropriate price signal, there is no dead-weight loss, and the producer recovers all of its costs.

TWO-PART TARIFF WITH DIFFERENT CUSTOMER CLASSES

An interesting twist occurs when a second class of customer is distinguished. In this case, the optimal two-part tariff would dictate that the entry fee (F) be allocated between the two classes and that each could pay a usage fee (P) that equals a common marginal cost (c) so that the firm recovers its total costs, that is,

$$TC = F + (Q_1 + Q_2)*c, \tag{8.18}$$

where

$$Q_2 < Q_1.$$

In other words, customers in the second group are much smaller users (consuming Q_2) than those in group 1 (consuming Q_1).

In addition, suppose that consumers in group 2 are not willing to pay the same entry fee as those in the first group; after all, they are not consuming as much output and it could be the case that the fixed fee more than offsets the gains from marginal cost pricing (compared to average cost pricing) so that a negative consumer surplus would be earned by the customers in group 1. That is, they could elect to drop out of the market and consume nothing whereby they would earn zero consumer surplus, which is clearly better than a negative surplus. In this case, how would a two-part tariff be constructed, especially if it were the case that customers in group 2 are clearly better off with group 1 in the market? Why? They might even be willing to pay a higher entry fee, thus subsidizing the customers in group 1, which is not an uncommon occurrence (see "Aside: The Issue of Cross-Subsidization in Utility Rate Making" later in this chapter).

Alternately, one of the groups may be willing to pay a usage fee that is above marginal cost but below what would be charged under average cost pricing, such as $P*$ in Figure 8.6.

---

[7]Often called an entry fee, as many types of businesses (e.g., amusement parks, fitness clubs, or even video stores) charge a fee to enter and then charge a usage fee (per ride, per fitness class or service, or lower rental price per video).

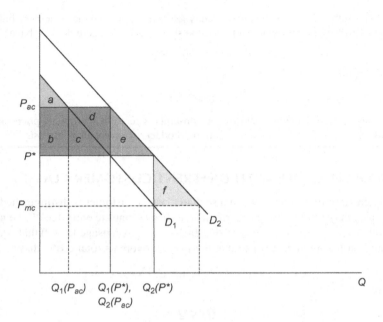

FIGURE 8.6   Two-part tariff with different customer classes. Group 1 customers are characterized by the demand curve labeled D_1. As smaller users, their willingness to pay is lower than that of the customers who are in group 2 (their demand curve is D_2).

It has been stated that the two-part tariff is better than the average cost pricing methodology for rate-making purposes. Given a choice, however, not all customers will choose the former. At P_{ac}, group 1 customers would consume Q_1 (P_{ac}) and gain the triangle "a" in surplus. Should they opt for the two-part tariff, they would pay a lower usage charge (P^*) and consume more [Q_1 (P^*)]. However, they would also have to pay the entry fee, which is equal to the shaded area "bcd," which appears to render a negative consumer surplus. This group is better off under the average cost pricing scheme.

Customers in group 2, however, present a different view. Under the two-part tariff, they pay P^* and consume Q_2 (P^*). However, unlike group 1, larger users maintain a positive surplus (area "e") even after paying the entry fee (area "bcd"). As such, this group of customers will select the two-part tariff.

What about the firm? Clearly the firm is better off with the two-part tariff, which we have seen before. In particular, the firm is able to charge P^*, which is above marginal cost. This translates into a profit equivalent to the area "f" in Figure 8.6. A seemingly pareto-optimal situation has occurred.

Note: It is often not the case that consumers have such a choice. Many utilities file two-part tariffs (or more than two parts). In the case of two such groups, what typically occurs is that the fixed fee is equivalent to the entire consumer surplus of the smaller users so they experience no gain at all, while larger consumers (and the producer) gain.

Example 8.3 (Rothwell and Gomez, Chapter 4, Exercise 4.2.4)

The demands of two distinct sets of 10 customers are given by:
Group 2 (four customers):

$$P_2 = 100 - 80*Q_2. \tag{8.19}$$

Group 1 (six customers):

$$P_1 = 100 - 6.3*Q_1. \tag{8.20}$$

The utility's cost is given by

$$TC = 500 + 20*Q, \tag{8.21}$$

where Q is $Q_1 + Q_2$. Given these conditions, what choice will each group of customers make?

The largest charge that group 2 consumers will pay is equal to their consumer surplus. In this case,

$$CS_2 = 0.5*(100 - 20)*1 = 40.$$

If the 10 customers divide the $500 fixed entry fee equally, each would have to pay $50. However, each customer earns a surplus of only $40, so group 2 customers would not be interested in the two-part tariff but would instead pay the higher average cost of $21.5 per kilowatt hour. The derivation of this is provided here.

What about group 1?

$$CS_1 = 0.5*(100 - 20)*12.7 = 508.$$

This group would be willing to go up to $508 to connect to the electric system.

Optimal Two-Part Tariff: The Solution

As stated, the optimal two-part tariff minimizes dead-weight loss by charging a connection fee equal to the consumer surplus of the smaller customer (group 2) and charges all customers the same usage fee (P^*). In this case,

$$CS_2 = 0.5*(100 - P^*)*[(100 - P^*)/80],$$

but a determination of P^* is required. Because total revenue is equal to 10 times the connect charge (in this case the consumer surplus of group 2) multiplied by price times output, which is equal to the quantity demanded of group one plus the quantity demanded by group two, we have

$$TR = 10*CS_2 + P*(Q_1 + Q_2)$$

or

$$TR = 10{*}CS_2 + P{*}[(100 - P{*})/6.3] + P{*}[(100 - P{*})/80],$$

and total cost is given by

$$TC = 500 + 20[(100 - P{*})/6.3] + 20[(100 - P{*})/80].$$

Because profit is equal to total revenue minus total cost, the solution (P^*) can be obtained by maximizing profit with respect to price. That is by setting the derivative of profit with respect to P^* equal to zero and then solving for P^*. That is:

$$\partial \pi / \partial P^* = 0, \tag{8.22}$$

which yields $P^* = 21.5$.

This implies that the access charge, which is equal to the consumer surplus of group 2, is given by

$$CS_2 = 0.5{*}(100 - P{*}){*}Q_2(P{*})$$

or

$$CS_2 = 38.5.$$

Aside: The Issue of Cross-Subsidization in Utility Rate Making

An excerpt from a 2004 article in *Public Utilities Fortnightly* (Casten and Meyer, 2004) makes this point well:

Indeed, the cross-subsidization concept is found throughout utility rates: From discounted rates to low income families to systems benefits charges, there are huge swathes of customers who pay less than their full cost of service, thus being subsidized by other customers who pay more to make up the difference. We tolerate and encourage such rate setting out of the belief that the social benefits created by such subsidization outweigh the resulting economic inefficiency.

The authors go on to cite the various forms that cross-subsidization, which are put in place to achieve certain economic, social, and political objectives. Included herein (see article for more details) are:

1. **Geographic diversity within the same rate class:** It is well known that urban customers subsidize those residing in rural cost areas. The latter, which are clearly more costly to serve due to lower density and more rugged terrain, typically pay the same rates and customer charges as those in more populated areas within each customer class.
2. **No price signal:** Instead of paying the actual cost of the power they consume at any given time, prices are based on average cost over the year (or during the time between rate cases). (This is obviated to some degree if the utility has some type of time-varying rates but, unless pricing is on a real-time basis, cross-subsidization still occurs.)

3. **Demand-side management or other energy-efficiency recovery charges:** These "below-the-line" items are charged to all customers in a particular class to fund a variety of energy-efficienct and renewable-power projects. However, not all ratepayers within the class benefit from these programs. (Note: Other below-the-line items include fuel adjustment clauses and cost recovery for environmental expenditures.)

4. **Interclass subsidization:** Representation at rate case proceedings on the behalf of residential and industrial customers often means that commercial customers are the most profitable to utilities. In the case of the former, the state's attorney general is the advocate on the consumer side, whose status as voters confers upon them certain benefits, which take the form of minimal changes in usage rates and customer charges. In the case of large industrial customers, the ability to leave the utility service territory [and even the state (or country)] confers similar benefits; in fact, it is often the case that industrial customers are represented by legal counsel who argue on their behalf at rate proceedings.

MULTIPART TARIFFS

What we have seen thus far is a variation of a *multipart* tariff, which differentiates customers based on the quantity of usage. (This is also known as nonuniform pricing.) However, what is seen more often is that the utility's tariff itself distinguishes usage levels by not only the usage charge but also the customer (or entry) charge and, in many cases, a demand (or capacity) charge. These are typically based on the class of customer (e.g., residential, commercial, industrial) and quite often a cross-subsidization among such classes exists; that is, there are different customer charges, energy charges, and demand charges, the latter of which only apply to industrial users (despite the fact that both residential and commercial users contribute toward the utility's peak demand and hence its capacity requirements).

NONUNIFORM PRICING: BLOCK RATES

Now we are starting to embark upon a path toward more efficient pricing of electricity; that is, by allowing (variable) usage charges to reflect the true marginal cost of providing service at the time (or level) required. In the simplest case, we recall Example 8.2, which exemplified the fact that the price of electricity in on-peak hours can differ quite vastly from prices that prevail in off-peak hours.[8]

It is often the case that utilities charge a different rate for different levels of usage, which may also vary depending on the season. For example, the first 1000 kilowatt hours of usage may be priced at one rate while all kilowatt

---

[8]On-peak hours are typically defined as weekday hours between 10:00 a.m. and 6:00 p.m. (EST).

hours above 1000 are charged a different rate. This is an example of a three-part tariff in that there is a fixed charge (or entry fee), and a breakpoint at a particular level of usage, or

$$P(Q_1) = P_1, \quad 0 < Q \le 1000$$

and

$$P(Q_2) = P_2, \quad 1000 < Q.$$

Such a pricing mechanism results in a "kinked" supply curve, such as that displayed in Figure 8.7.

Note: As displayed in Figure 8.7, block rates can be increasing or decreasing but only the former is a step toward pricing at marginal cost, as higher levels of output require more expensive generating units to come online to supply load. Not only do declining and flat block rates fail to yield a price signal to consumers, the former actually provides an incentive to overconsume, which is totally anathema to the ideas of conservation and efficiency currently being espoused.

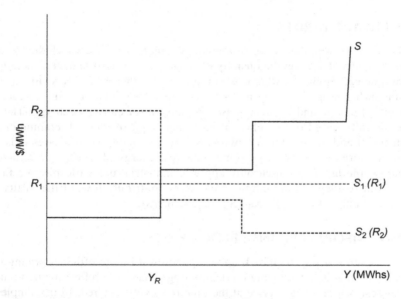

FIGURE 8.7 Supply curves for an electric utility with block rate pricing. Generating units are dispatched according to marginal cost, which is based predominantly on the cost of fuel. This is represented by S, the supply function. In the region of the supply curve where $Y < Y_R$, base load capacity is the first generation to be dispatched, which is typically pulverized coal, hydro (if in the northwestern part of the United States or if one of the power administrations is the supplier), or nuclear generation. The next units to come online may be natural gas fired, which are typically used for peaking capacity. Given the cost, renewable resources such as wind or solar may be the last to be dispatched *if there is no renewable portfolio standard*. Again, it depends on the marginal cost of the fuel at the time, which can affect the stacking order of the generation that is dispatched and whether market purchases supplement the utilities own generation.

In Mr. Kelly's article, which was quoted at the beginning of this chapter, there is a quote from the U.S. Department of Energy that specifically denigrates flat–block rate pricing, as it provides no price signal whatsoever. However, there are states whereby this is the standard methodology for setting usage charges. For example, in the state of Kentucky, the Kentucky Public Service Commission allows the utilities under its jurisdiction to offer flat rates to its customers. In fact, in a rate case filed in 2004, one of the largest utilities in the state went from offering seasonal rates for its weather-sensitive customer classes to a flat energy rate, which does not vary with usage. This is shown as R_1 in Figure 8.7.

Yet another type of block rate is the declining block rate in which the per-kilowatt-hour price of electricity actually declines with increased usage. The epitome of a perverse incentive (especially today when we are talking about conservation and energy efficiency as a means to reduce greenhouse gas emissions), there are numerous utilities in the United States of which the rates are fashioned in this respect. Duke Energy, Indiana, is an example of this, as are the Indianapolis Power and Light Company and the Northern Indiana Public Service Company. This rate structure is represented by R_2 in Figure 8.7.

S_1 (R_1) represents the supply curve of a producer charging a flat block rate (R_1), and S_2 (R_2) represents a supply curve that is a function of the declining block rate, R_2. The marginal cost curve is given by S, which increases with output, as higher cost generation must come online to serve load. For $Y < Y_R$ the utility is overcharging (i.e., it "overearns") for electricity, but for $Y > Y_R$ it does not charge enough to recover its costs and hence underearns, which typically triggers the filing of a rate case. Had the utility adopted a more reasonable approach to setting rates (i.e., an inclining block rate schedule), it could have recovered most of its incremental costs and not have to go through the time and the expense of a rate case. (Note: Overearning will *never* cause an investor-owned utility to file a rate case.)

Example 8.4: Block Rate Pricing

An electric utility offers the tariff shown in Table 8.1 to its residential customers. The pricing scheme in Table 8.1 yields the demand [and supply (or marginal cost)] curves in the summer season, that are displayed in Figure 8.8.

Consider the firm whose tariff is displayed in Table 8.1. What is the entry fee under an optimal tariff design?

TABLE 8.1

	Summer	Winter
Customer charge (monthly)	5.00	5.00
First 1000 kilowatt hours (cents per kilowatt hour)	0.08	0.04
Over 1000 kilowatt hours (cents per kilowatt hour)	0.05	0.01

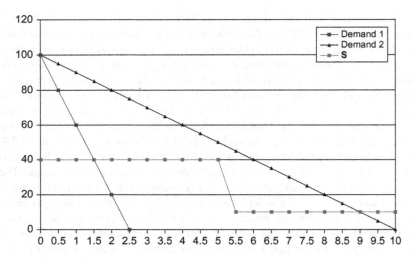

FIGURE 8.8   Seasonal two-part tariff.

Solution: The optimal entry fee is equal to the consumer surplus of the smaller consumer (group 1), which in this case is given by

$$CS = 0.5*(100 - 40)*1.5.$$

The entry fee is equal to $45 divided by the number of customers in group 1.

TIME-OF-USE RATES

A further extension of a multipart tariff is that of time-of-use rates, which includes a critical peak price in addition to a peak and off-peak price, the objective being to shift consumption to off peak when the cost of generating (or procuring) electricity is relatively low. In addition, extreme weather conditions may call for a critical peak price, which would reflect the cost of procuring power when demand is highest. For example, an electric utility may offer a tariff that distinguishes between different time periods throughout the weekday (that may be different on weekends), which likely depend on the season (but this may vary depending on the climate). Chapter 10 provides a case study on a time-of-use pricing pilot for a small midwestern utility.

A BRIEF HISTORY OF TIME-OF-USE PRICING

Until the 1970s there was little interest in pricing electricity efficiently. It was not until the twin energy crises of 1973–1974 and the late 1970s that there was a desire to set prices according to marginal cost, thus encouraging a more efficient use of energy and reducing the need for new generating capacity. During this time, numerous studies were funded by the Federal Energy Administration (the predecessor of the U.S. Department of Energy) to assess how customers would respond to time-varying rates, and in 1978 the Public Utility Regulatory Policies Act (PURPA) was passed by congress. In addition to other objectives (for more details, see Chapter 3), PURPA required state regulatory commissions to consider rates that varied by time, type of customer, and season.

A 1985 study authored by Dennis Aigner ("The Residential Electricity Time-of-Use Experiments: What Have We Learned") focused on a group of time-of-use pricing experiments for a group of utilities over a six-year period beginning in 1975. The main objective of these experiments was to determine whether time-of-use pricing would yield a change in the load shapes of residential customers and the impact of such pricing on utility revenues and consumer welfare. Depending on the results, important policy decisions could be made, which could influence the rate-making process and obviate the need for additional investment in utility infrastructure. One of the outcomes of the study was the estimation of own- and cross-price elasticities, which directly impact both the utilities revenues and the surplus of the consumer. The outcome is summarized from Miedema and White (1980, p. 4) as[9]:

All studies showed some reduction in usage during the peak period under TOU rates. However, reduction in usage during the peak period was not accompanied by statistically significant increases in baseperiod usage. Total usage seemed either to decline or remain the same in all projects. ... Peak-day usage shifts and average-day usage shifts appeared to be about the same.

Given the renewed interest in efficient pricing, time-of-use rates are clearly a step toward marginal cost pricing in that end-use customers are charged the cost of power during the time in which it is being consumed, for example, during peak usage times when more expensive peaking capacity [typically natural gas-fired combustion turbines (refer to Figure 8.7)] is being employed to generate electricity to meet demand. From an economic efficiency perspective, it is a necessary step in providing proper incentives to consumers to use energy wisely, which may also save them money by reducing their total bill. Numerous utilities in various states have adopted such a pricing structure, which varies by the time of day, day of the week, and month of the year (summer or winter, known as seasonal pricing). An excerpt from the Hydro Ottawa web site makes this point nicely:

Shifting electricity use to off peak periods reduces the need for investment in new electricity supply projects, which will help to moderate future rate increases. It also benefits the environment by reducing our reliance on additional power generation brought on-line or imported from other jurisdictions. Many create additional pollution and are more expensive to operate.

---

[9]It was duly noted that "A number of design considerations have an impact on the ultimate usefulness of the experimental data that have been forthcoming, not the least of which is the amount of variation available in peak, midpeak and off-peak prices. Many of the DOE experiments have but one set of TOU prices, and therefore the inferences available are limited to a single statistical comparison of control-group and experimental households" (Aigner, 1985).

This tariff went into effect on May 1, 2008 in various Ottawa neighborhoods and has four distinct periods on weekdays during summer months, five in winter months, and one on the weekend. Some companies, such as Arizona Power Service (APS), have a super critical peak price from 3:00 p.m. to 6 p.m. With 40% customer enrollment, APS leads the nation in time-of-use customer participation.[10] But the key is that there must be enough of a differential between peak and off-peak prices. Nevada Power's time-of-use rate, which went into effect on July 1, 2008, offers a nice example here whereby the residential on-peak rate, which is from 1:00 p.m. to 7:00 p.m., is almost three times the off-peak rate (23.08 cents per kWh vs. 7.097 cents per kWh).[11]

REAL-TIME PRICING

Taking this one step further still is the concept of real-time pricing, which has actually been around for some time. California was one of the first states to offer real-time prices in the mid-1980s. The price, which varied hourly, was quoted a day in advance for all energy consumed. As such, participants' entire load was exposed to the volatility that characterized the real-time prices they faced. These tariffs were designed to be revenue neutral over average climatic conditions, for the class of customers deemed likely to participate. However, because such a large portion of the revenues generated from these tariffs was related to actual hourly supply and/or weather conditions, revenue recovery could not be guaranteed.

Niagara Mohawk's Hourly Integrated Pricing Pilot, launched in 1988, introduced a new real-time tariff design: a two-part rate with a customer-specific access charge. A unique customer baseline load (CBL) profile, composed of a kilowatt-hour value for each hour of the year, was established for each participant from his historical interval billing data. The customer-specific access charge was calculated by applying the energy and billing demand rates from the customer's otherwise applicable tariff to their CBL load profile. Deviations between the customer's actual load and its CBL in each hour were settled at the prevailing real-time price. Because only marginal changes in usage were subject to real-time prices, participants had less exposure to price volatility, and the utility had greater revenue stability, compared to earlier real-time tariff designs.[12]

It was not until the mid- to late 1990s that real-time pricing became popular in states with retail choice. The real-time tariffs introduced in these states were generally based on a rate structure composed of hourly energy prices for the commodity component and unbundled transmission and distribution charges

[10]http://www.reuters.com/article/pressRelease/idUS155472+24-Mar-2008+BW20080324.

[11]http://www.nevadapower.com/conservation/home/home_rebates/time_of_use.cfm.

[12]An excerpt from "A Survey of Utility Experience with Real-Time Pricing," LBNL-54238 (http://www.osti.gov/energycitations/servlets/purl/836966-SZe2FO/native/836966.PDF).

assessed on the customer's billing demand and/or energy consumption. However, in most cases, it was predominantly larger commercial and industrial customers that participated.

In recent years, participation rates have declined significantly but that is now changing. With a renewed focus on energy efficiency and conservation, programs such as this are making a comeback, even offering residential customers the opportunity to participate. According to an article entitled "ComEd Pioneers Real-Time Pricing Program (RRTP) for Residential Customers,"[13] participants for all 12 months in 2007 experienced an annual savings between 7% and 12% compared to the fixed rate other residential customers received. RRTP participants are billed for the electricity they consume based on hourly wholesale market prices. They have access to hourly pricing information via the Internet and pricing alerts via text messaging and e-mail. Participants may choose to make adjustments in their electricity usage based on hourly prices. For instance, if pricing alerts indicate that electricity prices could reach or exceed 13 cents per kWh, about 2.5 cents higher than Commonwealth Edison's fixed rate, customers can save money by shifting their electricity usage to lower-priced hours later in the day.

UNDERSTANDING ELECTRIC UTILITY CUSTOMERS

The Electric Power Research Institute (EPRI) had included a new program in its 2012 research portfolio: Understanding Electric Utility Customers (Program 182). As a member of the advisory committee of this program, the author has been involved from day 1 helping to shape and direct this program. The excerpt that follows provides an overview of the program.

Understanding what customers want and how they perceive and realize value from electric services is becoming more important—in some cases, imminently so. There is a growing movement for electric utilities to better understand, as well as engage with, customers so that they may more fully realize the benefits of utility program activity and technology investments.

Other industries have been developing detailed knowledge of their customers' preferences and behaviors for decades. Obtaining customer intelligence has been an important element of utility operations as well, but the focus has been more directed to measuring customer satisfaction rather than developing an in-depth understanding of when and how they use electricity. Utilities today are grappling with the knowledge that the customer will play a pivotal role in ensuring a link between technology deployments and the investment benefits that are to be realized.

With new technologies being added to the grid to enable greater consumer participation in how they manage their electricity usage, there is an opportunity for the electric utility industry to get customers actively and sustainably involved

---

[13]http://www.reuters.com/article/pressRelease/idUS274507+31-Jan-2008+PRN20080131.

in electricity usage decisions. However, some fundamental research is first required to get to the root of various aspects of utility customer behavior, such as the effects of rate structure and information provision (or feedback) on customer response, response variation by customer segment, and other pertinent research questions.

Although research is already ongoing in some of these areas, an opportunity exists to make it more collaborative and probing, while at the same time more extensible to circumstances beyond just those of individual jurisdictions. Transparency and research rigor also may be bolstered by making data available to a wider research audience, a practice not typically followed in the industry. EPRI's collaborative research model is uniquely qualified to capitalize on these opportunities.

Approach

The Understanding Electric Utility Customers research program is composed of two project sets: (1) impacts of rates and pricing structures and (2) customer behavior. Because these areas complement EPRI's ongoing technology research, together the three areas of technology, pricing, and behavior will more fully encompass the range of utility research needs, and EPRI and its members will develop a foundation of knowledge that will help guide investment decisions that can help move the industry forward.

With this program, members will have access to information that can help them in these ways:

- A rate structure design manual that combines conceptual guidance from the disciplines of finance and economics with real-world experience about how customers respond to different rate structures.
- A framework for designing wholesale and retail markets to foster the optimal demand response.
- Facilitating the use of behavioral programs to tap into new sources of savings potential by assessing the appropriateness of using a deemed savings approach for estimating savings.
- New approaches to understanding customer diversity by building robust and reliable customer groupings and associations.

Of particular interest to the author and to the topic at hand is the assessment of rate structure options. This is described in more detail next.

ASSESSMENT OF RATE STRUCTURE OPTIONS

A wide assortment of alternative structures is available to price electricity. The relative effectiveness of many alternative pricing structures in achieving specific goals (e.g., efficiency, fairness, load shifting, conservations, and societal goals) has not been defined clearly and concisely. This is especially true of dynamic pricing and demand response structures that have rather narrowly constructed objectives, the attainment of which has not been established thoroughly or unequivocally. Utilities often are faced with needing to achieve a specified goal, or set of goals, without a way to determine what price structures are best suited to accomplish the desired result or a good understanding of the revenue implications of migrating customers from one rate structure to another.

Moreover, how customers respond to price structures is not well established, especially for dynamic pricing and demand response structures that are receiving more attention because they can address particular and vexing market imbalance situations.

The EPRI intends to develop the concepts, tools, protocols, and experience-based data needed by utilities to design rates that achieve predictable and desirable outcomes. The result will be a rate structure design manual that combines conceptual guidance from the disciplines of finance and economics with real-world experience about how customers respond to rate structures.

Approach

- Conceptual foundation: EPRI will use and expand upon the rate structure framework development undertaken in previous EPRI research (p. 170). That framework provides a comparison of how the higher-order classes of rate structures achieve various and often competing goals.
- Comparative framework: This project will develop a more detailed characterization of the expected impacts of alternative rate structures that can support designing rates for all classes of customers, applicable to at all levels of the sector (wholesale and retail), in all supply circumstances (short-run expediency and long-run sustainability).
- Design tools and protocols: Develop analytical methods and protocols for their application to specific rate structure design objectives that allow the designer to construct one or several rate structures, among other things.

Impact

A comprehensive design tool kit will help utilities implement a pricing structure, or portfolio of pricing structures, that is aligned with the goals that drive the initiative, undertake the initiatives with a high chance of success, and track achievements regularly and reliably.

Through the use of a standardized design process, other utilities gain from the collective knowledge that results about what works in rate structures, how well it works, and in what circumstances.

How to Apply Results

- Assess and quantify changes to existing rate structures.
- Design and evaluate purpose-specific rate structures, such as low income, economic development, peak shaving, and load shifting.

- Quantify the expected benefits from smart grid-enabled pricing structures.
- Develop and evaluate a portfolio of offerings made available through opt-in or opt-out implementation and administration.

From this research, a design manual will be produced in the form of a technical report, which should be completed by December 2012.

CONCLUSION

With all of this said, however, the need to price electricity efficiently is still at the forefront, especially given that emissions of greenhouse gases from (60%) coal-fired power plants are the worst offenders (emissions from a coal-fired generating plant are almost twice that from natural gas to fuel the generation of electricity). Given this, should not the customers of utilities with coal-fired generation pay more for their electricity (and not just via below-the-line items, such as cost-recovery mechanisms for environmental or automatic adjustments for fuel costs) than rate payers in the service territories of lower-emitting utilities? In other words, should not consumers pay more for electricity generated from "dirty sources" than that generated from "clean resources"? Also, in the case of investor-owned utilities, should there not be a higher return to the shareholders of firms that have made investments in renewable resources and other efficiency-improving investments?

It is often said that energy efficiency is the least cost option; however, energy efficiency requires investment in more energy-efficient appliances and equipment, which tends to command a relatively higher cost.[14] It is this author's opinion that conservation is clearly the least cost option; however, for this to occur there must be a price signal to which end users can respond. At the very least, this means that energy charges reflect marginal costs in that they increase with usage, as higher usage implies higher costs of generating (or procuring) electricity. (In other words, *at the very least* there must be an increasing block tariff; flat rates and declining block rates have no place in the pricing of energy, especially electricity!)

Also, while time-of-use and real-time pricing are attempts to emulate the marginal costs of supplying electric service, they also require a significant investment in infrastructure (smart meters, etc.) on the part of utilities (and hence consumers), which may be prohibitive.

To date, little work has been done in the estimation of the marginal costs of providing electricity for rate-setting purposes. As described in Chapter 4 (and elsewhere in this book), estimation of marginal costs would require an appropriately specified cubic cost model, which is not without its challenges. However, Chapter 6 presents such a model and estimation results as a case study.

---

[14]If you perform a life-cycle cost analysis you may be surprised; you may want to review the book entitled *Electricity Cost Modeling Calculations* (2010) by Greer for a discussion of discrete choice models and the theory underlying the trade-off between installation and operating costs in terms of appliance choices.

EXERCISES

1. In Example 8.2, it was stated that the pair of prices that yield the lowest dead-weight loss is $P_1 = \$0.09$ and $P_2 = \$0.034$.
   a. Why is this the most efficient outcome? Prove that this pair of prices minimizes the dead-weight loss.
   b. Does it make sense that the off-peak price is lower than the on-peak price? Why or why not?

2. Verify that the firm depicted in Example 8.1 (continued): Marginal Cost Pricing in the Presence of Fixed Costs suffers a loss of $1000.

3. For the firm depicted in Example 8.1 (continued):
   a. What is the profit-maximizing level of output under average cost pricing?
   b. What price corresponds to this level of output?
   c. What profit will be earned by this firm?
   d. What is the dead-weight loss associated with this price output combination?

4. An electric utility faces the following:

$$\text{Total cost} = 50 + 20*Q. \tag{8.23}$$

   The individual customer demand (inverse demand function) is given by

$$P = 100 - 6.25*Q. \tag{8.24}$$

   a. In the absence of price regulation:
      i. What price and output would prevail?
      ii. What is the profit earned by the firm?
   b. If the state regulatory commission were to impose average cost pricing:
      i. What price and output would prevail?
      ii. What is the firm's profit?
      iii. Is there a dead-weight loss? If so, calculate.
   c. Instead, assume the state regulatory commission requires marginal cost pricing:
      i. What price and output will prevail?
      ii. What is the firm's profit (loss)?
      iii. How can this be rectified?
      iv. Does a dead-weight loss result? If so, calculate.

5. The firm depicted in Exercise 2 has convinced the state regulatory commission that a two-part tariff would be better than any other pricing mechanism.
   a. What arguments might it have used to support its contention?
   b. What is the gain to the producer from adopting this pricing scheme?

6. A second type of customer is recognized by the state regulatory commission. This class of customer has a demand function given by

$$P = 100 - 8.0*Q. \tag{8.25}$$

a. If price is equal to marginal cost, what level of output will be consumed by this class of customer?
b. What will the firm's profit be?
c. What entry fee will be paid by each type of customer so that the firm covers all of its costs?

7. The public regulatory commission requires that all classes of customers be treated the same so that customers described in Exercises 3 and 4 must pay according to a two-part tariff.
a. What usage fee will be charged ($P*$)?
b. What will the access (or entry) charge be?
c. Is this an optimally designed tariff? Why or why not?

8. What type of tariff is represented in Table 8.1? Is it optimal? Why or why not?

Chapter 9

Price and Substitution Elasticities of Demand: How Are They Used and What Do They Measure?

A Review of Demand Models and the Relevant Literature

INTRODUCTION

Increasingly stringent environmental regulation has resulted in rising fossil fuel prices, which, of course, translates into higher electricity rates. Subsequently, many electric utilities are finding themselves filing rate cases more frequently than before. Figure 9.1 displays the recent trend in rate case filings by shareholder-owned utilities in the United States (13 in the first quarter of 2009). Reasons cited include decreased sales (which resulted from the economic recession), constrained liquidity, and higher costs of capital (thanks to the collapse of financial markets), along with regulatory requirements to implement tracking mechanisms and surcharges to recover energy efficiency and conservation expenses, construction work in progress, and implementation of a smart grid rider and related expenses.

It is expected that more frequent rate cases will ensue over the foreseeable future, which begs the question: What will the impact be on the demand for electricity and hence utility revenues?

PRICE ELASTICITY OF DEMAND

In the economics literature, the term *elasticity* is discussed frequently—often in reference to the effect on quantity demanded from a change in price, which is how it is used here. Often elusive—not to mention difficult to ascertain in practicality—the price elasticity of demand measures the percentage change in quantity demanded caused by a percent change in price. This elasticity is almost always negative and is usually expressed in terms of absolute value. If the elasticity is greater than one, demand is said to be elastic; between zero and one demand is inelastic; and if it equals one, demand is unit elastic. This is displayed in Figure 9.2.

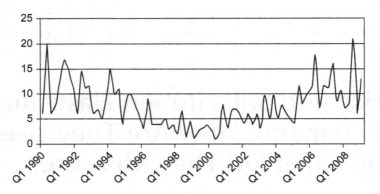

FIGURE 9.1   Number of rate cases filed: 1990—Q1 2009 (quarterly). Source: SNL Financial/ Regulatory Research Association and EEI Rate Department.

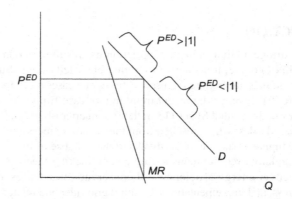

FIGURE 9.2   Price elasticity changes along a linear demand curve. Although the slope does not change along a linear demand curve, price elasticity does.

Formally, the price elasticity of demand is defined as

$$\varepsilon = \partial Q/\partial P \cdot P/Q. \tag{9.1}$$

A simple example follows.

Example 9.1 Price Elasticity of Demand

What price–output combination yields a price elasticity of demand equal to 1? A linear demand curve of the form

$$Q = A - 2P \tag{9.2}$$

yields

$$MR = A - 4P. \tag{9.3}$$

Solution:

Using Equation (9.1), it is straightforward to solve for the output and price combination that yields a price elasticity of demand equal to −1, which in the example coincides to the output at which the marginal revenue equals zero. In this case, the solution is

$$P = A/4. \tag{9.4}$$

This implies that

$$Q = A/2. \tag{9.5}$$

Thus, when demand = $A/2$, then the price elasticity of demand = −1. [Verify this using Equation (9.1).]

A BRIEF REVIEW OF THE LITERATURE: ENERGY DEMAND, ELASTICITIES OF DEMAND IN ENERGY MARKETS, AND FUNCTIONAL FORMS

The Demand for Energy: Econometric Models

Economic theory dictates that consumer demands for products are functions of the prices of goods (p) and income (Y) and producers consider the prices of inputs (w) and outputs (Q). In other words,
Demand:

$$Q = f(p, Y). \tag{9.6}$$

Supply:

$$C = f(w, Q). \tag{9.7}$$

In this section, much of which comes from Dahl (2006), we consider some of the econometric models that have been applied to estimate demands. It is reprinted with permission here.

The simplest models used are one equation models, which have the advantage of being simple and undemanding in terms of data requirements. The simplest of these models is a static model that regresses the quantity of the energy product (E) on the price of the fuel (P) and some measure of income (Y). For example,
Model (1):

$$E = \beta_o + \beta_1 P + \beta_2 Y. \tag{9.8}$$

These models can be made more complicated by adding other variables to represent demographics and weather and may include the prices of competing fuels to measure substitution across fuels. For aggregate data, the set of additional variables is usually rather limited.

However, in household survey data, the additional demographic and geographic data are often quite numerous. All such models that do not include any lagged variables and do not include the stock of energy using appliances we will call static models (Stat).

To the simple static model we could add some measure of the stock of energy using appliances or equipment (Sk):
Model (2):

$$E = \beta_o + \beta_1 P + \beta_2 Y + \beta_3 Sk. \tag{9.9}$$

These models, which include a stock of energy using equipment, will tend to capture short-run adjustments in energy demand and will be called static stock models (StatSk). Neither of these models is likely to capture total long-run adjustments, although Equation (9.8) might do so if adjustment time is very short or cross-sectional data are used, whereas Equation (9.9) might do so if adjustment time is short, the market is saturated, stock is measured as number of units owned, and all adjustment is in utilization or in changing the characteristics of the stock.

(Elasticities from model (1) are included in Table 9.1 in the column under P_{lr} and Y_{lr} to indicate their static nature, while elasticities from model (2) are under P_{sr} and Y_{sr}.) Their precise interpretation, however, may depend on the data and model type.

In the linear form, the price and income elasticities can be computed as [from Equation (9.8)]:

$$\varepsilon = \beta_1 P / E \tag{9.10}$$

$$\eta = \beta_1 Y / E, \tag{9.11}$$

where P is price, Y is income, and E is energy.

The expectation is, of course, that own-price elasticities will be negative and income elasticities will be positive. As expected, the price elasticity is negative and decreases in the longer run. Also, the income elasticity is positive and increases with time as it should.

TABLE 9.1 Price and Income Elasticities of Demand

Energy	P_{sr}	P_{lr}	Y_{sr}	Y_{lr}
Mean	−0.23	−0.72	0.60	1.21
Standard deviation	0.10	0.42	0.57	0.49
Minimum	−0.41	−1.20	0.06	0.91
Maximum	−0.13	−0.32	1.20	2.19

Functional Forms

Linear or linear in the logs forms are often employed in these simple models with the choice of the functional form being made by the researcher. On occasion the functional form is subject to testing, a practice I would urge on more researchers. A simple test that can be employed uses the Box Cox (BxCx) function. For example, the Box Cox formulation for Equation (9.1) would be

$$(E^\lambda - 1)/\lambda = \beta_o + \beta_1(P^\lambda - 1)/\lambda + \beta_2(Y^\lambda - 1)/\lambda. \qquad (9.12)$$

When $\lambda = 1$ we have a linear function.
When $\lambda = 0$ we have a log function.
When $\lambda = -1$ we have a function in the inverses of the variables.
When $\lambda =$ none of the above we get the Box Cox function.

Long-Run Elasticities versus Short-Run Elasticities

Distinguishing between short run and long run is typically done in three ways in the reduced-form framework. The first way is to associate strict cross-sectional with long-run adjustments, particularly if prices and incomes are very different across the cross-sections.

At this juncture, it is important to bring up the issue of omitted variable bias, which can occur when variables that influence the demand for energy are not included in the model (typically because data do not exist or are not available). If these variables are correlated with price or income, their effects will be attributed to price and income by the estimating model with the direction of the bias uncertain and depending on the relationships between included and excluded variables. Hartman (1979) believes that because of these locational and structural differences, cross-sectional data overstate elasticities, particularly for price.

Time-series data, particularly shorter series, are more likely to capture short-run effects. The disadvantage of short time series is that it is often the case that there is little change in the variables or that there are simply not enough observations. Longer time series may provide more changes in the variables and more observations, but may also suffer from structural change so that the entire series should not be used.

Given all of this, it would appear to be the case that panel data (cross-sectional time series) would give us the advantage of more variation across a much larger data set, which would measure some mix of long- and short-run effects. However, the issues described earlier would also apply here.

Data can be divided further into their periodicity. Annual data are by far the most commonly used. Quarterly and monthly data can dramatically increase the sample size. However, many series are not available this often and problems of seasonality need to be taken into consideration. A second way of distinguishing long run from short run using panel data is described in Baltagi and Griffin (1983). Using their methodology, the estimation equation is

$$E_{it} = \beta_0 + \beta_1 P_{it} + \beta_2 Y_{it}. \tag{9.13}$$

The across-country variation will be associated with the long run and would be obtained by regressing the means of each country's energy demand on the means of each country's prices and incomes and any other variables that are in the model, or

$$E_i = \beta_0 + \beta_1 P_i + \beta_2 Y_i. \tag{9.14}$$

Within-country variation would be associated with the short run and captured by a pooled regression where each country's demand and explanatory variables are deviated from their respective means, or

$$E_{it} - E_i = \beta_1 (P_{it} - P_i) + \beta_2 (Y_{it} - Y_i). \tag{9.15}$$

This model is equivalent to running a pooled regression in which a dummy variable is allowed for each country.

The third, and probably most ubiquitous, technique for separating out short- and long-run effects on reduced-form single-equation models is to make the model dynamic by adding lagged values to the model. The simplest and most common way of doing this is to use lagged endogenous variables, which are also known as stock adjustment, partial adjustment, adaptive expectations, Koyck, or geometric lag models after the economic process represented, the originator, or the shape of the lag. The simplest lagged endogenous (LE) model is

$$E_t = \beta_0 + \beta_1 P_t + \beta_2 Y_t + \beta_3 E_{t-1}. \tag{9.16}$$

The advantage of this model is that it is simple and flexible to use with an intuitively appealing lag shape. The disadvantages include a fairly restrictive shape for the lag constrained to be the same for all variables. Furthermore, collinearity between lagged endogenous variable and included current values of other variables can render rather erratic estimates.

There are more flexible forms that nest the lagged endogenous model within a form that allows an inverted V lag as well. The two standard procedures for doing this are (LE^1),

$$E_t = \beta_0 + \beta_1 P_t + \beta_2 P_{t-1} + \delta_1 Y_t + \delta_2 Y_{t-1} + \sigma_1 E_{t-1}, \tag{9.17}$$

and (LE^2),

$$E_t = \beta_0 + \beta_1 P_t + \delta_1 Y_t + \sigma_1 E_{t-1} + \sigma_2 E_{t-2}. \tag{9.18}$$

Although these lags are less restrictive than the LE model, they seem to suffer an even greater tendency toward multicollinearity.

A more general way to make a simple model dynamic is to add lags to some or all of the independent variables. These models will be called distributed lag models and can be represented as

$$E_t = \beta_0 + \sum_i \beta_i P_{t-i} + \sum_i \delta_i Y_{t-i}. \tag{9.19}$$

This model has the advantage of being flexible and allowing different lags on different variables. In practice, however, there is often so much collinearity across time for the variables that the model does not perform very well, and lags as long as adjustment might reasonably be expected to occur and can rapidly chew up our degrees of freedom. If lags are constrained to be on a polynomial to help deal with problems of collinearity and loss of degrees of freedom, the model will be a polynomial distributed lag.

Each of the preceding dynamic approaches only indirectly accounts for the fact that energy is an indirect demand that is always consumed with energy using equipment. Often this equipment is very long lived and therefore complete adjustment can take a considerable amount of time. However, information on the stock may be unavailable and expensive to collect. Two early approaches to deal with a nonavailable stock of appliances are those by Houthakker and Taylor (1970) and by Balestra and Nerlove (1966).

The Houthakker and Taylor model (1970) is designed to deal with demand for a durable good where purchases of the good add to an existing stock or to a nondurable, where the existing stock of the good is considered the habit of using the good with additional purchases adding to the stock of habits. It is not well designed to deal with a good where the purchased good is used with a stock of another good, as in the case of energy. Nevertheless, the model is used occasionally in the energy context where we must remember that the stock variable in the initial model is not the stock of energy using appliances but the habit formation variable. In their model, the demand for energy source E is a function of price, income, and the stock of habitual energy use, and is given by

$$E = \beta_o + \beta_p P + \beta_y Y + \beta_{Sk} Sk. \tag{9.20}$$

This implies that

$$Sk = (E - \beta_o - \beta_p P - \beta_y Y)/\beta_{Sk}. \tag{9.21}$$

Then the change in the habitual energy stock is equal to E minus depreciation of the habit (rSk), or

$$\Delta Sk = E - rSk, \tag{9.22}$$

so that changes in energy consumption are given by

$$\Delta E = \beta_p \Delta P + \beta_y \Delta Y + \beta_{Sk} \Delta Sk. \tag{9.23}$$

Plugging in for ΔSk from Equation (9.22) into Equation (9.23) yields

$$\Delta E = \beta_p \Delta P + \beta_y \Delta Y + \beta_{Sk}(E - rSk), \tag{9.24}$$

and plugging Sk from Equation (9.21) into Equation (9.24), we get

$$\Delta E = \beta_p \Delta P + \beta_y \Delta Y + \beta_{Sk}(E - r(E - \beta_o - \beta_p P - \beta_y Y))/\beta_{Sk}. \tag{9.25}$$

Rearranging and solving for E yields the equation to be estimated:

$$E = \beta_o r/(r - \beta_{Sk}) + \beta_p/(r - \beta_{Sk})\Delta P + \beta_y/(r - \beta_{Sk})\Delta Y \\ + r\beta_p/(r - \beta_{Sk})P + r\beta_y/(r - \beta_{Sk})Y + 1/(r - \beta_{Sk})\Delta E. \tag{9.26}$$

This model can be estimated by ordinary least squares (OLS), but to recover the coefficient β_{Sk} it will have to be estimated by a nonlinear approach as the model is overidentified (see later for details). The interpretation of β_{Sk} is as follows. In the long run, E and Sk are both constant. Let these constant long-run values be E^* and Sk^*. Then

$$E^* = \beta_o + \beta_p P + \beta_y Y + \beta_{Sk} Sk^* \tag{9.27}$$

and the change in the stock is zero, or

$$\Delta Sk = E^* - rSk^* = 0, \tag{9.28}$$

which implies that

$$E^* = rk^*. \tag{9.29}$$

Deviation of current purchases of the good from the long-term equilibrium equals

$$E - E^* = \beta_o + \beta_p P + \beta_y Y + \beta_{Sk} Sk - (\beta_o + \beta_p P + \beta_y Y + \beta_{Sk} Sk^*), \tag{9.30}$$

which is equal to

$$\beta_{Sk}(Sk - Sk^*), \tag{9.31}$$

and is proportional to the difference of the current stock or habit of using the good from its long-term level. If β_{Sk} is negative, then current purchases are above the long-term level when the stock is below its long-term level, which is the case of the stock adjustment model. If β_{Sk} is positive, then current purchases are above the long-term level when the stock is above its long-term level, which is the case of habit formation. The variable r is the rate of depreciation if the good is a consumer durable. In the nondurable case, as in an energy context, it has a more nebulous interpretation.

Equations using this modeling approach will be designated as HT, but as mentioned above, this model is not well designed for nondurable goods used in conjunction with durable goods, as is the case for energy, although it has been used on occasion. Further, there is often a lot of correlation between current variables and change variables so econometric results are poor as well.

Expenditure Share Models

In the simplest share models either a linear or a log-linear form was chosen for each of the equations. But soon the most popular approach to estimating these types of systems was to use a flexible functional form such as the translog, where the share is the share of expenditures (*Ex*) on the *i*th energy form. To model consumers in this approach, we begin as in Pindyck (1980) with the indirect utility function:

$$LnV = \alpha_o + \sum_k \alpha_k \ln P_k / Ex + (1/2) \sum_k \sum_j \beta_{kj} \ln(P_k) / Ex \ln(P_j / Ex). \quad (9.32)$$

This function is considered to be a second-order Taylor approximation of any indirect utility function. The estimating equations become the budget shares of goods, which for the *j*th good is equal to (summations always run to *m*, the number of goods)

$$Sh_j = P_j X_j / Ex. \quad (9.33)$$

Typically, shares are most often included on energy subchoices, such as oil, coal, gas, and electricity. From this formulation partial own-price, cross-price, and income elasticities can be computed [for more detail, see Dahl (2006)].

These elasticities are made assuming that expenditure stays constant or, if applied to a fuel subaggregate, like energy, which is composed of various fuels, it assumes that the expenditure on energy is constant.

For the firm, the translog model becomes somewhat simpler. Following Griffin (1979), the indirect cost function *C* is the following function of input prices (P_i):

$$\ln C = \beta_o + \sum_i \beta_i \ln P_i + (1/2) \sum_i \sum_j \beta_{ij} \ln P_i \ln P_j. \quad (9.34)$$

From the cost function, the share equation to be estimated can be derived as

$$Sh_i = \partial \ln C / \partial \ln P_i = \beta_i + \sum_j \beta_{ij} \ln P_j. \quad (9.35)$$

Elasticities can then be computed from estimated equations and

$$\sigma_{ij} = (\beta_{ij} + Sh_i Sh_j) / Sh_i Sh_j \quad (9.36)$$

for all $i \neq j$, and

$$\sigma_{ii} = (\beta_{ii} + Sh_i^2 - Sh_i) / Sh_i^2. \quad (9.37)$$

The own- and cross-price elasticities are

$$\varepsilon_{ij} = \sigma_{ij} Sh_j \quad (9.38)$$

for all i, j.

A second function that has been used for share equations is the logit equation. This model has been used for shares of energy, but its most popular application has been in structural models of appliance choice. The relevant equations in the logit model are developed and discussed in Considine (1989). He begins with share equations as

$$Sh_i = \exp(\beta_i + \sum_j \beta_{ij} \ln P_j) / \sum_i \exp(\beta_i + \sum_j \beta_{ij} \ln P_j), \quad (9.39)$$

where the own elasticity is given by

$$\varepsilon_{ii} = \partial \ln(Sh_i)/\partial \ln(P_i) + Sh_i - 1 \tag{9.40}$$

and the cross-price elasticity is given by

$$\varepsilon_{ik} = \partial \ln(Sh_i)/\partial \ln(P_k) + Sh_j. \tag{9.41}$$

The logit cost-share model collapses to a constant elasticity of substitution (CES) model for two inputs or when the elasticities of substitution are all equal. Producer theory constraints cannot hold globally but are constrained to hold around specific shares, $Sh_i{}^*$, which are typically chosen to be the shares at the mean of data.

The three approaches just described look at interfuel substitution from share equations. An alternative formulation that has been used is the generalized Leontief. In this model, following Dowd et al. (1986) for the linear homogeneous case, the cost function conditional upon output Q is given by

$$C = Q\left(\sum_i \sum_j \beta_{ij} P_i^{0.5} P_j^{0.5}\right). \tag{9.42}$$

From Shepherd's lemma,

$$X_i = \partial C/\partial P_i \tag{9.43}$$

or

$$X_i = Q\sum_j \beta_{ij}(P_j/P_i)^{0.5}, \tag{9.44}$$

which can be estimated using

$$a_i = X_i/Q = \sum_j \beta_{ij}(P_j/P_i)^{0.5}, \tag{9.45}$$

where a_i denotes that these are the input–output demand equations.

PRICE AND SUBSTITUTION ELASTICITIES

One nice feature of flexible functional forms such as the generalized Leontief is that they place no a priori restrictions on the substitution elasticities. In this case, the Hicks–Allen partial elasticities of substitution for a general dual cost function with n inputs are computed as

$$\sigma_{ij} = C^* C_{ij}/C_i^* C_j, \tag{9.46}$$

where i, j are the first and second partial derivatives of the cost function with respect to input prices (p_i, p_j) and $i, j = 1, \ldots, n$.

More specifically, the substitution elasticities are given by

$$\sigma_{ij} = \frac{1}{2} {}^* C\beta_{ij}(P_j P_i)^{-0.5}/Q\, a_i a_j \tag{9.47}$$

for all $i, j = 1, \ldots, n$ but $i \neq j$, and the own Hicks–Allen elasticities by

$$\sigma_{ii} = -\frac{1}{2} * C \sum_j \beta_{ij} P_j^{0.5} P_i^{-1.5} / Q\, a_i^2 \qquad (9.48)$$

for $i = 1, \ldots, n$.

Cross-Price Elasticities

Cross-price elasticities, which measure the degree to which one input is substituted for another in response to a change in price, are given by

$$\varepsilon_{ij} = \partial X_i / \partial P_j * P_j / X_i. \qquad (9.49)$$

More specifically, for the generalized Leontief cost function, we have

$$\varepsilon_{ij} = -\frac{1}{2} * \beta_{ij} (P_i / P_i)^{-0.5} / a_i \qquad (9.50)$$

for all $i, j = 1, \ldots, n$ but $i \neq j$.

However, and in the case of electric usage, own-price elasticities measure the change in usage that occurs in response to a change in own price and are computed as

$$\varepsilon_{ii} = \partial X_i / \partial P_i * P_i / X_i. \qquad (9.51)$$

In the case of the generalized Leontief cost function, we have

$$\varepsilon_{ii} = -\frac{1}{2} * \beta_{ij} (P_i / P_i)^{-0.5} / a_i \qquad (9.52)$$

for all $i = 1, \ldots, n$.

Remarks

As this particular form will be the subject of further analysis and a case study, several observations are worth mentioning (Berndt, 1992).

1. Caution needs to be maintained in assuming that, like the Hicks–Allen substitution elasticities, price elasticities are also symmetric. In general, they are not. In other words, while substitution elasticities, which are given by Equation (9.47), satisfy

$$\sigma_{ij} = \sigma_{ji},$$

cross-price elasticities, which are displayed in Equation (9.50), do not:

$$\varepsilon_{ij} \neq \varepsilon_{ji}.$$

(You will prove this in an exercise.)

2. The sign of the estimated coefficient β_{ij} allows one to ascertain whether inputs are substitutes, not related, or complements as β_{ij} is positive, zero, or negative. In the case of electricity, shifting household usage to lower-priced periods implies that these are substitutes for those hours in which

price is the highest; again, mid- to late afternoon on a hot summer day, for example.

3. As noted in Berndt (1991, p. 465):

Since C and a_i, a_j appear in the elasticity expressions, computations of these estimated elasticities are based on the estimated parameters and the predicted or fitted values of C and the a_i and a_j, not on the observed values. It is quite easy to make computational errors when using the above equations. One useful check on elasticity calculations involves the following summation of elasticities, which must always hold:

$$\sum_j \varepsilon_{ij} = 0 \qquad (9.53)$$

for all i = 1, ..., n.

4. As required by theory, it is necessary to make sure that the estimated cost function is monotonically increasing and strictly quasi-concave in input prices[1]; that is, one must verify that fitted values for all input–output equations are positive and that the $n \times n$ matrix of substitution elasticities is negative semidefinite at every observation.

5. Finally, it is important to keep in mind that computed elasticities are based on estimated parameters, which implies that they are stochastic and, as such, have variances and covariances. The rather substantial nonlinearities inherent in elasticity calculations have required approximation techniques to calculate these variances. In addition, because distributional properties of such elasticity estimates have not yet been derived (as of 1991), the basis for employing statistical inference on them does not yet exist.

DYNAMIC MODELS

Moving on, various approaches have been used to make these more flexible multi-equation approaches dynamic. Berndt et al. (1981) consider three generations of dynamic models. In the simplest first-generation cases, lagged endogenous or other variables have been included as in the aforementioned reduced-form models. In later generation models, outputs are separated into fixed and variable.

They discuss two second-generation models. In the first, by Nadiri and Rosen (1973), the Koyck model is generalized to multi-equations in which disequilibrium in one factor market is related to disequilibrium in another factor market. In this approach, if x_t is a vector of inputs and x_t^* is a vector of desired inputs, then

$$x_t - x_{t-1} = \beta x_{t*} - x_{t-1}, \qquad (9.54)$$

---

[1]You may recall from Chapter 4 that this requires that diagonal terms of the Hessian matrix, which is the matrix of second derivatives of the cost function with respect to input prices, must be nonpositive so that the factor-demand equations are negative semidefinite.

where β is an $n \times n$ partial adjustment matrix and x_t^* is chosen to be some function of the prices of factors and can be one of the functional forms such as the translogarithmic or generalized Leontief. In a slight modification of this model, Lucas (1967) assumes an adjustment matrix only for quasi-fixed factors such as capital and makes the β matrix a function of variables such as the discount rate and a technology parameter.

Another second-generation approach discussed by Berndt et al. (1981) is to estimate a restricted cost function with variable factors represented by price (P), quasi-fixed factors represented by quantities (X) and with output (Y) included, or

$$C = C(P, X, Y). \tag{9.55}$$

Estimating this cost or the related demand or share equations yields short-run elasticities.

From the restricted cost function one can also derive long-run elasticities from the long-run relationship that the negative of the price of the fixed factor $-P_x$ equals the partial derivative of the restricted cost function, or

$$-P_x = \partial C(P, X, Y) / \partial X. \tag{9.56}$$

By solving this equation for the desired fixed factor (X^*), long-run elasticities can be obtained.

In Dahl (2006), this type of second-generation model is indicated by the number 2. For example, in the translog formulation, this model would be designated Tl2. This model allows us to capture short-run, long-run, and capacity utilization, but does not allow us to capture an adjustment path. In a third-generation model, the change in the fixed variable ($X\cdot$) is added to the restricted cost function to represent the cost of adjustment, or

$$C = C(P, X, X\cdot, Y). \tag{9.57}$$

This function can be estimated directly or factor-demand or factor-share equations can be estimated depending on the functional form.

From dynamic cost minimization, the time path of capital accumulation must satisfy

$$-C_X - rC_{X\cdot} - P_X + C_{X\cdot, X}X\cdot\cdot + C_{X, X}X\cdot = 0, \tag{9.58}$$

where C is the estimated restricted cost function, r is the interest rate, and $X\cdot$ and $X\cdot\cdot$ are the first and second derivatives of the fixed factor with respect to time, which will be zero in long-run equilibrium. [Dahl (2006) designates these third-generation models by 3. For example, in the generalized Leontief case the model would be Gl3.] Examples of this last approach can be seen in Berndt and Watkins (1977), Pindyck and Rotemberg (1983), Walfridson (1987), Morrison (1988), and Kolstad and Lee (1992). The author refers the interested reader to these papers for a more complete discussion of this technique.

STRUCTURAL MODELS

Structural models are theoretically pleasing because they provide more detailed information on adjustment and hold promise for microanalysis. Because they tend to find rather different results than reduced-form models on aggregate data, their usefulness for aggregate forecasting and policy analysis at the macro level needs to be investigated. In structural models, the short-run decision is on the use of the ith appliance stock U_i, while the long-run decision is to decide on what the appliance stock is to be Sk_i. As such, total demand for energy at any point in time is given by

$$E = \sum_i U_i Sk_i, \tag{9.59}$$

where use of the ith stock of equipment might be represented as

$$U_i = E_i / Sk_i \tag{9.60}$$

or

$$U_i = F(P_i, Y, Sk_i, X), \tag{9.61}$$

where X represents other relevant variables. The purchase decision of the ith piece of equipment might be

$$Sk_i = F(P_i, P_s, Pk_i, Pk_s, Y, X), \tag{9.62}$$

where s represents the price or prices of substitute energy products, Pk_i is the price of the stock of ith energy using equipment, and Pk_s is the price of substitute energy using equipment. In the case of consumer appliances, a popular approach has been to model the appliance choice using a logit or other discrete choice model [for a comprehensive discussion of discrete choice models, see Greer (2011)].

EXPENDITURE SYSTEM MODELS

Expenditure system models look at all consumer expenditures as a system. The simplest of these models is the linear expenditure system (Ex-l) where the estimating equations for the jth product take the form

$$p_j q_j = p_j \gamma_j + \beta_j (Ex - \sum_k p_k \gamma_k), \tag{9.63}$$

where p_j is the price of good j, q_j is the quantity consumed of good j, and Ex is total expenditure.

The first expression on the right is considered the base expenditure, perhaps representing the basic necessity, and the second amount is the portion of income above subsistence that the person consumes on this good. (Note: Linear homogeneity in input prices requires that $\sum \beta_j = 1$.) Dividing through by p_j gives the representative estimating equation for the system as

$$q_j = \gamma_j + (\beta_j/p_j)(Ex - \sum_k p_k \gamma_k), \qquad (9.64)$$

where β_j and γ_j are estimated parameters.

Desirable properties of this technique are that it satisfies all the theoretical restrictions on systems of demand equations and it can be derived from a specific utility function. A disadvantage is that the restrictions are imposed and hence cannot be tested. A model developed to test some of the restrictions is the Rotterdam model Ex-Rot discussed in Deaton and Muellbauer (1980b). In this model, the estimating equation becomes the following difference equation in the natural logs:

$$w_i d\ln q_i = \beta_j \sum_k w_k d\ln q_k - \sum_j \beta_{ij} d\ln p_j, \qquad (9.65)$$

where w_i represents the budget share of good q_i and d represents the total differential, which is represented by the first difference, for estimation purposes.

With development of the translog and other flexible functional forms, Deaton and Muellbauer (1980a) wanted a model with the flexibility of the translog and the Rotterdam model.

The expenditure system model they developed (Almost Ideal Demand System Ex-AIDS), with its rather unfortunate acronym, is estimated using the following equation:

$$w_i = \beta_o + \sum_j \beta_{ij} \ln P_j + \beta_{yi} \ln(Ex/P), \qquad (9.66)$$

where P is the translog price index for all goods, defined as

$$\ln P = \alpha_o + \sum_k \alpha_k \ln Pk + \sum_k \sum_m \beta_k m \ln P_k \ln P_m. \qquad (9.67)$$

See the original article for restrictions testing in the context of this model.

As with modeling approaches, both functional forms and estimation techniques have taken on more sophistication over time. The most popular functional forms, early on, were log linear (ln) and linear (l), but increasingly translog (Tl), logit (Lg), and other more complicated models have been used. With estimation techniques, ordinary least squares gave way to techniques that paid more specific attention to econometric problems and included generalized least squares with corrections for serial correlation (-s) or heteroskedasticity (-h) or an error components model. Other techniques reported include maximum likelihood, two-stage least squares, seemingly unrelated (SUR), three-stage least squares, nonlinear least regressions, and full information maximum likelihood (FIML).

ECONOMETRIC ISSUES: IDENTIFICATION AND SYSTEMS BIAS

Although many econometric models of demand ignore supply, we know that prices in markets are typically determined by the interaction of both supply and demand, with a variety of other variables affecting the demand and supply equations. To see what problems ignoring supply might cause us, consider Figure 9.3, which represents the market for gasoline at different times. Over time, both demand and supply shift. In each panel, the intersection of demand and supply at various market equilibriums creates a data point. In Figure 9.3(a), if only the supply shifts and we fit a line through the three data points, we will

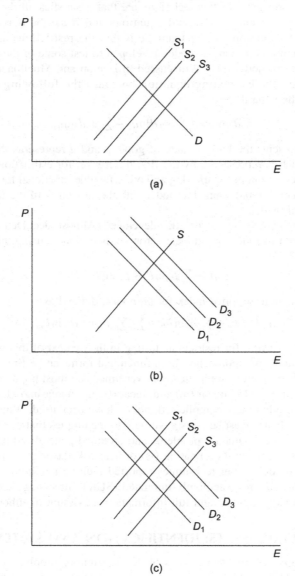

(a)

(b)

(c)

FIGURE 9.3 Changes in demand and supply over time: (a) supply, (b) demand, and (c) supply and demand. The identification problem is illustrated.

estimate the demand equation. In Figure 9.3(b), if only the demand curve shifts and we fit a line through the three data points, we will estimate the supply equation. In Figure 9.3(c), if both the demand and supply curves shift and we fit a line through the nine data points, we will get neither the demand nor the supply curve. Thus, when evaluating demand models we need to determine whether a demand curve was estimated as in Figure 9.3(a). The statistical term for whether we really have a demand curve is called *identification*, and mathematical properties for whether a demand curve is identified can be found in Pindyck and Rubinfeld (1991).

Aside: The Identification Problem

Just because one knows the reduced form of a system of equations does not mean that the parameters in the original equation can be estimated. Consider the following example of a supply–demand time-series model in which there are no predetermined, or exogenous, variables:

Supply:

$$Q_t = \alpha_1 + \alpha_2 P_t + \varepsilon_t \tag{9.68}$$

Demand:

$$Q_t = \beta_1 + \beta_2 P_t + \mu_t, \tag{9.69}$$

where P_t and Q_t are price and quantity in period t, respectively.

Assuming that the market is in equilibrium, then quantity demanded equals quantity supplied so that at each period of time there is only one value for price (P^*) and one for quantity sold (Q^*). In other words, only data available are P^* and Q^* in any given time period. (Note: It is likely the case that error values will render the values of P^* and Q^* in different time periods to be different; they will, however, lie close to the values for P^* and Q^* that would be obtained by direct solution of the equations of the model.) This is depicted in Figure 9.4.

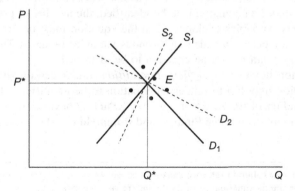

FIGURE 9.4 A market in equilibrium. Despite being in equilibrium, supply and demand curves are unidentified.

An attempt to estimate separate supply and demand equations using market data yields meaningless results, as there is no way to ascertain the true supply and demand slopes given on equilibrium data (it is only because of the errors that any estimation is possible; otherwise, all data points would be at point E).

The model described here is one in which both supply and demand curves are *unidentified*. That is, there is no way to obtain values of the structural parameters from the reduced-form equations [i.e., the reduced from is the equation that describes the point of intersection of the demand and supply curve with errors attached obtained by solving Equations (9.68) (supply) and (9.69) (demand) for P_t^* and Q^*]. In deviations form, these are given by

$$P_t = (\mu_t - \varepsilon_t)/(\alpha_2 - \beta_2) \tag{9.70}$$

$$Q_t = (\alpha_2\mu_t - \beta_2\varepsilon_t)/(\alpha_2 - \beta_2). \tag{9.71}$$

Clearly, additional information is required.

Consider the following supply–demand system of equations:
Supply:

$$Q_t = \alpha_1 + \alpha_2 P_t + \varepsilon_t \tag{9.72}$$

Demand:

$$Q_t = \beta_1 + \beta_2 P_t + \beta_3 Y_t + \mu_t, \tag{9.73}$$

where Y_t is income in period t.

In this case, as income varies over time it is possible to determine the supply curve (because income is an exogenous variable). However, because there is no such variable in the supply equation, the demand equation remains unidentified. Of course, by including an exogenous variable in the supply curve (such as technological change), the demand curve would be identified as well.

This brings us to the point in which it is necessary to determine the conditions under which identification of an equation is guaranteed; in other words, a formal "rule of thumb," which is known as the *order condition*. The order condition states that if an equation is to be identified, the number of predetermined (or exogenous) variables excluded from the equation must be greater than or equal to the number of included endogenous variables minus 1. This is known as a *necessary condition* for an equation to be identified.

This is not, however, a *sufficient condition*; under certain situations, the order condition may fail to render an equation to be identified. This involves an understanding of the *rank*[2] of a matrix, which is beyond the scope of this chapter [for more details, see Pindyck and Rubinfeld (1991) or Greene (1993)].

---

[2]In linear algebra, the **column rank** (**row rank** respectively) of a Matrix A with entries in some field is defined to be the maximal number of columns (rows respectively) of A, which are linearly independent.

SIMULTANEOUS EQUATIONS

Even if the identification problem has been dealt with adequately, simultaneous system bias is a second problem that can be encountered with simultaneous systems, such as the supply–demand model in Equations (9.68) and (9.69). Also known as an *endogeneity* problem (as both price and quantity are contained in both equations and are endogenous variables), it renders inconsistent estimation by OLS (and biased estimates) so that an alternative method must be employed. This is discussed in more detail later. For now, let us understand this issue further.

Basic economic theory dictates that both price (P) and output (Q) are determined simultaneously in equilibrium. That is, as displayed in Equations (9.68) and (9.69), the system of supply and demand equations constitutes an interdependent system of equations in which price and quantity are determined endogenously—within the system of equations. In a nutshell, price affects quantity and quantity affects price.

Consider the following system of equations:
Supply:

$$Q_t^S = \alpha_1 + \alpha_2 P_t + \alpha_3 P_{t-1} + \varepsilon_t \tag{9.74}$$

Demand:

$$Q_t^D = \beta_1 + \beta_2 P_t + \beta_3 Y_t + \mu_t \tag{9.75}$$

Equilibrium:

$$Q_t^S = Q_t^D. \tag{9.76}$$

Recalling the discussion on identification, prior-period price (P_{t-1}) and current-period income (Y_t) are predetermined variables. The important distinction between the two is that income is exogenous to the system, while prior-period price is determined within the system by past values of the variables.

To demonstrate that estimation by ordinary least squares yields inconsistent and biased estimates, consider the following system in equilibrium:
Supply:

$$Q_t = \alpha_1 + \alpha_2 P_t + \varepsilon_t \tag{9.77}$$

Demand:

$$Q_t = \beta_1 + \beta_2 P_t + \beta_3 Y_t + \mu_t. \tag{9.78}$$

In deviations (from mean values) form, we have
Supply:

$$q_t = a_2 p_t + \varepsilon_t \tag{9.79}$$

Demand:

$$q_t = \beta_2 p_t + \beta_3 y_t + \mu_t, \tag{9.80}$$

where q_t is $Q_t - \underline{Q}$ (mean of output), p_t is $P_t - \underline{P}$ (mean of price), and y_t is $Y_t - \underline{Y}$ (mean of income).

This is known as a structural model. Solving for endogenous variables yields reduced-form equations, which are given by

$$q_t = [\alpha_2\beta_3/(\alpha_2 - \beta_2)]y_t + [(\alpha_2\mu_t - \beta_2\varepsilon_t)/(\alpha_2 - \beta_2)] = \pi_{12}y_t + v_{1t} \qquad (9.81)$$

$$p_t = [\beta_3/(\alpha_2 - \beta_2)]y_t + [(\mu_t - \varepsilon_t)/(\alpha_2 - \beta_2)] = \pi_{22}y_t + v_{2t}. \qquad (9.82)$$

Estimation of Equation (9.79) by OLS would yield

$$a_2 = \sum p_t q_t / \sum p_t^2. \qquad (9.83)$$

Substitution for q_t in Equations (9.79) and (9.80) yields

$$a_2 = \sum p_t(\alpha_2 p_t + \varepsilon_t)/\sum p_t^2 \qquad (9.84)$$

or

$$a_2 = \alpha_2 + \sum p_t \varepsilon_t / \sum p_t^2. \qquad (9.85)$$

It is only when the price-related term on the right side of the equation equals zero that the estimate of α_2 is unbiased, that is,

$$\sum p_t \varepsilon_t / \sum p_t^2 = 0. \qquad (9.86)$$

Similarly, if this term approached zero as the sample size became large, then OLS estimation would be consistent. However, it is typically the case that neither of these is true because the error terms in such models tend to be correlated (because the endogenous variables in one equation are inputs into the variables in another equation in the system), which yields biased and inconsistent estimators. A further discussion on the direction of this bias can be found in Pindyck and Rubinfeld (1991).

CONSISTENT PARAMETER ESTIMATION

So how does one obtain consistent estimates of parameters in simultaneous equation models? One methodology is to use the *instrumental variables* (IV) approach in which predetermined variables serve as *instruments* (or appropriate regressors), as they are correlated with the endogenous variables but uncorrelated with the error term (because they are predetermined). In the supply–demand example described earlier [Equations (9.81 and 9.82)], y_t serves as a suitable instrument so that the estimated value of α_2 is given by

$$a_2^* = \sum y_t q_t / \sum p_t y_t. \qquad (9.87)$$

Another method is *two-stage least squares* (2SLS):

1. In the first stage, the reduced-form equation for p_t (9.82) is estimated using ordinary least squares by regressing p_t on all of the predetermined variables

in the system of equations. From here, fitted (i.e., predicted) values of p_t are determined and are, by construct, independent of the error terms e_t and u_t. As such, a variable has been created that is related to the model variables but is uncorrelated with the error term in the supply equation.

2. In the second stage, the supply equation of the structural model is estimated by replacing variable p_t with the first-stage fitted variable p_t^*. Ordinary least squares will yield a consistent estimator of the supply parameter α_2, as well as other parameters on any other predetermined variables that might appear in the equation.

MORE ADVANCED ESTIMATION METHODS

Thus far, we have examined two straightforward methods for estimating parameters in a system of simultaneous equations. Now we will delve into more sophisticated techniques that are more efficient to estimate, including *seemingly unrelated regression, three-stage least squares*, and *full information maximum likelihood*.

One of the most commonly employed estimation procedures is the seemingly unrelated regression[3] model in which the entire system of equations is estimated jointly, which yields more efficient estimates than the 2SLS and IV methods, because these estimate only a single equation within the system. (Inefficiency arises because single-equation estimation does not take into account cross-equation correlation among errors.) The SUR method achieves an improvement in efficiency by taking into explicit account the fact that cross-equation error correlations may not be zero. Generalized least squares is the estimation technique that is typically employed. The following excerpt from SAS Help and Documentation provides a description of this:

When you have a system of several regression equations, the random errors of the equations can be correlated. In this case, the large-sample efficiency of the estimation can be improved by using a joint generalized least squares method that takes the cross-equation correlations into account. If the equations are not simultaneous (no dependent regressors), then seemingly unrelated regression (SUR) can be used. This method requires an estimate of the cross-equation error covariance matrix, Σ. The usual approach is to first fit the equations by using OLS, compute an estimate Σ^* from the OLS residuals, and then perform the SUR estimation based on Σ^*. The MODEL procedure estimates Σ by default, or you can supply your own estimate of Σ.

If the equation system is simultaneous, you can combine the 2SLS and SUR methods to take into account both simultaneous equation bias and cross-equation correlation of the errors. This is called three-stage least squares or 3SLS.

A different approach to the simultaneous equation bias problem is the full information maximum likelihood estimation method. The FIML does not require

---

[3]For more details, see Zellner (1962).

instrumental variables, but it assumes that the equation errors have a multivariate normal distribution. 2SLS and 3SLS estimation do not assume a particular distribution for the errors.

In the case study presented on the real-time pricing of electricity, FIML is used to ensure symmetry of a GL model.

An example in Pindyck and Rubinfeld (1991) illustrates this concept nicely. It is shown here with permission.

Example 9.2 The Demand for Electricity

Price elasticity of demand is quickly becoming an important topic in the era of rising prices due to increasingly stringent environmental regulations. A study by Halvorsen (1975) provides a nice example of a simultaneous equations model, which is given by

$$\text{Log } Q = \alpha_1 + \alpha_2 \log P + \alpha_3 \log Y + \alpha_4 \log G + \alpha_5 \log D + \alpha_6 \log J + \alpha_7 \log R + \alpha_8 \log H + \varepsilon \tag{9.88}$$

$$\text{Log } P = \beta_1 + \beta_2 \log Q + \beta_3 \log L + \beta_4 \log K + \beta_5 \log F + \beta_6 \log R + \beta_7 \log I + \beta_8 \log T + \varepsilon, \tag{9.89}$$

where Q is average annual residential electricity sales per customer, P is marginal price of residential electricity (real), Y is annual income per capita (real), D is heating degree days, J is average July temperature, R is percentage of population living in rural areas, H is average size of households, T is time, L is cost of labor, K is percentage of generation produced by publicly-owned utilities, F is cost of fuel per kilowatt hour of generation, and I is the ratio of total industrial sales to total residential sales.

Equation (9.88) is a residential demand equation in which quantity demanded is a function of price and other demand-related variables. Equation (9.89) is the supply equation in which the supply of electricity is assumed to be fixed and electricity is sold at block rates with the price declining as the volume of sales increases.

Viewed as a system of equations, the model has two endogenous variables, P and Q, and is simultaneous since Q appears on the right side of the supply Equation (9.89) and P appears on the right side of the demand equation. Furthermore, both equations are identified since there are five exogenous variables in each that do not appear in the other equation.

The data are pooled time-series cross-section (also known as panel) data for 48 states for the years 1961–1969. Note that because the model variables are in logarithms all of the coefficients are elasticities. The equations are estimated via 2SLS and yield a price elasticity of demand that is negative (as expected) and statistically different from zero (standard error in parentheses):

Price elasticity of demand $= -1.15 \, (0.3)$.

(Only the relevant price variable is reported here.) This is the long-run direct elasticity of demand, which is greater than unity in absolute value and higher than one would have anticipated in the short run [see Dahl (1993), Wade (2003)].

Not surprising, the decline in the real price of electricity has led to an increase in the residential demand for electricity.

Next, assume that one wanted to examine the effect of a tax on the use of electricity in the long run. The initial impact would be measured by the demand equation (since a tax affects the price and price is a regressor in that equation). But the supply equation also comes into play since it measures the secondary effect of a change in the quantity sold on price. Thus, to measure the total long-run elasticity of demand, the elasticity must be calculated from the reduced-form equation, which yields a much larger result (in absolute value):

$$\text{Elasticity from reduced}-\text{form equation} = -3.70.$$

ADDITIONAL ECONOMETRIC ISSUES (BERNDT, 1991)

In *Electricity Cost Modeling Calculations*, the author emphasized how important data are in obtaining reasonable parameter estimates that can be used to attain appropriate conclusions in empirical work. Later in this chapter you will see examples that yield results contrary to a priori expectations, which may be due to a lack of appropriate data. With this said, there are other econometric issues that one must consider in attempting to estimate the demand for electricity and hence price and income elasticities. Berndt (1991) provides a thorough discussion of these so they are just summarized here.

1. Serial correlation of disturbances: Time-series data often yield serially correlated errors. Also known as autocorrelation, it violates the ordinary least-squares assumption that error terms are uncorrelated. While it does not bias OLS coefficient estimates, standard errors tend to be underestimated (and hence the t statistics overestimated) when autocorrelations of errors at low lags are positive.
2. Omitting the intramarginal price of electricity in the regression equation: Taylor (1975) argued that both marginal price and some measure of the intramarginal price should be included as regressors in the demand equation if it is to be consistent with economic theory. Failure to include the intramarginal price variable biases the parameter estimates.[4] As quoted in Berndt (1991),

---

[4]From Wikipedia, the free encyclopedia: In *statistics*, omitted-variable bias (OVB) is the bias that appears in estimates of *parameters* in a *regression analysis* when the assumed *specification* is incorrect, in that it omits an independent variable that should be in the model. Two conditions must hold true for omitted-variable bias to exist in *linear regression*: the omitted variable must be a determinant of the dependent variable (i.e., its true regression coefficient is not zero) and the omitted variable must be correlated with one or more of the included *independent variables*. As an example, consider a *linear model* of the form

$$y_i = x_i\beta + z_i\delta + u_i, \quad i = 1, ..., n, \tag{9.90}$$

where x_i is a $1 \times p$ row vector and is part of observed data; β is a $p \times 1$ column vector of unobservable parameters to be estimated; z_i is a scalar and is part of observed data; δ is a scalar and is an

[If] average and marginal prices are positively correlated, then using one in the absence of the other will lead, in general, to an upward bias in the estimate of the price elasticity. That this is so follows from the theorem on the impact of an omitted variable.

3. Simultaneity between electricity price and quantity: Also known as simultaneous equation bias, given a downward sloping demand curve, it is evident that

$$P = f(Q) \tag{9.96}$$

and

$$Q = f(P). \tag{9.97}$$

As such, ordinary least-squares estimation will yield upward-biased estimates of the demand response.

Note: In the case of rate-regulated utilities, this is not likely to be an issue, as the price (or rate) of electricity is determined exogenously by public regulatory commissions and not endogenously by the model.

---

unobservable parameter to be estimated; *error terms* u_i are unobservable *random variables* having *expected value* 0 (conditionally on x_i and z_i); and dependent variables y_i are part of observed data. We let

$$X = \begin{bmatrix} x_1 \\ \vdots \\ x_n \end{bmatrix} \in \mathbb{R}^{n \times p}, \tag{9.91}$$

and

$$Y = \begin{bmatrix} y_1 \\ \vdots \\ y_n \end{bmatrix}, \quad Z = \begin{bmatrix} z_1 \\ \vdots \\ z_n \end{bmatrix}, \quad U = \begin{bmatrix} u_1 \\ \vdots \\ u_n \end{bmatrix} \in \mathbb{R}^{n \times 1}. \tag{9.92}$$

Then through the usual least-squares calculation, the estimated parameter vector $\hat{\beta}$ based only on observed x values but omitting observed z values, is given by

$$\hat{\beta} = (X'X)^{-1}X'Y, \tag{9.93}$$

where the "prime" notation means the transpose of a matrix. Substituting for Y based on the assumed linear model,

$$\begin{aligned} \hat{\beta} &= (X'X)^{-1}X'(X\beta + Z\delta + U) \\ &= (X'X)^{-1}X'X\beta + (X'X)^{-1}X'Z\delta + (X'X)^{-1}X'U \\ &= \beta + (X'X)^{-1}X'Z\delta + (X'X)^{-1}X'U. \end{aligned} \tag{9.94}$$

On taking expectations, the contribution of the final term is zero; this follows from the assumption that U has zero expectation. On simplifying the remaining terms:

$$\begin{aligned} E[\hat{\beta}] &= \beta + (X'X)^{-1}X'Z\delta \\ &= \beta + \text{bias}. \end{aligned} \tag{9.95}$$

The second term is the omitted variable bias in this case. Note that the bias is equal to the weighted portion of z_i, which is "explained" by x_i.

4. The choice of functional form: Much of the literature on econometric estimation of the demand for electricity employs the log-linear functional form, as the parameter estimate of the price-related variable is the elasticity of demand.

Example 9.3 A Log-Linear Demand Model for Hourly Electricity Usage

Equation (9.97) represents a model in which hourly demand (load) is a function of price, weather (temp), and a binary variable that distinguishes demand between weekdays and weekends, which are likely to be different. Note that load is a function of the temperature in the hours leading up to the current period (t), which will likely lead to serially correlated errors as described earlier:

$$\ln \text{Load}_t = \beta_0 + \beta_1 \ln \text{Price}_t + \Sigma_i \alpha_i \text{Temp}_{t-i} + \delta_1 \text{weekend}_t + \varepsilon_t. \quad (9.98)$$

However, there are some problems with this form. The first is that β_1 [the estimated elasticity of demand in Equation (9.98)] is constant; in other words, it does not vary with price.

The second issue is that the estimated model should be consistent with the theory of utility maximization, which means that the underlying utility function is well defined and exhibits positive but declining marginal utility. In addition, the resulting demand equations, which can be derived from a well-behaved utility function, must be homogeneous of degree zero in prices and income, and the share-weighted sum of the latter must be unity.[5]

A BRIEF SURVEY OF THE LITERATURE: PRICE ELASTICITY OF DEMAND

Over the past few decades, numerous studies have been performed that attempt to estimate price elasticities of electricity consumption. One of the earliest is that of Houthakker (1951), who estimated residential customer demand on a cross-section of 42 towns in the United Kingdom and found that price elasticity was close to unity. Houthakker was among the first to recognize the importance of distinguishing between short- and long-run responses not only to price but also from the stock of equipment. This differed radically from what was typical of industry practitioners, who for many years merely used extrapolation techniques, which worked well until the 1970s when energy prices increased dramatically and actual sales fell far short of forecasts. The implications of this were grave: New generating assets that were built based on forecasted demand that never materialized were (at least some portion) disallowed in the rate base so that regulated utilities were not able to recover all of the costs of these

---

[5]It is often the case in studies of this nature that there are no data on income, at least at the individual customer level. Because survey respondents tend to be sensitive when asked questions of this nature, such questions are often left out of the survey instrument.

investments in rates; as such, the utilities themselves or shareholders (in the case of investor-owned firms) bore the financial consequences of these inaccurate forecasts.

One of the first models that explicitly included equipment stock was that of Fisher and Kaysen (1962). Their model of the short-run demand for electricity of residential customers used engineering information on kilowatts used per hour of normal consumption by each appliance in a typical household, which was then aggregated to form a composite variable, W_{it}, for the ith household in time t; that is, the model they specified was given by

$$Q_{it} = P_{it}^{\alpha} Y_{it}^{\beta} W_{it}, \tag{9.99}$$

where P is price, Y is income, and α, β are parameters to be estimated.

Nonlinear in parameters, a logarithmic transformation yields

$$\ln Q_{it} = \alpha \ln P_{it} + \ln Y_{it} + W_{it}. \tag{9.100}$$

Attempting to estimate W_{it} by states and years with "any kind of reliability is simply out of the question" (Berndt, 1991). Rather, they assumed that the stock of appliances in a given household in the ith state grew at a constant rate of γ percent per year, or

$$W_{it}/W_{it-1} = \exp(\gamma_i) \tag{9.101}$$

or

$$\ln W_{it} - \ln W_{it-1} = \gamma_i. \tag{9.102}$$

Lagging Equation (9.100) by one time period and subtracting it from Equation (9.100) yields

$$\ln Q_{it} - \ln Q_{it-1} = \gamma_i + \alpha_i(\ln P_{it} - \ln P_{it-1}) + \beta_i(\ln Y_{it} - \ln Y_{it-1}). \tag{9.103}$$

A random disturbance term (assumed to be independently and identically normally distributed, or i.i.d.) was then added to reflect the effects of stochastic elements and omitted variables, which were assumed to be uncorrelated with the regressors. Using data from 1946 to 1957, Fisher and Kaysen (1962) estimated the parameters α_i, β_i, and γ_i for each state in the United States using ordinary least squares.

Given the logarithmic specification of their model, the estimates of α_i and β_i are estimated short-run price and income elasticities of demand for electricity, conditional on the stock of equipment in the household. In most cases, they found that these elasticities were close to zero except in states of which the economies were less developed; in these cases, the elasticities were much larger but less than unity (in absolute value).

At this juncture, it is important to reiterate how important underlying data are in obtaining results. As is often the case, and as was conceded by Fisher and Kaysen in the estimation of a long-run model, the poor quality of data yielded positive estimates of the coefficient of the price variable, and often those of the income variable were not statistically different from zero in many instances.

According to Berndt (1991), one important lesson to be learned from the Fisher–Kaysen study is that while in principle it may be desirable to include measures of equipment stocks directly into short-run electricity demand equations, data problems can be severe, which can result in very imprecise parameter estimates.

This data issue was also emphasized by Taylor et al. (1984), who concluded that (Berndt, 1991, p. 315)

The results ... are better than might have been realistically expected, but they are clearly much poorer than might have been hoped for. In general, the utilization equations are very good, whereas the capital stock equations leave much to be desired.

Aside: The Necessity to Separate Short-Run from Long-Run Effects and Methodologies

The energy price increases in the 1970s and early 1980s spawned a plethora of studies. The demand (or quantity demanded) of electricity is considered to be inelastic (at least in the short run). According to several studies, the price elasticity of demand for electricity is between −0.1 and −0.56. A survey of 21 residential, 7 commercial, and 18 industrial studies of electric price elasticity by Dahl (1993) showed wide variances in price elasticity estimates. Econometric in nature, these studies show a long-run price elasticity of aggregate demand to be near −1.0, with elasticity for residential customers being between −0.91 and −0.71. Later studies indicate that for residential customers, the short-run elasticity is likely between −0.20 and −0.34 (Dahl, 2006; Wade, 2003).

In the short run, changes in energy prices elicit behavioral responses that affect the utilization of energy-using equipment (e.g., adjusting thermostats and turning off lights and appliances when not in use). In the long run (the EIA assumes 25 years), the price elasticity of demand is more elastic, declining to between −0.43 and −0.49, as price responses tend to occur via changes in the stock of energy-consuming equipment, which incorporates improving efficiency standards in addition to short-run behavioral responses to changes in price. Commercial customers tend to be more inelastic than residential customers in the short run, averaging between −0.1 and −0.2. In the long run, however, they tend to be similar to their residential counterparts with a price elasticity of −0.45.

SUBSTITUTION ELASTICITIES

Another topic, which is especially relevant when time-of-use rates are in effect, is that of elasticities of substitution. These measure the degree to which consumers shift usage from one period to another in response to changes in price.

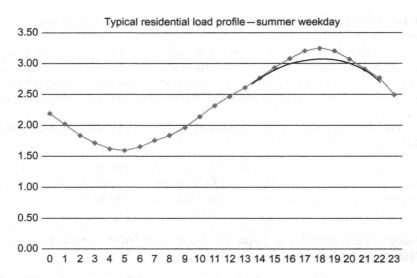

FIGURE 9.5   The impact of time-of-use pricing on peak demand.

One feature of time-of-use rates is that prices to end-user customers reflect the cost of power at the time that it is demanded. Typically, weekday afternoons during summer months command the highest hourly prices, especially in the case of residential customers.

Figure 9.5 displays a typical residential customer's summer hourly load profile. The peak, which occurs at hour 18, could be reduced by a time-of-use rate, which is significantly higher in the midafternoon-to-early-evening period than it is during other periods of the weekday. This can yield a flattening of the peak, which is shown by the smooth line below the load curve. In Figure 9.5, the load is reduced from 3.3 kW to just over 3 kW, which may not seem like much. However, multiplying this by the number of residential customers can lead to a rather substantial reduction in peak demand, which can delay the construction of new generating capacity, a move that typically triggers a rate case and raises rates for all end users.

Also known as peak load pricing, the optimality of this practice is well established in the literature on public utility economics.[6] Substitution elasticities help determine the extent to which usage is shifted to later hours or even to the weekend. These are examined in more detail later.

Faruqui and Sergici (2009) provide a survey of experimental evidence on household response to time-of-use pricing using 15 experiments conducted over the 1996–2007 timeframe, which is also reviewed in more detail later.

---

[6]Crew et al. (1995) provide a nice survey of the literature on this topic.

But first, an excerpt from Faruqui and Sergici (2009) provides a nice introduction. It is included here with permission from the authors.

The California Debacle: A Decade Later

In *Electricity Cost Modeling Calculations* (Greer, 2011), the author points out some of the flaws in Order AB 1890, which legislated the restructuring of California's electric markets and mandated how this would be facilitated. One fatal flaw, which is also mentioned by Faruqui and Sergici (2009), is that the energy crisis that transpired in the western United States in 2000–2001 could have been obviated (at least in part) if the prices paid for electricity by retail customers were allowed to vary in response to supply conditions. To wit: Despite the fact that wholesale prices rose precipitously during this time, prices paid by end users remained at or below prices charged before California's electric markets were restructured. Clearly, retail end-use customers had no incentive to reduce demand so they did not alter their consumption behavior, and the crisis ensued. More specifically (Greer, 2011, p. 321),

In fact, under the guise of protecting ratepayers, not only were rates frozen but also a mandatory rate reduction was imposed—rates were reduced by 10% for residential and small commercial customers from 1996 levels in 1998 with the intention of reaching 20% by 2002. However, the crisis that ensued put an end to retail access in 2001. (Again, no price signal to consumers!)

Furthermore, and as Greer (2011) argued, a group of economists put forth a manifesto on the crisis, which argued (Bandt et al., 2003, p. 9):

Any structural model for the industry should include a mechanism for charging consumers for the cost of the production and delivery of electricity at the time of its consumption. Electricity at midnight in April is completely different from electricity at noon on a hot August day. ... Prices to most end users don't signal when electricity is cheap or dear for the industry to produce. Nor are consumers offered the true economic benefit of their conservation efforts at times of peak demand. Customers suffer further when unchecked peak demands grow too fast, pushing up costs for all. Wholesale electricity markets also become more volatile and subject to manipulation when rising prices have no impact on demand. Indeed, a functioning demand side to the electricity market in California would have greatly reduced the likely private benefits, and consequent social cost, of any strategic behavior engaged in during the crisis.... Regardless of other reform efforts that are pursued in California, real-time pricing or other forms of flexible pricing is a key to enhanced conservation, more efficient use of electricity, and the avoidance of both unnecessary new power plants as well as concerns about the competitiveness of wholesale electricity markets.

As is pointed out in Faruqui and Sergici (2009), there were two unanswered questions left by this manifesto: (1) whether customers would respond to higher prices by reducing demand and (2) whether, from an economics perspective, it would make sense to equip end-use customers with the technologies that would allow them to respond to dynamic price signals.[7]

The answer to the first question, which emanates from several programs that have been implemented and assessed throughout the United States, Canada, and elsewhere around the globe [see Faruqui and Sergici (2009) for a study that includes results from studies in France and Australia], is a resounding YES; customers do respond to higher prices by reducing demand. The result of California's experiment with dynamic pricing showed that customers responded to high prices by lowering peak usage by 13% (Faruqui and George, 2005; Herter, 2007; Herter et al., 2007). In addition to benefits from demand response, there can be operational benefits to the distribution system from advanced meter infrastructure (AMI) as well (Faruqui and George, 2005).

However, the answer to the second question is more difficult; again appealing to Faruqui and Sergici (2009),

The cost of upgrading all residential meters in the US would be staggering. Using the California cost estimates as a proxy, the nationwide cost would be around $40 billion.

(Note: This does not include upgrades to the meters of commercial and industrial customers!)

The authors go on to describe the two conditions under which it would be worthwhile to pursue the investment in AMI technology. Clearly, it is necessary that dynamic pricing be adopted by providers of electricity, which represents a major change in the pricing (or regulatory) paradigm. This requires a departure from traditional rate making on the part of state regulatory commissions, which may be controversial to say the least. (Please see Chapter 8 for more detail.) This is a subject of much debate within many state regulatory commissions, especially those in states where rates are deemed "low" and tend to have flat or declining block rates.

FEDERAL LEGISLATION

In recent years, three federal laws have, in fact, requested that state commissions consider the deployment of smart meters to elicit demand response. More specifically, the Energy Policy Act of 2005 (EPACT), the Energy Independence and Security Act of 2007 (EISA), and the American Reinvestment and Recovery Act of 2009 (ARRA) suggested the deployment of smart meters to accomplish this goal. These are described in more detail elsewhere. A synopsis is given here.

EPACT and EISA

The Energy Policy Act of 2005 and the Energy Independence and Security Act of 2007 both include specific language asking state utility commissions to

---

[7]This is more relevant for residential and small commercial customers than it is for large commercial and industrial customers.

consider the deployment of smart meters and demand response. Subsequently, a 2008 survey of state regulatory activity found that 38 state commissions had initiated regulatory consideration of smart meters and demand response in response to federal legislation and 32 had completed their consideration (Faruqui and Sergici, 2009; see Chapter 3 for more detail).

American Recovery and Reinvestment Act[8]

The U.S. Department of Energy (DOE) has been charged with leading national efforts for modernization of the electric grid. Within the DOE, the Office of Electricity Delivery and Energy Reliability (OE) is responsible for heading this effort. In recent years, DOE's research and energy policy programs have been responsible for coordinating standards development, guiding research and development, and convening industry stakeholders involved in implementation of the smart grid. The ARRA has placed an unprecedented funding resource in the hands of the DOE, resulting in the Smart Grid Investment Grant program and the Smart Grid Demonstration program ("Smart Grid Programs").

As part of the Smart Grid Programs, DOE will award approximately $4 billion to utilities, equipment suppliers, regional transmission organizations, states, and research organizations to jump start smart grid deployment and demonstration on a massive scale. Projects to be undertaken by these recipients will support critical national objectives, including:

- Number and percentage of customers using smart grid–enabled energy management systems.
- Number and percentage of distribution system feeders with distribution automation.
- Number and percentage of transmission lines instrumented with networked sensors used to assess and respond to real-time grid disturbances (i.e., time-synchronized situational awareness capability).

Smart grid assets improve the ability to automate and control grid operations remotely and also provide customers with real-time data so that they can make informed decisions about their energy consumption. For instance, AMI gives the utility the ability to conduct real-time load measurement and management and it gives the customer real-time data required to optimize their electricity use to reduce cost.

Furthermore, assets can include energy resources that either deliver electricity or contribute to load reduction. Energy resources that interact with the grid include distributed generation, stationary electricity storage, plug-in electric vehicles, and smart appliances. These resources can communicate and make operating decisions based on signals from the grid or customers.

---

[8]Sources: http://apps1.eere.energy.gov/news/news_detail.cfm/news_id=12209 and http://www.oe .energy.gov/DocumentsandMedia/final-smart-grid-report.pdf.

The DOE is particularly interested in six areas, including:

1. Job creation and marketplace innovation
2. Peak demand and electricity consumption
3. Operational efficiency
4. Grid reliability and resilience
5. Distributed energy resources and renewable energy
6. Carbon dioxide emissions

Certain metrics will be used to evaluate the impact of these programs on the six areas listed.

The first biennial report was issued in December 2008 and the findings were substantial:

1. There has been a fivefold increase in the number of end-use customers with AMI technology (up from 1% in 2006).
2. Eight percent of all energy customers in the United States participate in a demand response program.
3. The total electrical demand that can be shed through demand response programs is now at 5.8% of U.S. peak demand, or close to 41,000 megawatts, a more than ten fold increase from the 2006 estimate of about 3400 megawatts.

Policies and Programs

Policies and programs may also be implemented along with smart grid assets to obtain the maximum benefits possible. For example, customers who have access to dynamic pricing programs have an incentive to use the information provided by advanced metering infrastructure. The DOE will be collecting information regarding the deployment and adoption of these types of policies and programs. Examples of these policies and programs include, but are not limited to:

- Demand response
- Dynamic pricing
- Critical peak pricing
- Distributed resource interconnection policy
- Policy/regulatory progress for rate recovery

Many of these are discussed in more detail in Chapter 8.

Energy resources include:

- Distributed generation
- Renewable energy
- Storage
- Plug-in electric vehicles
- Smart appliances/devices

There are benefits to smart grid technology to utilities as well, including

1. Improved reliability
2. Deferred capital spending for generation, transmission, and distribution
3. Reduced operations and management costs
4. Increased efficiency in power delivery
5. Integration of renewable energy and distributed resources
6. Improved system security

TECHNOLOGY—THE "SMART GRID"—HOW IT WORKS

Also known as advanced metering infrastructure, end-use customers receive signals when prices are about to change so that they can respond accordingly. For example, when prices increase, one might respond by curtailing unnecessary activities such as laundry or running the dishwasher until later when prices are reduced (i.e., the concept of substitution elasticity, which is discussed in more detail later). Subsequently, the California Public Utilities Commission (CPUC) initiated proceedings on advanced metering, demand response, and dynamic pricing (see http://docs.cpuc.ca.gov/published/proceedings/R0206001.htm).

As cited in Faruqui and Sergici (2009, p. 4), which is quoted with permission here,

As part of the proceeding, the state carried out an elaborate experiment with dynamic pricing. It showed conclusively that customers responded to high prices by lowering peak usage by 13 percent. The three investor-owned utilities in the state relied on the results from the experiment to develop their AMI business cases. They showed that while AMI yielded many operational benefits to the distribution system, such benefits only covered about 60 percent of the total investment. The remaining 40 percent had to be covered through demand response.

Based on these results, the CPUC approved the deployment of almost 12 million smart meters for electricity to occur within the next five years. In addition, all customers who have such meters will be placed on critical peak pricing, where the price of electricity consumed during peak hours will be significantly higher than that consumed outside of peak hours.

For example, a typical weekday residential hourly load profile on a summer day is displayed in Figure 9.6. Critical price hours are those beginning midafternoon and extending to early evening, which would correspond to hours 13–18 in Figure 9.6, which displays an average load profile of a residential customer not on a smart rate (as such assigned to the "Control" group) under two scenarios: one in which no critical peak price (CPP) event is called (which indicates that heat is not extreme and, as such, demand is lower—0-Control), and the other where a CPP event is called (most likely due to extreme heat, which yields higher demand—1-Control). A case study on real-time pricing of electricity is presented in Chapter 10.

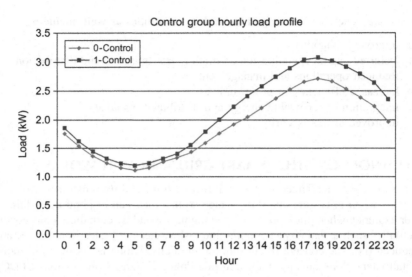

FIGURE 9.6   Hourly load profile of control group under different weather scenarios.

ELASTICITIES

The beauty of time-varying prices is that the elasticities of price and substitution can be calculated, which provide a plethora of information about consumer behavior. The next section reviews one of the models used to estimate the elasticity of substitution for electricity between peak and off-peak periods.

Models for Estimating Price and/or Substitution Elasticities: The Constant Elasticity of Substitution Functional Form

Cost Models

In *Electricity Cost Modeling Calculations*, the author spent some time detailing the constant elasticity of substitution function as a choice for modeling the cost of electricity. In recent experimental work on price and substitution elasticities, the CES form is a popular choice. More specifically, the substitution equation (named this because it measures the degree to which electricity usage is moved to lower-priced (i.e., off-peak) periods is defined as follows.

Formally, the elasticity of substitution between inputs, which measures the degree of substitutability between inputs (x_i, x_j), is given by

$$\sigma_{ij} = \partial \ln(x_i/x_j)/\partial \ln(P_i/P_j), \qquad (9.104)$$

where P_i, P_j are the marginal products of x_i, x_j.[9]

---

[9]The marginal product (P_i) of an input is equal to the partial derivative of y with respect to that input (x_i):

$$\partial y/\partial x_i = P_i. \qquad (9.105)$$

Kenneth Arrow, Hollis Chenery, Bagicha Minhas, and Robert Solow (1961) show that solving for $\partial \ln (x_i/x_j)$ and then integrating Equation (9.104) yield

$$\ln(x_i/x_j) = \text{constant} + \sigma \ln(F_i/F_j), \tag{9.106}$$

where F_i/F_j is the marginal rate of technical substitution (MRTS) between x_i and x_j.

The integral of the MRTS yielded the implied production function, which is known as the constant elasticity of substitution (CES) production function.

Duality: The CES Demand Function

From consumer demand theory, we know that when consumers maximize utility with respect to a budget constraint, the indifference curve is tangent to the budget line so that slopes of the two are equal. This is shown in Figure 9.7, which displays an indifference curve and a budget line.

$$U(\text{max}) = U(X^*, Y^*). \tag{9.107}$$

Letting m denote the slopes of each, we have

$$m_{\text{indifference}} = m_{\text{budget}}. \tag{9.108}$$

This implies that the marginal rate of substitution between x and y equals the ratio of the prices of x and y, or

$$\text{MRS}_{xy} = P_x/P_y. \tag{9.109}$$

An example of peak versus off-peak electricity serves nicely here.

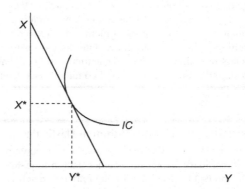

FIGURE 9.7 The consumer's utility maximization problem. The consumer's optimal consumption bundle, which is denoted by X^*, Y^*.

Example 9.3 Peak versus Off-Peak Electricity

In this case, substituting Q_p/Q_{op} (ratio of the consumer's demand for electricity peak vs. off-peak) for x_i/x_j [the producer's demand for inputs in Equation (9.105) and using Equation (9.109) (the MRS)] yields a basic substitution equation:

$$\ln(Q_p/Q_{op}) = \alpha + \sigma \ln(P_p/P_{op}), \qquad (9.110)$$

where Q_p is the average energy use per hour in the peak period for the average day, Q_{op} is the average energy use per hour in the off-peak period for the average day, σ is the elasticity of substitution between peak and off-peak energy use (following convention, this is taken to be a positive number for substitutes and a negative number for complements), P_p is average price during the peak pricing period, and P_{op} is average price during the off-peak pricing period.

In addition, other regressors are typically included in the substitution equation to increase the explanatory power of the model. For example, Faruqui and Sergici add the difference in cooling degree hours between peak and off-peak periods and a fixed-effects variable for each customer so that the final form of their model is given by

$$\ln(Q_p/Q_{op}) = \alpha + \sigma \ln(P_p/P_{op}) + \delta(CDH_p - CDH_{op}) + \Sigma_i \theta_i D_i + \varepsilon, \qquad (9.111)$$

where δ is measure of weather sensitivity, CDH_p is cooling degree hours per hour during the peak pricing period, CDH_{op} is cooling degree hours per hour during the off-peak pricing period, θ_i is a fixed-effect coefficient for customer i, D_i is a binary variable equal to 1 for the ith customer, 0 otherwise, where there are a total of N customers, and ε is a random error term.

Next, a price elasticity equation provides a second summary measure in the CES demand system. Again appealing to Faruqui and Sergici, a typical log-linear specification is employed of the form

$$\ln(Q_d) = \alpha + \eta \ln(P_d) + \delta(CDH_d) + \Sigma_i \theta_i D_i + \varepsilon, \qquad (9.112)$$

where Q_d is the average daily energy use per hour, η_d is the price elasticity of demand for daily energy (defined later), P_d is the average daily price (e.g., a usage weighted average of peak and off-peak prices for the day), CDH_d is cooling degree hours per hour during the day, and ε is the regression error term.

The authors include variations on these themes. For example, the substitution and price elasticity equations would likely differ across customers in different socioeconomic classes, in the saturation of electricity-using appliances (i.e., heating type in homes), and even weather-related conditions (see later for details).

Example 9.4 Estimating Price and Substitution Elasticities

Using data in "RRP for Elasticity Calculations.csv," which contains a cross-section of individual customer load, price, and weather data in June–July 2010, estimate Equations (9.111) and (9.112) to obtain implied price and substitution elasticities (for off-peak electricity consumption vs. peak electricity consumption). Include any indicator variables that may be appropriate. Estimation results are displayed in Table 9.2.

TABLE 9.2 Estimated Coefficients for Price and Substitution Equations

Variable	Sub Ed	t Statistics	Price Ed	t Statistics
Intercept	−0.02723	−7.59	0.03872	12.38
Price	−0.01199	−19.41	−0.01200	−19.55
Weather	−0.00026	−4.92	0.00132	14.31

Discussion of Estimation Results

The second and third columns of Table 9.2 pertain to the substitution elasticity, or the degree that electricity is shifted from peak to off-peak when prices are lower. As can be seen, the coefficient is negative, which implies that electricity between the two periods is a complement rather than a substitute. Another aspect to note is that regression equations constitute a system of equations, that is, one for each participant. As such, the typical goodness-of-fit statistic, the adjusted R^2, is not valid when there is a system of equations.[10]

Next, regarding the price elasticity equation, we see that the coefficient of price is of the expected (negative) sign and statistically different from zero. Again estimated as a system of equations, the coefficient of the price variable is negative as required by economic theory. In addition, the weather-related

---

[10]In the single-equation context, in which the regression model is given by

$$Y = X\beta + e. \qquad (9.113)$$

Most computer software programs compute R^2 as

$$R^2 = 1 - e'e/(Y - Y^*)'(Y - Y^*), \qquad (9.114)$$

where Y^* is the sample mean of Y. Note that in the single-equation context,

$$e'e = (e - e^*)'(e - e^*). \qquad (9.115)$$

This is the case because least-squares estimation ensures that the sum of residuals equals zero and therefore their mean also equals zero. There are two reasons that this measure of R^2 is not appropriate in equation systems.

1. It is possible that the R^2 from a particular equation could be negative, since with system estimation it is not necessarily the case that within each equation the sum of the residuals is zero. As such, it is possible that the numerator of Equation (9.114) is larger than the denominator, which results in a negative R^2.
2. While single-equation least squares minimize $e'e$ and therefore maximize R^2, in general, system estimation methods do not minimize $e'e$.

variable (cooling degree days) is of the appropriate (positive) sign and statistically different from zero.

MORE COMPLEX MODELS: PRICE AND SUBSTITUTION ELASTICITIES USING CONSTANT ELASTICITY OF SUBSTITUTION MODEL

It was stated earlier that modifications can be made to the CES model substitution and price elasticity equations. For example, in addition to fixed effects, these elasticities could vary with weather and other factors, such as cooling degree days and the presence of central air conditioning in the dwelling. The author's concern is that the price variable in the elasticity of substitution Equation (9.111) is negative; a priori expectations would be that a higher price in peak hours would cause a shift toward consumption in off-peak hours (i.e., a positive coefficient, indicating that participants are substituting away from higher-priced hours toward lower-priced hours—shifting tasks such as laundry, running the dishwasher, and so on to off-peak hours). The author's suspicion is that the model is misspecified, as described in Chapter 4, which means that it likely suffers from omitted variable bias, an incorrect functional form, or some other specification error.

Adding other explanatory variables, such as (1) the cross-product between the difference in cooling degree hours is being multiplied by the ratio of price between peak and off-peak and (2) the presence of central air conditioning, Equation (9.111) (substitution elasticity) becomes

$$
\begin{aligned}
\ln(Q_p/Q_{op}) = {} & \alpha + \sigma \ln(P_p/P_{op}) + \delta(CDH_p - CDH_{op}) \\
& + \lambda(CDH_p - CDH_{op}) \ln(P_p/P_{op}) + \varphi(CAC)\ln(P_p/P_{op}) \\
& + \sum_i \theta_i D_i + \varepsilon,
\end{aligned}
\tag{9.116}
$$

and Equation (9.112) becomes

$$
\begin{aligned}
\ln(Q_d) = {} & \alpha + \eta \ln(P_d) + \delta(CDH_d) + \rho(CDH_d) \ln(P_d) \\
& + \psi(CAC) \ln(P_p) + \sum_i \theta_i D_i + \varepsilon.
\end{aligned}
\tag{9.117}
$$

Note: Elasticities are computed by taking the derivative of each with respect to the price variable. As is evident, elasticities do not change with price or load.

OVERVIEW OF RESULTS—FARUQUI AND SERGICI STUDY

In response to rising energy prices in the 1970s, various electricity pricing experiments were conducted under the auspices of the Federal Energy Administration. Experiments focused on measuring customer response to time-varying and seasonal rates. Data from the top five were analyzed by Christensen Associates for the Electric Power Research Institute. Results were conclusive: customers responded to higher prices by reducing usage and by shifting to less-expensive off-peak hours. As stated in Faruqui and Sergici (2009),

The results were consistent around the country once weather conditions and appliance holdings were held constant. Customer response was higher in warmer climates and for customers with all-electric homes. The elasticity of substitution for the average customer was 0.14. Over the entire set of customers, it ranged between 0.07 and 0.21.

The Faruqui and Sergici study followed 15 experiments, detailing for each the rate design and subsequent results. Across the range of experiments, time-of-use rates induced a drop in peak demand between 3% and 6%, and critical peak pricing tariffs led to a drop in peak demand of 13% to 20%. Furthermore, when coupled with enabling technologies, CPP tariffs led to a reduction in peak demand between 27% and 44%. Needless to say, this has important implications for the future of the energy industry. But first, pricing electricity efficiently is key and was expounded upon in Chapter 8.

CONCLUSION

This chapter contains numerous demand-related models that can be used to compute price and substitution elasticities. Incorporating prices into such demand models allows us to estimate the impact on electric loads and peak demand. As prices continue to rise in response to pending legislation (carbon, mercury, and other hazardous air pollutants) and other uncertainties (regulatory, fuel prices, etc.), it is more critical than ever that the impact of changes in price be factored into sales forecasts and other analyses used by utilities for resource planning purposes. In addition, results of such models have important implications for reliability and least cost rate-making. On this point, it is even more critical that energy prices reflect the underlying costs so that a customer response can be elicited, which can reduce the need for additional generating capacity, thus reducing energy costs going forward. In addition, this could also result in a reduction of greenhouse gas emissions, which is also the objective of pending legislation (for more details, see Chapter 3).

FOR INTERESTED READERS

There is a set of exercises at the end of the chapter designed to provide hands-on experience in working with some of the models presented in this chapter. A word of caution: Some are rather advanced in that they require the use of rather sophisticated econometric analysis and SAS programming skills. Nonetheless, give them a try and see how you do. (You can always contact the author for assistance.)

RECOMMENDED READING

You also may want to review the following, which will give guidance on the more advanced exercises: "Customer Response to RTP in Competitive Markets: A Study of Niagara Mohawk's Standard Offer Tariff" (Boisvert et al., 2006, available on the Internet),"Household Response to Dynamic Pricing of Electricity: A Survey of the

Experimental Evidence" (Faruqui and Sergici, 2009, available on the Internet), and "Industrial and Commercial Customer Response to Real-Time Electricity Prices" (Boisvert et al., 2004, available at www.bneenan.com).

EXERCISES

1. In this chapter it was stated that unlike Hicks–Allen substitution elasticities, price elasticities are not symmetric. More specifically, cross-price elasticities, which are displayed in Equation (9.50), are not, so that

$$\varepsilon_{ij} \neq \varepsilon_{ji}.$$

   Why is this the case?

2. Using data in "RRP for Elasticity of Substitution calculations.csv," estimate Equation (9.97).
   a. How many lags of temperature are required?
   b. Calculate the elasticity of substitution between usage on peak and off-peak. Are they substitutes or complements?
   c. Does the type of day (i.e., weekday vs. weekend) affect price and substitution elasticities? If so, how?

3. **Advanced:** Using data in "RRP for Elasticity of Substitution calculations. csv," estimate Equations (9.111) and (9.112).
   a. Do your results match those in Table 9.2? If not, how are they different? What do you think is causing the difference?
   b. What are the price and substitution elasticities that result from estimation of these equations?
   c. Examine the goodness-of-fit statistics. Are the signs of the adjusted R^2 values as expected? Why or why not?
   d. Does the addition of any/all of the following improve model results?
      i. An electric water heater
      ii. Central air conditioning
      iii. Large home (hint: define this as home size >4 bedrooms and people >4)
      iv. Other

4. **Advanced:** Using data in "RRP for Elasticity of Substitution calculations. csv," estimate Equations (9.112) and (9.113) and compute price and substitution elasticities.
   a. How do your results differ from those obtained in Exercise 3?
   b. Has the addition of the cross-product between the difference in cooling degree hours is being multiplied by the ratio of price between peak and off-peak *and* the presence of central air conditioning improve model results?
   c. What are the price and substitution elasticities associated with the addition of these variables?

Chapter 10

Time-of-Use Case Study

INTRODUCTION

In 1996, Federal Energy Regulatory Commission (FERC) Orders 888 and 889 were passed to facilitate wholesale competition in the bulk power supply market. More specifically, Order 888 addresses the issues of open access to the transmission network giving FERC the jurisdiction over all transmission issues, especially pricing. Order 889 requires utilities to establish electronic systems to share information about available transmission capacity. In addition, as of June 30, 1996, 44 states and the District of Columbia (more than 88% of the nation's regulatory commissions) had started activities related to retail competition in one form or another.

California was the first state to pass restructuring legislation that allowed retail choice among consumers beginning in 1998. Prior to the passage of deregulation legislation, the electric industry in California was composed of both publicly and investor-owned vertically integrated utilities (IOUs), the latter of which supplied 75% of California's retail load (Rothwell and Gomez, 2003). The remainder was served by a mix of publicly and cooperatively-owned entities, and two of the largest cities in California were served by the former. In Los Angeles, the Los Angeles Department of Water and Power provided electricity to 9.6% of California's total native load customers. In Sacramento, the Sacramento Municipal Utility District provided electric service to 4.0% of the total load in California.

At this time, rates paid by ultimate consumers were among the highest in the nation and, like most states, were distinguished by the type of customer (i.e., residential, commercial, industrial, or other). Figure 10.1 displays average rates paid by California native load customers by customer class from 1990 to 2007. Note the rather precipitous increase after 1999, which is due to several factors that are discussed in this case study.

Prior to passing restructuring legislation, utilities were regulated by three separate and distinct entities: the California Public Utilities Commission (CPUC), which had jurisdiction over the rates and operations of the utilities; the California Energy Commission, which oversees new plant siting and construction activities; and the FERC, which regulates wholesale electricity trading and interstate transmission for all energy suppliers in every state.

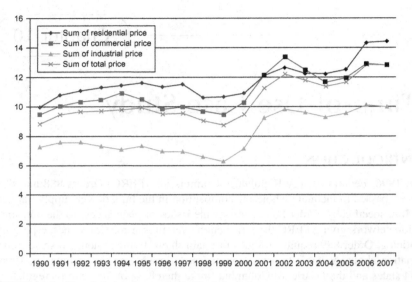

FIGURE 10.1   Historical electric rates in California by end-use sector. Rates paid by end users in California were among the highest in the nation, averaging 9 cents per kilowatt hour in 1998. By 2001, the confluence of events that transpired to lead to what is known as the California debacle resulted in even higher rates (a 24% increase over 1998 rates; by 2002, the increase was over 40%). Not surprising, the state of California has subsequently rescinded restructuring legislation.

THE ELECTRICITY CRISIS: SUMMER 2000

Factors Precipitating the Crisis

A confluence of events and poor decisions ultimately led to what is known as the California debacle. Included are:

1. No new generation in years.
2. Utilities were forced to divest generating assets and buy back from a newly created power exchange at spot prices.
3. No bilateral trading.
4. Retail rates were frozen (even reduced by 10%).
5. Reliability was compromised: blackouts ensued.
6. Ability to game system (Cal ISO problems)—EnronOnline.

Each is discussed in turn.

Chronology of the California Electricity Crisis: Lessons Learned (or What Not to Do)

Instead of building new generating assets in the years leading up to the restructuring of California's electric market to keep up with double-digit population growth (13% throughout the 1990s), the state relied on imported power from the north; the Bonneville Power Administration had been producing ample hydroelectric generation to supply much of the load in Washington and Oregon and had enough electricity to export power to California to supply the load that was not being met by California's own utilities. Despite this, California had among the highest rates in the country, which precipitated the passage of California Assembly Bill 1890 (AB 1890), which introduced competition into California's electricity market. Key features of AB 1890 included establishment of the California Independent System Operator to operate the transmission facilities of California's investor-owned utilities, which encouraged (read: mandated) investor-owned utilities to sell off their generation assets, thus requiring them to buy all of their power in a newly created "spot" market run by the California Power Exchange (PX). In addition, IOUs in the state were forbidden from entering into long-term, "bilateral" contracts, which can serve as a hedge against future price volatility. Furthermore, AB 1890 capped retail rates that the utilities could charge customers below the then-current cost of electricity.

From April 1998 when the California market commenced until late May 2000, the plan worked relatively well. But in May 2000, spot prices began to rise notably: From 1999 to 2000 the total wholesale cost of supplying electricity to meet load in California nearly quadrupled, rising from $7.4 to $28 billion. Average wholesale electricity prices typically exceeded $100 per megawatt hour and, in some cases, exceeded $300 per MWh. In summer 2000, unusually warm temperatures, coupled with other factors delineated earlier, resulted in prices reaching $500/MWh (Palo Verde, summer 2000—Chronology, p. 15), which was the then-imposed price cap at the time. This is displayed in Figure 10.2.

The drought conditions in the northwestern United States in 2000 resulted in a lower-than-anticipated water runoff to fuel the hydroelectric power generation upon which California had become dependent. In addition, disruptions in the supply of natural gas (California utilities were predominantly gas fired due to environmental standards) and transmission constraints further exacerbated the imbalance between supply and demand. Finally, the rules that accompanied California's restructuring legislation led to a situation that was ripe for market manipulation, which is exactly what transpired.

Unable to pass on the higher costs of purchasing electricity (due to the rate freezes that accompanied AB 1890), the utilities financial well-being was severely jeopardized. In June 2000, rolling blackouts occurred in the San Francisco area. Two months later, San Diego Gas & Electric filed a complaint against the utilities that were selling into the California ISO and to the PX markets requesting that the price cap that had been imposed at $500/MWh be lowered to $250/MWh. A formal investigation was launched, and in November the FERC issued its report on the bulk power market in California:

The electric market structure and market rules for wholesale sales of electric energy in California are seriously flawed and that these structures and rules, in conjunction with an imbalance of supply and demand in California, have caused, and continue to

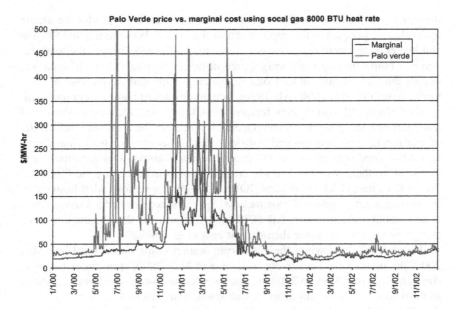

FIGURE 10.2 Wholesale prices at Palo Verde. In the summer of 2000, unusually warm temperatures, coupled with other factors, resulted in prices reaching $500/MWh upon several occasions, which was the price cap imposed at the time.

have the potential to cause, unjust and unreasonable rates for short-term energy (Day-Ahead, Day-of, Ancillary Services, and real-time energy sales) under certain conditions. While this record does not support findings of specific exercises of market power, and while we are not able to reach definite conclusions about the actions of individual sellers, there is clear evidence that the California market structure and rules provide the opportunity for sellers to exercise market power when supply is tight and can result in unjust and unreasonable rates under the FPA.

(For more details, see http://www.ferc.gov/industries/electric/indus-act/wec/chron/chronology.pdf.)

A number of remedies were proposed and on December 15, 2000 many were adopted. Nonetheless, it is informative to review the mistakes that were made that contributed to the energy crisis that occurred. The mistakes that were made in the adoption of restructuring legislation are enumerated here.

Mistake #1

As part of the restructuring process, AB 1890 required that IOUs divest themselves of their generating assets and then required them to purchase the power required to serve native load (or load, as very few retail customers chose a different supplier) from the power exchange (this was to reduce the possibility of horizontal market power). (Note: IOUs served 75% of California's native load in 1998.) In addition, no buy-back provisions (or long-term power contracts, also known as bilateral trades) left utilities at the mercy of skyrocketing whole-sale power prices (which reached $500/MWh in Palo Verde in the summer 2000, see Figure 10.2).

Mistake #2

Under the guise of protecting ratepayers, not only were rates frozen but also a mandatory rate reduction was imposed—rates were reduced by 10% for residen-tial and small commercial customers from 1996 levels in 1998 with the intention of reaching 20% by 2002. However, the crisis that ensued put an end to retail access in 2001. (Again, no price signal to consumers!) This, coupled with mis-take #1, eventually led to huge losses for IOUs and the almost financial ruin and subsequent bankruptcy of Pacific Gas & Electric, one of California's largest electric suppliers, which supplied 33.5% of the total native load in 1998.

Mistake #3

Reliance on imported power (20% in 1999) to serve native load left Califor-nians even more exposed to the vagrancies of the wholesale power market and to manipulation by the players in the game (see next section).

Mistake #4

The Ability to Game the System

Enron (and others) proved to be masterminds in figuring out ways to game the California market. Enron was instrumental in the development of a proprietary trading platform, known as EnronOnline, which became an industry-trading platform used by many of the players in the industry to view bid and ask prices for electricity (and natural gas), to make trades, and to value portfolios, which were done via the mark-to-market accounting methodology.[1] Despite the fact

---

[1] According to Wikepedia (http://en.wikipedia.org/wiki/Mark-to-market_accounting), mark-to-market or fair value accounting refers to the accounting standards of assigning a value to a position held in a financial instrument based on the current fair market price for the instrument or similar instruments. Fair value accounting has been a part of the U.S. Generally Accepted Accounting Prin-ciples (GAAP) since the early 1990s, and investor demand for the use of fair value when estimating the value of assets and liabilities has increased steadily since then as investors desire a more realistic appraisal of an institution's or company's current financial situation. Mark-to-market is a measure of the fair value of accounts that can change over time, such as assets and liabilities. It is the act of recording the price or value of a security, portfolio, or account to reflect its current market value rather than its book value. The practice of *mark-to-market* as an accounting device first developed among traders on futures exchanges in the 20th century. It was not until the 1980s that the practice spread to big banks and corporations far from the traditional exchange trading pits, and beginning

that Enron traders spoke in codes[2] (using the gaming nomenclature described), the FERC eventually caught on and Enron and several others (including El Paso Natural Gas Trading Company, Williams Energy Marketing & Trading Company, and AES Southland, Inc.) were eventually brought down. Unfortunately, in the interim, the havoc wreaked upon ratepayers, shareholders, and taxpayers was in the billions of dollars.

One such scheme was called "Fat Boy" and involved scheduling power to the California ISO to nonexistent or exacerbated loads. (Computer systems at the California ISO were apparently unable to recognize a systematic pattern of abuse.) A number of market participants engaged in this practice until the Powerex eventually caught on. In the meantime, however, it has been estimated that this scheme alone cost consumers over $3.5 billion (McCullough Memorandum, 2003).

Another scheme was known as "wash trades." According to the Federal Energy Regulatory Commission's final report on Price Manipulation in Western Markets: Fact-Finding Investigation of Potential Manipulation of Electric and Natural Gas Prices, the term "wash trade" is generally defined as a *prearranged* pair of trades of the same good between the same parties, involving no economic risk and no net change in beneficial ownership. These trades expose the parties to no monetary risk and serve no legitimate business purpose. Potential motives for wash trading are numerous. Wash trades may be used to create the illusion that a market is liquid and active or to increase reported trading revenue figures. Wash trades might be arranged at prices that diverge from the prevailing market in an attempt to send false signals to other market participants. Alternatively, the intent might be to affect the average or index price reported for a market, which in turn could benefit a derivatives position or affect the magnitude of payments on a contract linked to the index price.

---

in the 1990s, mark-to-market accounting began to give rise to scandals. To understand the original practice, consider that a futures trader, when taking a position, deposits money with the exchange, called a "margin." This is intended to protect the exchange against loss. At the end of every trading day, the contract is marked to its present market value. If the trader is on the winning side of a deal, his contract has increased in value that day and the exchange pays this profit into his account. However, if the market price of his contract has declined, the exchange charges his account that holds the deposited margin. If the balance of this account falls below the deposit required to maintain the position, the trader must immediately pay additional margin into the account to maintain his position (a "margin call"). As an example, the Chicago Mercantile Exchange, taking the process one step further, marks positions to market *twice* a day, at 10:00 a.m. and 2:00 p.m. As the practice of marking to market caught on in corporations and banks, some of them seem to have discovered that this was a tempting way to commit accounting fraud, especially when the market price could not be determined objectively (because there was no real day-to-day market available or the asset value was derived from other traded commodities, such as crude oil futures), so assets were being "marked to model" in a hypothetical or synthetic manner using estimated valuations derived from financial modeling, and sometimes marked in a manipulative way to achieve spurious valuations.

[2]"Death Star"—This game involved … (Congestion Manipulation, FERC Part 2, "Price Manipulation in Western Markets). Selling nonfirm energy as firm (FERC Part 2, "Price Manipulation in Western Markets)."Get Shorty" (FERC Part 2, "Price Manipulation in Western Markets).

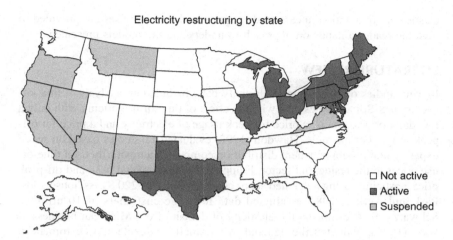

FIGURE 10.3 Status of electric restructuring (May 2009). To date, electric industry restructuring has been something of a failure; it only remains in a handful of states.

Subsequent FERC findings were that:

In general, wash trading is viewed as damaging to the integrity of a market and has the potential to mislead a host of market stakeholders (including competitors, regulators, analysts, and investors) through the various forms of manipulation outlined above. Although the Commission has no regulations on wash trading, wash trades are prohibited in markets regulated by the Commodity Futures Trading Commission. This chapter provides statistical evidence indicating that wash trading in energy products was more common than previously recognized.

And concerning EnronOnline:

Staff concludes that EnronOnline (EOL), which gave Enron proprietary knowledge of market conditions not available to other market participants, was a key enabler of wash trading (FERC Findings Part 1).

This confluence of mistakes proved to be fatal flaws in the deregulation of the industry and would eventually lead to subsequent reregulation not only in California but also in seven other states, which are indicated in gold and displayed in Figure 10.3 (http://www.eia.doe.gov/cneaf/electricity/page/restructuring/restructure_elect.html).

CHAPTER OVERVIEW

This chapter explores the literature on such studies and shows how changing price schedules can be used to calculate price and substitution elasticities for electricity, along with various models that can be estimated to calculate these elasticities. In addition, a case study is presented and exercises are provided to give the reader a hands-on approach to understand the models presented.

LITERATURE REVIEW

Recent studies of electric price elasticity have included an analysis of time-of-use pricing. Some studies deal with the effect of time-of-use pricing with only a few daily prices, that is, a price for peak usage of electricity and a price for off-peak usage. For example, Hawdon (1992) evaluated 11 studies based on seven experimental programs where distribution companies temporarily used time-of-use prices to the residential sector. Filippini (1995) estimated peak and off-peak price elasticities using a detailed set of characteristics for 220 Swiss households. Park and Action (1984) evaluated data for large customers of 10 utilities. Schwarz (1984) estimated the elasticity of demand from Michigan businesses. Woo (1985) evaluated the demand of 64 small nonresidential customers in southern California. Tishler (1991) evaluated the demand of two industrial firms in southern California under time-of-use pricing. Tishler (1998) estimated electric price elasticities using data from 500 small- to medium-size Israeli businesses. The sample evaluated in a number of these studies is described, while the range of price elasticities estimated in each of these studies is shown in Table 10.1.

TABLE 10.1 Ranges of Price Elasticity Estimates for Time-of-Use Studies[*]

| | Residential | | | | Business | | | |
| | Peak | | Off Peak | | Peak | | Off Peak | |
	Most Elastic	Least Elastic	Most Elastic	Least Elastic	Most Elastic	Least Elastic	Most Elastic	Least Elastic
Hawdon (1992)	+	−0.80	+	−0.90				
Filippini (1995)	−1.50	−1.25	−2.57	−2.30				
Park and Action (1984)					−0.25	−0.00		
Schwarz (1984)					+	−0.07		
Woo (1985)					−0.04	−0.03	−0.05	−0.03
Tishler (1991)					−0.09	−0.04	−0.06	−0.02
Tishler (1998)					−0.47	−0.01	−0.38	−0.02

[*] Some studies provided results such as the price elasticity of the midpeak that are not reported here. Results are not reported for methodologies that the study determined to be deficient. A plus sign indicates that the estimation produced the theoretically invalid result of a positive elasticity.

Other studies deal with the effect of more complex dynamic pricing. Dynamic pricing can involve the supplier providing a set of 24 hourly prices in periods of either one day ahead or one to two hours ahead. Real-time pricing (RTP) refers to a form of dynamic pricing based on current market conditions.

There have been numerous dynamic pricing programs. In the mid-1980s some electric utilities adopted small RTP programs. A study done by Newcomb and Byrne (1995) identified 29 North American utilities with real-time pricing programs. These programs were aimed primarily at large industrial customers with some large commercial customer participation. The hourly programs were complicated, as they were run by integrated utilities under cost-of-service regulation. In a review by Mak and Chapman (1993), four of these utilities conducted customer surveys and found that high prices indeed induce customers to reduce loads significantly, despite the fact that the hourly pricing programs differed significantly. Patrick and Wolak (1997) analyzed customer data from medium and large industrial customers in five industries participating in purchasing electricity on the basis of half-hourly prices in England and Wales during the period 1991–1995. They found that price elasticities varied considerably across industries and the pattern of within-day substitution in electricity consumption. During high price periods, they found that, despite small elasticities, significant load reduction occured for these participants. Price elasticities were reported only for the most price elastic industry—the water supply industry; these price elasticities ranged from −0.142 to −0.27. King and Shatrawka (1994) found that dynamic pricing in England produced more significant interday load shifting, with price elasticities ranging from 0.1 to 0.2, than intraday load shifting. They found that between 33% and 50% of participating customers responded to time-varying prices.

In 2001, Georgia Power initiated one of the largest dynamic pricing programs in the United States, where 1600 industrial customers with about 5000 MW of load participated. At the time, commercial customers constituted about 15% of the program load. Georgia Power offered day-ahead or hour-ahead firm price notices and would post the day-ahead price, based on its expected hourly marginal cost plus $4/MWh at 4 p.m. by e-mail or the Internet. The hour-ahead price was set at the next hour's expected marginal cost plus $3 because the utility's risk that the price will not be accurate is lower for these prices than the day-ahead prices. The price response of this program has been significant: The utility reported much greater load reduction on peak compared with on-peak-load reductions under its historical fixed-price method. Georgia Power also reported that as wholesale price volatility increased, customers did not abandon the day-ahead or hour-ahead firm price options. Instead, many purchased risk management tools such as contracts for differences, caps, or collars protecting certain loads during specific time periods.

Not surprisingly, participants in the hour-ahead program exhibited greater demand elasticity than day-ahead program participants: hour-ahead customers had a demand elasticity of −0.154, −0.171, and −0.189 for prices of $0.25, $0.50, and $1.00/kWh, respectively. For very high prices, participating industrial customers in the hour-ahead program reduced their electric load by 29% and 60% for increases in the maximum daily price to the $300–$350 and $1500–$3000/MWh range, respectively.

Day-ahead participants exhibited significantly smaller demand elasticities, ranging from one-half to one-tenth of the demand elasticities of day-ahead participants across the price levels of $0.25 to $1.00/kWh. For very high prices, day-ahead participants reduced their load by 8% and 20% for price increases to the $300–$350 and $1500–$2000/MWh range, respectively.

Some utilities have also offered residential customers dynamic pricing options. Braithwait (2000) reported that GPU initiated a time-of-use pilot program for residential customers where the utility communicated price signals to home thermostats that could be set to adjust to these signals automatically. During the summer months, these residential participants reduced their weekday electricity consumption by 7% on average. The overall elasticity for these residential customers was about 0.30, much higher than the elasticities reported in other residential time-of-use programs.

More recently, The New York ISO implemented an Emergency Demand Response Program in the Albany area during the summer of 2001. About 70 MW of retail load participated in the program on an hourly basis. During peak demand periods, participants would be notified to reduce load and receive the higher of $500/MWh or locational marginal price times the amount of load reduction. The load reduction program produced significant savings, particularly during a four-day heat wave in early August. Participants reduced total load by 3.5%. A consultant determined that wholesale electricity prices would have been 32% higher absent this load reduction.

Later in this chapter, RTP programs at Niagara Mohawk and Central and Southwest (now a part of AEP) will be used to introduce the generalized Leontief model to calculate substitution elasticities.

Aside: Difficulties in Price Elasticity Estimations

Experience in estimating price elasticities has highlighted two major difficulties. First, the estimate of price elasticity will be biased if substitution of other inputs for the use of electricity occurs but is ignored by the model used to make the price elasticity estimation. (Of course, such inclusion requires a more complex model with the corresponding difficulties involved.) The theoretical possibility of such bias has long been known.[3] For example, empirically

---

[3]For example, this issue was debated in six articles in the *Journal of Business and Economic Statistics*, July 1983 (1), pp. 202–228.

modeling the substitution of labor for electricity raised the elasticity estimates for the Tishler studies from the least elastic to the most elastic estimations reported in Table 10.1.

We will see that the elasticity estimate can vary widely depending on the selection of the functional form of the equation to be estimated. Second, price elasticity seems to vary widely across different industries and, needless to say, the level of the price itself. Regional variations occur as well depending on the mix of industries.

EFFECT OF TIME-OF-USE PRICING ON PEAK UTILITY LOAD

According to Levesque (2001), Georgia Power's time-of-use pricing has been reported to have 1600 customers and reduced peak load by 17% on the most expensive days. While other factors that may have affected demand have not been accounted for, these results indicate that a utility's peak load will be decreased if time-of-use pricing is implemented.

In Wisconsin, Caves and Christensen (1983) evaluated how customers altered their loads in response to a retail peak-load pricing program. The customers involved were found to be quite responsive in altering their load. The program confronted different customers with peak/off-peak price ratios ranging from 2:1 to 8:1, while the length of the peak period varied from 6 to 12 hours in duration. Even though customers were told that a participant's bill would not change if they continued their prior consumption pattern, they altered their consumption considerably. Customers facing a 2:1 peak/off-peak ratio reduced their consumption of electricity during summer months by 11% to 13%, while those customers facing an 8:1 price ratio reduced their consumption by 15% to 20% during summer peak periods. In addition, this study found there was greater shifting of load on critical days to off-peak periods during heat waves with reduction in consumption up to 31% compared with nonpeak days. Caves and Christensen (1983) found that customers reduced their consumption of electricity more during the time of the system peak than during the remainder of the peak pricing period in the summer compared to the winter. Customers were more willing and able to shift their midday usage (the time of the summer peak period) than their evening usage (the time of the winter peak period). They also found that the time-of-use pricing encouraged conservation as customers reduced their overall consumption of electricity over the combined summer and winter pricing periods.

RECENT EXPERIENCE IN CALIFORNIA

Although the focus on demand response is usually related to price changes, electricity demand is a function of other things in addition to prices. A recent striking example of this is the significant drop in electricity demand in California. Demand from February to May 2001 averaged over 4.9% less per month than the corresponding months of 2000 (weather adjusted data). As the renowned economist Paul Joskow (2001) aptly states:

Since retail electricity prices did not rise significantly until June 2001 we must attribute the decline in demand, in part, to the effects of all the publicity about electricity on consumer behavior and the formal energy conservation programs implemented by the state.

While such reductions due to public awareness deserve attention, the fact remains that pricing signals are the most effective approach to rationing supply shortages. In fact, on February 16, 2001, then California Governor Gray Davis conceded: "Believe me, if I wanted to raise rates, I could have solved this problem in 20 minutes."

In 2009, Ahmad Faruqui and Sanem Sergici, economists with the Brattle Group, produced a survey of studies on household response to dynamic pricing. More specifically, they reviewed 15 recent dynamic pricing experiments. As is typical in such experiments, a control group faces the "normal" rate, while the treatment group (or smart rate group) is subject to dynamic prices and is given enabling technology so that they may be able to respond by curtailing demand. It is important to point out that certain conditions need apply to maximize the accuracy and benefit of the experiment:

1. The control group should be chosen randomly.
2. Measurement should me made during the pretreatment period to allow detection of self-selection bias that may exist.
3. Multiple price points should be part of the design so that demand models and price/substitution elasticities can be computed.

These have important implications for the case study presented later in the chapter.

THE CALIFORNIA DEBACLE: A DECADE LATER

Some of the flaws in the policy that ordered California's electric market to be restructured in 1996 were pointed out earlier in the chapter. This is also mentioned by Faruqui and Sergici (2009); that is, it is clearly the case that the energy crisis that transpired in the western United States in 2000–2001 could have been obviated (at least in part) if prices paid for electricity by retail customers were allowed to vary in response to supply conditions. Despite the fact that wholesale prices rose precipitously, those prices paid by end users remained at or below prices charged before California's electric markets were restructured via Order AB 1890. Retail customers had no incentive to reduce demand so they did not alter their consumption behavior, and the crisis ensued. More specifically (Greer, 2011, p. 321).

In fact, under the guise of protecting ratepayers, not only were rates frozen but also a mandatory rate reduction was imposed—rates were reduced by 10% for residential and small commercial customers from 1996 levels in 1998 with the intention of reaching 20% by 2002. However, the crisis that ensued put an end to retail access in 2001. (Again, no price signal to consumers!)

Recently, and as Greer (2011) argued, a group of economists put forth a manifesto on the crisis. It is quoted here with permission from Bandt et al. (2003) in Faruqui and Sergici (2009):

Any structural model for the industry should include a mechanism for charging consumers for the cost of the production and delivery of electricity at the time of its consumption. Electricity at midnight in April is completely different from electricity at noon on a hot August day. ... Prices to most end users don't signal when electricity is cheap or dear for the industry to produce. Nor are consumers offered the true economic benefit of their conservation efforts at times of peak demand. Customers suffer further when unchecked peak demands grow too fast, pushing up costs for all. Wholesale electricity markets also become more volatile and subject to manipulation when rising prices have no impact on demand. Indeed, a functioning demand side to the electricity market in California would have greatly reduced the likely private benefits, and consequent social cost, of any strategic behavior engaged in during the crisis. ... Regardless of other reform efforts that are pursued in California, real-time pricing or other forms of flexible pricing is a key to enhanced conservation, more efficient use of electricity, and the avoidance of both unnecessary new power plants as well as concerns about the competitiveness of wholesale electricity markets.

As pointed out in Faruqui and Sergici (2009), two unanswered questions were left by this manifesto: (1) whether customers would respond to higher prices by reducing demand and (2) whether, from an economics perspective, it would make sense to equip end-use customers with the technologies that would allow them to respond to dynamic price signals.[4] Also known as advanced metering infrastructure (AMI), end-use customers receive signals when prices are about to change so that they can respond accordingly. For example, when prices increase, one might respond by curtailing unnecessary activities such as laundry or running the dishwasher until later when prices are reduced, which gives rise to the concept of elasticity of substitution. Subsequently, the CPUC initiated proceedings on advanced metering, demand response, and dynamic pricing (see http://docs.cpuc.ca.gov/published/proceedings/R0206001.htm).

As cited in Faruqui and Sergici (2009),

As part of the proceeding, the state carried out an elaborate experiment with dynamic pricing. It showed conclusively that customers responded to high prices by lowering peak usage by 13 percent. The three investor-owned utilities in the state relied on the

---

[4]This is more relevant for residential and small commercial customers than it is for large commercial and industrial customers.

results from the experiment to develop their AMI business cases. They showed that while AMI yielded many operational benefits to the distribution system, such benefits only covered about sixty percent of the total investment. The remaining forty percent had to be covered through demand response.

Based on these results, the CPUC approved the deployment of almost 12 million smart meters for electricity to occur within the next five years. In addition, all customers who have such meters will be placed on critical peak pricing, whereby the price of electricity consumed during peak hours will be significantly higher than that consumed outside of peak hours.

For example, a typical weekday residential hourly load profile on a summer day is displayed in Figure 10.4. The critical price hours are those beginning midafternoon and extending to early evening, which would correspond to hours 13–18 in Figure 10.4, which displays an average load profile of a residential customer not on a smart rate (as such assigned to the control group) under two scenarios: one in which no critical peak price (CPP) event is called (lower demand—0-Control) and the other where a CPP event is called (most likely due to extreme weather, which yields higher demand). Later, a case study is presented that evaluates the impact on the profile from critical peak pricing where the CPP is roughly three times the price in the hour before the CPP event begins.

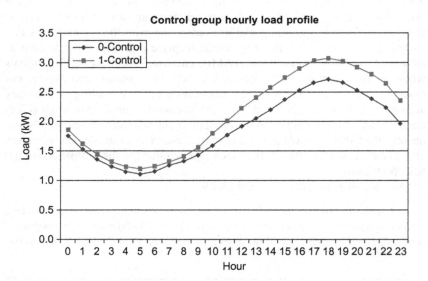

FIGURE 10.4 Hourly profiles of control group under different weather scenarios. Summer 2010— a typical load profile and an increase in response due to extreme heat.

USING REAL-TIME PRICING TO ESTIMATE PRICE ELASTICITY OF DEMAND

Real-time pricing of electricity has made it much easier to estimate price elasticity over various time periods (i.e., hourly, daily). After all, there can be no price response if prices do not vary with changes in marginal cost. During the summer of 2010, a midwestern utility ran a residential responsive pricing program, which is used in the case study presented in this chapter. Data are contained in a data set called "RTP." Using these data, we can estimate a very basic model, which is given by

$$Load_t = \beta_0 + \beta_1 Price_t + \varepsilon_t, \tag{10.1}$$

where $Load_t$ is the average hourly load at time t and $Price_t$ is the per-kWh price of electricity at time t.

In this case, the price elasticity of demand is given by

$$\eta_t = \beta_1 * Price_t / Load_t, \tag{10.2}$$

where β_1 is the estimated coefficient of $Price_t$.

Alternately, one could estimate an equation that is similar except that both price and output (load) are in natural logarithms, which implies that the coefficient of price is the elasticity, that is, estimate

$$\ln Load_t = \beta_0 + \beta_1 \ln Price_t + \varepsilon_t. \tag{10.3}$$

This implies that

$$\eta_t = \beta_1. \tag{10.4}$$

Example 10.1 Estimating Price Elasticity of Demand

Estimate the parameters of the models given earlier. What are the a priori expectations of the sign of β_1, the coefficient of price, in the two equations that follow?
　　Estimation results are

1. $Load_t = 2.656 + 0.327\ Price_t$

　　　64.17　　　　6.33

　　Durbin – Watson statistic $= 0.148$

　　Adjusted $R^2 = 0.276$.
$$\tag{10.5}$$

2. $\ln Load_t = 0.773 + 0.153\ \ln Price_t$

　　　19.35　　　　7.39

　　Durbin – Watson statistic $= 0.155$

　　Adjusted $R^2 = 0.372$.
$$\tag{10.6}$$

Note: t statistics are below coefficient estimates.
Question: Do these results make sense? Why or why not?

Clearly, they do not make sense; basic economics tells us that price is inversely related to quantity and the statistically significant coefficient estimate of the price variable is of the incorrect sign. This is most likely due to the fact that the model is clearly misspecified (i.e., omitted variable bias and/or the incorrect functional form, which is discussed in Chapter 4). There are other issues as well; the Durbin–Watson test statistic indicates that serial correlation is an issue, which has also been discussed in other chapters in this book. In addition, the overall fit of the model is poor (i.e., a low adjusted R^2).

One obvious way to improve the model fit is to add another relevant variable. Clearly, temperature affects load so possibly that is a choice variable. Including this in the model yields the following:

$$1.\ Load_t = -3.95 - 0.019\ Price_t + 0.086\ Temp_t$$
$$\qquad\qquad -17.48 \qquad -4.34 \qquad\quad 29.53$$

Durbin $-$ Watson statistic $= 0.203$

Adjusted $R^2 = 0.399$.

$$(10.7)$$

$$2.\ \ln Load_t = -10.71 - 0.12\ \ln Price_t + 2.72\ \ln Temp_t$$
$$\qquad\qquad\quad -30.12 \qquad\quad -6.79 \qquad\qquad 32.28$$

Durbin Watson statistic $= 0.209$

Adjusted $R^2 = 0.446$.

$$(10.8)$$

Note: t statistics are below coefficient estimates.

Adding the temperature in the current period improved both models. The signs of the estimated coefficients of price are now negative and statistically significant. The temperature variable has the expected (positive) sign and is also relevant to the model. However, the explanatory power of the model is still not up to par, as evidenced by the Durbin–Watson statistic and the adjusted R^2.

Unlike other commodities, the demand for electricity is also affected by temperatures in the hours (days) leading up to the current period. This is the case here, including several lags of hourly temperatures results in significant improvement in both models. The adjusted R^2 has increased and indicates that the models explain about 80% of the variation in load. Results are displayed in Table 10.2. (Note: Hourly lags of temperature variable are indicated by the variable name, for example, Temp1 is the temperature in the previous hour and Temp24 is the temperature 24 hours prior.)

The implied elasticity of demand in the log-linear model is −0.177, which is of the correct sign and appears reasonable; the price elasticity of demand for electricity in the short run is thought to be about −0.2 (Bohi, 1984; Dahl, 2010; Taylor, 1975; Wade, 2003). As stated, the implied price elasticity of demand that results from the linear model varies; it must be computed for every price–load combination. The average is −0.084 with a standard deviation of 0.05, which is less elastic than what might have been expected. Clearly, a review of the model results is in order.

One notable issue is that the sign of the estimated coefficient of Temp (lnTemp) is incorrect; it should be positive. In addition, variables Temp3 and Temp6 have rather low t statistics, which may or may not be true (there could be an issue with heteroskedasticity, which inflates standard errors, thus lowering t statistics). Also, there is still the issue of serial correlation for which there may or may not

TABLE 10.2 Linear and Log-Linear Estimation Results with Lagged
Temperature Variables: Discussion of Results

	Linear			Logarithmic	
Variable	Coefficient Estimate	t Statistic	Variable	Coefficient Estimate	t Statistic
Intercept	−8.777	−38.42	Intercept	−18.267	−54.53
Price	−0.034	−12.55	lnPrice	−0.177	−15.88
Temp	−0.016	−2.47	lnTemp	−0.222	−1.27
Temp1	0.032	3.12	lnTemp1	0.860	3.14
Temp2	0.028	2.79	lnTemp2	0.731	2.66
Temp3	0.012	1.16	lnTemp3	0.439	1.59
Temp4	0.020	1.96	lnTemp4	0.542	1.97
Temp5	0.015	1.46	lnTemp5	0.395	1.44
Temp6	0.013	1.24	lnTemp6	0.255	0.93
Temp7	0.013	1.96	lnTemp7	0.400	2.27
Temp24	0.009	3.23	lnTemp24	0.274	3.69
Temp48	0.007	2.60	lnTemp48	0.229	3.09
Temp72	0.016	6.18	lnTemp72	0.561	8.29
Adjusted R^2 = 0.79		DW = 0.356	Adjusted R^2 = 0.81		DW = 0.359

be a panacea [except the addition of an autoregressive disturbance term to correct
for first-order serial correlation, or an AR(1) term].

There is a variable that could be included in the models: a binary variable indicating hours in which the critical peak price is in effect. Despite the fact that the
coefficient is statistically significant and of the correct (positive) sign, the aforementioned issues are not remedied. At this point, the inclination is to add an AR(1)
term, which yields the results in Table 10.3.

The results have improved dramatically. The previously mentioned issues have
been remedied and the price elasticities are reasonable and in accord with other
studies of this nature. For the linear model, the average price elasticity is −0.15
(standard deviation = 0.09) versus −0.23 for the log-linear model. Both are reasonable and in line with a priori expectations.[5]

---

[5]A colleague of the author had a concern about this model. He felt that including the binary variable
for a CPP event was possibly introducing some type of bias into the results. Creating a new data set
in which no CPP events occurred, the author reran the log-linear model (of course deleting the CPP
variable) and found that the price elasticity of demand was −0.25, again reasonable and in line with
a priori expectations.

TABLE 10.3 Linear and Log-Linear Estimation Results with AR(1) Correction: Discussion of Results

	Linear			Logarithmic	
Variable	Coefficient Estimate	t Statistic	Variable	Coefficient Estimate	t Statistic
Intercept	−9.089	−23.39	Intercept	−18.848	−32.46
Price	−0.061	−17.4	lnPrice	−0.233	−19.89
Temp	0.008	2.24	lnTemp	0.365	3.67
Temp1	0.025	6.77	lnTemp1	0.692	6.64
Temp2	0.027	7.31	lnTemp2	0.693	6.65
Temp3	0.013	3.54	lnTemp3	0.471	4.55
Temp4	0.019	5.05	lnTEmp4	0.515	5.00
Temp5	0.016	4.18	lnTemp5	0.405	3.95
Temp6	0.012	3.27	lnTemp6	0.260	2.57
Temp7	0.014	4.00	lnTemp7	0.406	4.33
Temp24	0.006	1.88	lnTemp24	0.200	2.09
Temp48	0.006	1.88	lnTemp48	0.291	3.04
Temp72	0.006	1.76	lnTemp72	0.320	3.38
CPP	0.992	11.07	CPP	0.205	8.20
Adjusted R^2 = 0.94		DW = 1.55	Adjusted R^2 = 0.94		DW = 1.58

FURTHER IMPLICATIONS OF THE RESIDENTIAL RESPONSIVE PRICING PILOT

An interesting and unexpected thing occurred during the summer of 2010: The peak demand for customers on the responsive pricing program actually increased (as opposed to just shifting or being reduced) when prices rose. Contrary to the purpose of the program, which was to shift electricity usage from higher-cost hours, this did not transpire. Figure 10.5 shows a typical load profile for a residential customer during June 2010 who is not on the time-of-use rate but pays a flat block rate of just under 7.0 cents per kWh.

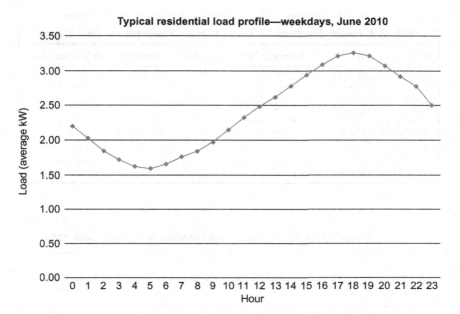

FIGURE 10.5   Load profile for a residential customer on a flat block rate (summer 2010).

PEAK-LOAD PRICING

The objective in the residential responsive pricing program would be to reduce demand in the midafternoon to early evening hours (i.e., the peak period when electricity is the most costly to generate). This could be affected by charging a higher price during those hours (hours 13–19 or 20) and a lower price in other hours to encourage customers to shift load to less-costly hours. However, caution must be exercised in pricing hours just after the peak price, which is one of the lessons to be learned here. In this case, the pricing structure adopted is displayed in Table 10.4.

Figure 10.6 displays the result and shows the control group (3) versus the treatment group (2). Due to the extremely hot weather, there were several CPP events during this period.

Discussion of Results

Clearly, the results are not as expected. Despite paying a higher price in the afternoon hours, the treatment group exceeded the demand of the control group in hours 15–18 (after this, the price paid by the treatment group falls to below that paid by the control group; again, see Table 10.4). The question is: Why did the treatment group respond to a price signal at 1 p.m. but not later, especially on a CPP day?

TABLE 10.4 Pricing Structure for Customers on Residential Pricing Program

Pricing Period[a]	Summer 2010 Weekdays	Summer 2010 Weekends
9:00 p.m.–10:00 a.m.	$P_1 = \$0.05$	$P_1 = \$0.05$
10:00 a.m.–1:00 p.m., 6:00 p.m.–9:00 p.m.	$P_2 = \$0.06$	$P_1 = \$0.05$
1:00 p.m.–6:00 p.m.	$P_3 = \$0.11$	$P_2 = \$0.06$
Critical peak:		
3:00 p.m.–7:00 p.m.	$P_4 = \$0.30$	

[a]Hour beginning.

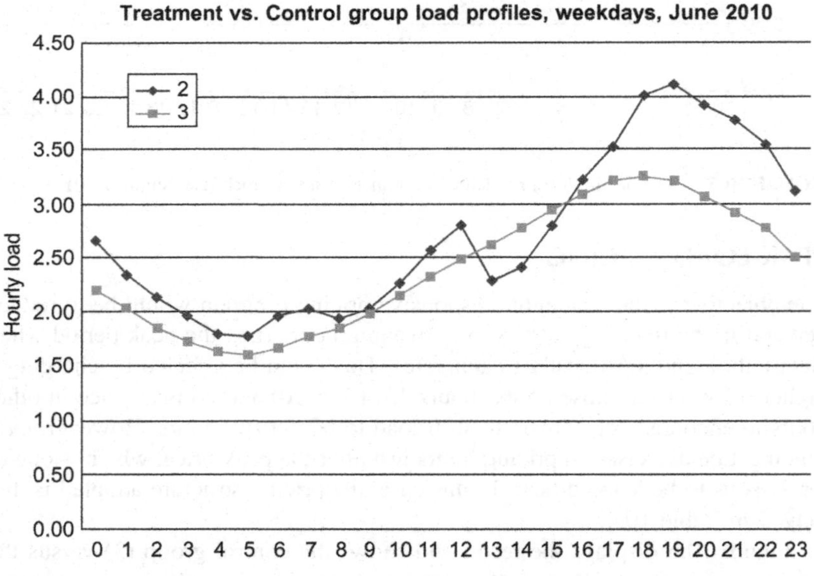

Treatment vs. Control group load profiles, weekdays, June 2010

FIGURE 10.6   Treatment versus control group load profiles (summer 2010). Not only was the peak not reduced for Residential Responsive Pricing (RRP) participants, but also it actually increased, shifting from hour ending 18 to hour 19. This brings up a critical point: What is the optimal length of a CPP event? And begs the question: How different would results be if it lasted until hour ending 19?

Alternately, we might show CPP event days separately, which is also informative. This is displayed in Figure 10.7. The hourly demand on CPP events days is labeled "1" and is lower than the demand on non-CPP days until hour 19, when loads appear to be equal. Oddly enough, the demand on CPP event days continues to increase (the peak has, in fact, shifted), which is not a desired effect.

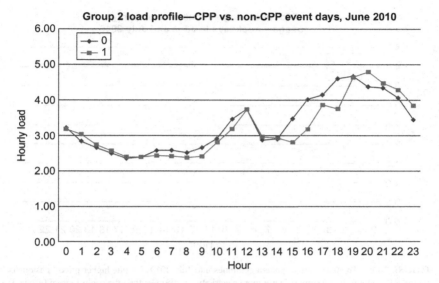

FIGURE 10.7 Treatment group load profiles under different weather scenarios.

What Is Going on Here?

Referring to Figure 10.8, we see that this same pattern continued into July 2010: Despite higher prices, customers in the RRP pilot program (group 2) use more electricity, on average, than the control group (group 3) *under the same weather conditions.*

So, exactly what *is* going on here?

Economic Theory

Given these results, the author believes that there is something else happening here. The author suspects that there is, in fact, a response to the changes in price. However, it is also necessary to control for other variables, such as income and home size. To wit: The increase in price does elicit a reduction in the *quantity demanded* of electricity (i.e., a movement along the demand curve). However, other factors, such as disparities in incomes and house sizes, likely result in a rightward shift of the demand curve so that, on net, the demand in a given hour is higher (i.e., the demand curve shifts right as the result of income and weather effects, etc.[6] so that usage increases. Unfortunately, it is difficult to separate the effects of price versus other effects; this would require a properly specified econometric demand model with data on income, home sizes, and other relevant variables to separate such effects. Figure 10.9 displays what the author has just described.

---

[6]Results are not weather normalized—the summer of 2010 was warmer than normal, which contributed to increased usage.

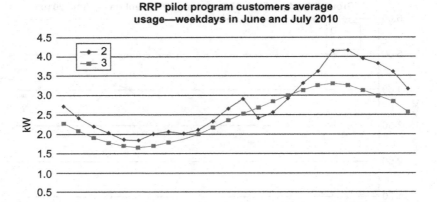

FIGURE 10.8 Treatment group pattern continues into July 2010. Despite higher prices, customers on the RRP pilot program (group 2) use more electricity, on average, than the control group (group 3) *under the same weather conditions*. This can be attributed to a variety of things, including flaws in the design of the program, which includes a disparity in average incomes, house sizes, and so on among the groups (recall the earlier discussion about not introducing bias into the sample selected for the treatment group).

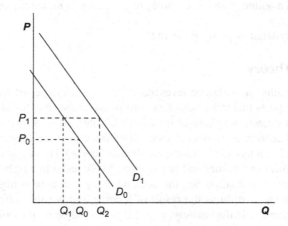

FIGURE 10.9 Economic theory behind contrary results obtained in the real-time pricing pilot program. The original demand curve is given by D at price = P_1 and quantity = Q_1. When price increases (P_0 to P_1), there is a decline in quantity demanded (Q_1). However, other factors, such as income, weather, and house size, shift the demand curve to the right so that the demand for electricity has increased (to D_1).

SUBSTITUTION ELASTICITIES

Now let us shift to another topic that has been discussed throughout this book and is relevant for time-of-use pricing studies. This section uses the generalized Leontief to calculate substitution elasticities for customers in the "RRP for elasticity of substitution calculations" data set.

Recalling from the KLEM chapter that the GL is superior to the CES model because it places no a priori restrictions on the substitution elasticities, let's explore this model in more detail. However, unlike the KLEM chapter, the author chose to estimate the substitution elasticities by using expenditure-share equations, which are detailed here.

A Generalized Leontief Model to Calculate Substitution Elasticities

Defining P_p and P_o as prices of peak and off-peak electricity and ES_p and ES_o as shares of electricity expenditure in peak and off-peak hours, then the two-commodity homothetic generalized Leontief (GL) indirect utility function and budget shares for electricity can be written as

$$V = Q/[\beta_{pp}P_p + 2\beta_{po}(P_pP_o)^{0.5} + \beta_{oo}P_o] \tag{10.9}$$

$$ES_p = [\beta_{pp}P_p + \beta_{po}(P_pP_o)^{0.5}]/[\beta_{pp}P_p + 2\beta_{po}(P_pP_o)^{0.5} + \beta_{oo}P_o] \tag{10.10}$$

$$ES_o = [\beta_{oo}P_o + \beta_{po}(P_pP_o)^{0.5}]/[\beta_{pp}P_p + 2\beta_{po}(P_pP_o)^{0.5} + \beta_{oo}P_o]. \tag{10.11}$$

The budget-share equations are homogeneous of degree zero in the parameters β_{pp}, β_{po}, and β_{oo}. As such, the equations can be normalized as with the following adoption:

$$\beta_{pp} + 2\beta_{po} + \beta_{oo} = 1. \tag{10.12}$$

Thus, for this two-commodity case, the elasticity of substitution between the peak and the off-peak is given by

$$\sigma_{po} = [\beta_{po}(P_pP_o)^{0.5}]/\{2ES_p(1 - ES_p)[\beta_{pp}P_p + 2\beta_{po}(P_pP_o)^{0.5} + \beta_{oo}P_o]\}. \tag{10.13}$$

Note: It is clearly the case that σ_{po}, ES_p, and ES_o can be nonnegative only if β_{po} is nonnegative.

To estimate the elasticities of substitution, it is necessary to have estimates of the parameters of the indirect utility function in Equation (10.9). Because there is no knowledge about the individual's utility levels, V, it is necessary to use only the expenditure-share equations. More specifically, we can first compute the ratio of the expenditure-share equations:

$$ES_p/ES_o = [\beta_{pp}P_p + \beta_{po}(P_pP_o)^{0.5}]/[\beta_{oo}P_o + \beta_{po}(P_pP_o)^{0.5}]. \qquad (10.14)$$

Thus, the set of equations are nonlinear in parameters and, as such, require use of a nonlinear estimation procedure. However, they have been reduced to a single equation, 10.14, and the only restriction required is the symmetry condition, which is given by

$$\beta_{pp} + 2\beta_{po} + \beta_{oo} = 1, \qquad (10.15)$$

and, of course, as required by Young's theorem, that

$$\beta_{po} = \beta_{op}. \qquad (10.16)$$

As stated, these equations are nonlinear in parameters, which complicates the estimation procedure. One technique that could simplify the procedure to some degree is to perform a logarithmic transformation on Equation (10.14) (you may recall that this is a procedure that the author employed in simplifying, at least to some degree, her quadratic and cubic cost models). However, because the equation is still nonlinear in parameters, it still requires a nonlinear estimation procedure to estimate the coefficients of the model presented in Equation (10.17).

The logarithmic transformation of Equation (10.14) yields

$$\ln(ES_p/ES_o) = \ln[\beta_{pp}P_p + \beta_{po}(P_pP_o)^{0.5}] - \ln[\beta_{oo}P_o + \beta_{po}(P_pP_o)^{0.5}] + e_t,$$
$$(10.17)$$

where e_t is an additive disturbance term.

Other variables are important—weather clearly drives consumption for residential and even commercial customers. Including such variables yields the final equation to be estimated, which is given by

$$\ln(ES_{p,i}/ES_{o,i}) = \alpha_0 + \gamma_{1,i} CD_i + \{\ln[\gamma_{p,i} H_p + \beta_{pp}P_p + \beta_{po}(P_pP_o)^{0.5}]\}$$
$$- \{\ln[\gamma_{o,i} H_{po} + \beta_{oo}P_o + \beta_{po}(P_pP_o)^{0.5}]\} + e_t, \qquad (10.18)$$

where i is an individual RRP participant, W is on-peak cooling-degree days, and H is a binary variable that equals 1 when the average temperature heat index >85°F.

Empirical Estimation

As you may suspect, empirical estimation of this model requires a nonlinear system-of-equation estimation technique as each respondent has his or her own equation to be estimated (denoted by i).[7]

---

[7]Similar to Boisvert et al. (2006), the author employed full information maximum likelihood as the estimation method within PROC MODEL (in SAS) to ensure symmetry.

TABLE 10.5 Estimated Coefficients of Generalized Leontief Equation: Discussion of Estimation Results

Variable	Parameter	Estimate	t Statistic
	α_0	−0.05214	−3.81
CDDs	γ	0.009672	3.28
Central air conditioning	α_1	0.000339	3.60
Electric water heater	α_2	0.004544	2.11
Freezer	α_3	−0.0067	−3.86
	$\beta_p = \beta_o$	−0.25845	−1.41
	β_{pp}	0.01129	4.55
	β_{oo}	0.001365	0.30
	$\beta_{po} = \beta_{op}$	0.493672	436.72
Restriction 1	$\beta_{pp} + 2\beta_{po} + \beta_{oo} = 1$	−300.293	−2.25E+06
Restriction 2	$\beta_p = \beta_o$	−4.72968	−0.07
Restriction 3	$\beta_{po} = \beta_{op}$	−253.912	−7.58

It is possible to include other relevant variables. Three other binary variables are included, which denote:

1. Whether the participant's home had central air conditioning.
2. Whether the participant had an electric water heater.
3. Whether the participant had a separate freezer.

Estimation results are displayed in Table 10.5.

Thirty different equations were estimated with varying degrees of success as indicated by the adjusted R^2 values, which fell between 0.16 and 0.78. More specifically, referring to Table 10.5, we can see that the estimated coefficients are not all of the expected sign. To wit: It is expected that each of the binary variables would result in an increase in the consumption of electricity, ceteris paribus. However, results indicate that the presence of a separate freezer in a home decreases the electricity consumption in the household.

ELASTICITIES OF SUBSTITUTION

The next step is to use results obtained to calculate the elasticities of substitution by inserting the estimated parameters into Equation (10.13). (Note: These will differ by individual participant and month and day during the summer of 2010.) Upon doing this, results are displayed in Table 10.6.

TABLE 10.6 Load-Weighted Daily Substitution Elasticities
for RRP (Summer 2010): Discussion of Results

	Peak to Off-Peak Substitution Elasticity
Maximum	0.78
Average	0.45
Minimum	0.23

By definition, elasticities of substitution should be between 0 and 1. In many studies, they are lower than those displayed in Table 10.6, especially maximums. However, the complexity of the equations involved often yield econometric results that can be misleading; with this said, however, it is informative to engage in these types of exercises. Quite possibly, inclusion of the additional three binary variables has caused such large substitution elasticities. In an exercise, you will be asked to reestimate Equation (10.18) without including these three variables.

CONCLUSION

This chapter presented a rather detailed case study on how real-time pricing can be used to make inferences not only about price elasticity of demand, but also how consumers shift usage from one period to another based on the price of electricity. Several studies are detailed, many of which demonstrate that consumers do respond to a price signal. Providing customers with such a price signal can reduce peak loads, thus delaying the building of power plants, which raises costs to all end-use customers. In addition, such pricing can help reduce the very greenhouse gases that may be responsible for some of the natural disasters that we have been experiencing in recent years.

The exercises contained in the next section provide a hands-on experience with the models and concepts that are explored in this chapter. The author encourages you to give them a try and see how you do. Part of the fun of being an economist is exploring models and possibly discovering something new and providing a valuable contribution to the economics/econometrics literature. Enjoy!

RECOMMENDED READING

"Customer Response to RTP in Competitive Markets: A Study of Niagara Mohawk's Standard Offer Tariff" (Boisvert et al., 2006, available on the Internet).

"Household Response to Dynamic Pricing of Electricity—A Survey of the Experimental Evidence" (Faruqui and Sergici, 2009, available on the Internet).

"Industrial and Commercial Customer Response to Real-Time Electricity Prices" (Boisvert et al., 2004, available at www.bneenan.com).

EXERCISES

1. Using data in "RTP," estimate Equations (10.7) and (10.8) and calculate the elasticity of demand for each.
   a. Are they different?
   b. If so, how and why?

2. To obtain the price elasticity of demand (hourly), an econometric model was specified in which hourly load (the dependent variable) is a function of hourly price, hourly temperature, and other variables. Using data in "RTP," estimate the following equation:

$$Load_t = \beta_0 + \beta_1 Price_t + \alpha_2 Temp_{t-1} + \alpha_3 Temp_{t-2} + \alpha_4 Temp_{t-3}$$
$$+ \alpha_5 Temp_{t-4} + \alpha_6 Temp_{t-5} + \alpha_7 Temp_{t-6} + \alpha_8 Temp_{t-24}$$
$$+ \alpha_9 Temp_{t-48} + \alpha_{10} Temp_{t-72} + \delta_1 Day21 + \delta_2 CPP_t \quad (10.19)$$
$$+ \delta_3 Hour0_t + \varepsilon_t,$$

   where $Load_t$ is the average hourly load at time t, $Price_t$ is the per-kWh price of electricity at time t, $Temp_t$ is the temperature at time t, where $t = 1,...,$ 72 (hours), $Day21$ is a binary variable that equals unity if the 21st of the month, CPP is a binary variable that equals unity if critical peak pricing was in effect, $Hour0$ is the hour ending 12:59 a.m.
   a. Why are binary variables $Day21$ and $Hour0$ included? Should any other variables of this nature be included? Why or why not? (Hint: Always look at data!)
   b. In the linear form of the equation, how is the price elasticity of demand calculated?
   c. What is the price elasticity of demand for this equation? Does it seem reasonable? Why or why not? (Hint: It is not a constant.)
   d. Breaking down this equation further, we can see if elasticity results differ by weekend. Add a binary variable for weekend and rerun the estimation. Are results materially different on the weekend than on weekdays?
   e. Breaking data into pricing periods, how do the calculated elasticities differ between pricing period 2 and pricing period 3?

3. From results obtained in Exercise 2, answer the following:
   a. Do the estimated elasticities accord with other studies of this nature during low and medium pricing periods?
   b. How does a CPP event change the magnitude of the price elasticity?
   c. In general, what other observations can be made about the price elasticities obtained (e.g., by month, and weekend vs. weekday)?
   d. Refer to the Faruqui and Sergici (2009) report entitled "Household response to dynamic pricing of electricity—A survey of experimental evidence." Do your findings accord with theirs when they state that

   We find conclusive evidence that households (residential customers) respond to higher prices by lowering usage. The magnitude of price response depends on several factors, such as the magnitude of the price increase, the presence of central air conditioning and the availability of enabling technologies such as two-way programmable communicating thermostats and always-on gateway systems that allow multiple end-uses to be controlled remotely. ... When accompanied with enabling technologies, the latter set of tariffs (CPP) lead to a drop in peak demand in the 27 to 44 percent range.

4. **Advanced:** Using data in "RRP for Elasticity of Substitution Calculations," estimate Equation (10.18) exactly as it is, that is, without including the three additional binary variables *Central AC*, *Electric Water Heater*, and separate *Freezer*. (Hint: You will likely need to obtain the Boisvert publications listed in the recommended reading section to perform this exercise.)
   a. How do the estimation results compare to those displayed in Table 10.5? (Hint: You may want to add an autoregressive disturbance term.)
   b. How are the goodness-of-fit statistics—are they better or worse than results obtained from inclusion of the three binary variables, which are displayed in Table 10.5?

5. **Advanced:** (Hint: You will likely need to obtain the Boisvert publications listed in the recommended reading section for this.) Using Equation (10.13), calculate the elasticities of substitution implied by the estimation results obtained in Exercise 4 (again, there will be a value for each of the participants in the RRP program).
   a. How do your results accord with those displayed in Table 10.6?
   b. Do they appear to be reasonable in light of what has been published in the literature? Why or why not?

References

Aigner, D., 1985. The residential electricity time-of-use pricing experiments: what have we learned? In: Hausman, J.A., Wise, D.A. (Eds.), Social Experimentation. University of Chicago Press, Chicago, pp. 11–54.

Alchian, A., Demsetz, H., 1972. Production, information costs, and economic organization. Am. Econ. Rev. 62 (December), 777–795.

Amemiya, T., 1985. Advanced Econometrics, Basil Blackwell, 1985.

Arrow, K.J., Chenery, H.B., Minhas, B., Solow, R., 1961. Capital-labor substitution and economic efficiency. Rev. Econ. Stat. 43 (5), 225–254.

Atkinson, S.E., Halvorsen, R., 1984. Parametric efficiency tests, economies of scale, and input demand in the U.S. electric power generation. Int. Econ. Rev. 25 (3), 647–662.

Balestra, P., 1967. The Demand for Natural Gas in the United States. North Holland Publishing Co., Amsterdam.

Balestra, P., Nerlove, M., 1966. Pooling cross section and time series data in the estimation of a dynamic model: the demand for natural gas. Econometrica 34 (3), 585–611.

Baltagi, B.H., 1995. Econometric Analysis of Panel Data. John Wiley and Son, New York.

Baltagi, B.H., Griffin, J.M., 1983. Gasoline demand in the OECD: an application of pooling and testing procedures. Eur. Econ. Rev. 22 (2), 117–137.

Bandt, W.D., Campbell, T., Danner, C., Demsetz, H., Faruqui, A., Kleindorfer, P.R. et al., 2003. 2003 Manifesto on the California Electricity Crisis. The manifesto can be accessed at this web site. http://www.aei-brookings.org/publications/abstract.php?pid=341.

Barbose, G., Goldman, C., 2004. A survey of utility experience with real time pricing. Ernest Lawrence Berkeley National Laboratory, LBNL-54238.

Baumol, W., Panzar, J., Willig, R., 1982. Contestable Markets and the Theory of Industrial Structure. Harcourt, Brace, and Jovanovich, New York.

Begosso, A., Azagury, J., Porter, T., 2011. Navigating in the age of uncertainty. Pub. Utilities Fortn. 149 (9), 44–48.

Bergson, A., 1936. Real income, expenditure proportionality, and frisch's 'new method of measuring marginal utility'. Rev. Econ. Stud. 4 (1), 33–52.

Berndt, E., 1991. The Practice of Econometrics, Classic and Contemporary. Addison Wesley Publishing Co., Reading, MA, 1991.

Berndt, E.R., Wood, D., 1975. Technology, prices and the derived demand for energy. Rev. Econ. Stat. LVII (3), 259–268.

Berndt, E.R., Morrison, C., Watkins, G.C., 1981. Dynamic models of energy demand: an assessment and comparison. In: Berndt, E.R., Fields, B. (Eds.), Modelling and Measuring Natural Resources Substitution, MIT Press, Cambridge, MA, pp. 259–289.

Berndt, E.R., Watkins, G.C., 1977. Demand for natural gas: residential and commercial markets in Ontario and British Columbia. Can. J. Econ. 10 (1), 97–111.

Berry, D., 1994. Private ownership form and productive efficiency: electric cooperatives versus investor-owned utilities. J. Regul. Econ. 6, 399–420.

Bickel, P.J., Klaassen, C.A.J., Ritov, Y., Wellner, J.A., 1998. Efficient and Adaptive Estimation for Semiparametric Models. Springer-Verlag, New York.

Boisvert, R.N., Cappers, P., Goldman, C., Neenan, B., Hopper, N., 2006. Customer Response to RTP in Competitive Markets: A Study of Niagara Mohawk's Standard Offer Tariff. Funded by the California Energy Commission and by the U.S. Department of Energy.

Boisvert, R.N., Cappers, P., Neenan, B., Scott, B., 2004. Industrial and commercial customer response to real-time electricity prices. www.bneenan.com.

Borenstein, S., 2005. The long-run efficiency of real-time pricing. Energy J. 26 (3), 93–116.

Braithwait, S.D., 2000. Residential TOU price response in the presence of interactive communication equipment. In: Faruqui, A., Eakin, K. (Eds.), Pricing in Competitive Electricity Markets, Kluwer Academic Publishers, Amsterdam, pp. 349–358.

Brockway, N., 1997. Electric deregulation may leave poor in dark. Forum Appl. Res. Public Policy 11, 13–17.

Brown, S.J., David, S.S., 1986. The Theory of Public Utility Pricing. Cambridge University Press, New York.

California Energy Commission, 2007. Integrated Energy Policy Report. www.energy.ca.gov/2007_energypolicy/.

California Public Utilities Commission, 2006 (CPUC R. 02-06-001). http://docs.cpuc.ca.gov/published/proceedings/R0206001.htm.

Casten, S., Meyer, J., 2004. Cross subsidies: getting the signals right: should regulators care about the inefficiencies? Pub. Utilities Fortn. 46–50.

Caves, D.W., Christensen, L.R., Tretheway, M.W., 1984. Economies of density versus economies of scale: why trunk and local service airline costs differ. Rand J. Econ. 15, 471–489.

Caves, D.W., Christensen, L., 1983. Time-of-use rates for residential electric service: results from the wisconsin experiment. Pub. Utilities Fortn. 111, 30–35.

Chiang, A., 1984. Fundamental Methods of Mathematical Economics. McGraw-Hill, New York.

Christensen, L., Greene, W., 1976. Economies of scale in U.S. electric power generation. J. Polit. Econ. 84, 655–676.

Claggett Jr., E.T., 1987. Cooperative distributors of electrical power: operations and scale economies. Q. J. Bus. Econ. 26 (3), 3.

Claggett Jr., E.T., 1994. A Cost function study of the providers of TVA power. Manag. Decis. Econ. 15, 63–72.

Claggett, E., Hollas, D., Stansell, S., 1995. The effects of ownership form on profit maximization and cost minimization behavior within municipal and cooperative electric distribution utilities. Q. Rev. Econ. Finance 35 (Special Issue), 533–550.

Cobb, C., Douglas, P.H., 1928. A theory of production. Am. Econ. Rev. 18 (Suppl.), 139–165.

Conkling, R., 1999. The marginal cost pricing doctrine. Contemp. Econ. Policy 17 (1), 20–32.

Considine, T.J., 1989a. Estimating the demand for energy and natural resource inputs: trade-offs in global properties. Appl. Econ. 21 (7), 931–945.

Considine, T.J., 1989b. Separability, functional form, and regulatory policy in models of interfuel substitution. Energy Econ. 11 (2), 82–94.

Creedy, J., Johnson, D., Valenzuela, M., 2003. A cost function for higher education in Australia. Aust. J. Labour Econ. 6 (1), 117–134.

Crew, M.A., Fernando, C.S., Kleindorfer, P.R., 1995. The theory of peak load pricing: a survey. J. Regul. Econ. 8, 215–248.

Dahl, C., 1993. A Survey of Energy Demand Elasticities in Support of the Development of the NEMS. Contract No. DE-AP01-93EI23499, Washington, DC.

Dahl, C., 2006. Survey of Econometric Energy Demand Elasticities Progress Report. Colorado School of Mines, Golden, CO.

De Alessi, T., 1974. An economic analysis of government ownership and regulation: theory and the evidence from the electric power industry. Pub. Choice 19 (1), 1–42.

Deaton, A., Muellbauer, J.S., 1980a. An almost ideal demand system. Am. Econ. Rev. 70, 312–326.

Deaton, A., Muellbauer, J.S., 1980b. Economics and Consumer Behaviour. Press Syndicate of the University of Cambridge, New York.

Dowd, J., Dye, R., Kaufmann, R., 1986. Do flexible functional forms generate accurate elasticities? In: Wood, D.O. (Ed.), The Changing World Economy: Papers and Proceedings of the Eighth Annual North American Conference of the International Association of Energy Economists, MIT, Cambridge, MA, pp. 456–460.

Dudley, R.M., 2002. Real Analysis and Probability. Cambridge University Press, Cambridge, UK.

Evans, D.S., 1983. Ed. Breaking up Bell: Essays on Industrial Organization and Regulation. North Holland, New York.

Evans, D.S., Heckman, J.J., 1984. A Test for subadditivity of the cost function with an application to the bell system. Am. Econ. Rev. 74 (4), 615–623.

Fare, R., Grosskopf, S., Logan, J., 1985. The relative performance of publicly-owned and privately-owned electric utilities. J. Pub. Econ. 26 (11), 89–106.

Farsi, M., Fetz, A., Filippini, M., 2007. Economies of Scale and Scope in the Swiss Multi-Utilities Sector. Center for Energy Policy and Economics, Swiss Federal Institutes of Technology, Working Paper No. 59, September.

Faruqui, A., George, S., 2005. Quantifying customer response to dynamic pricing. Electricity J. 18 (4), 53–63.

Faruqui, A., Sergici, S., 2009. Household Response to Dynamic Pricing of Electricity – A Survey of the Experimental Evidence. http://www.hks.harvard.edu/hepg/Papers/2009/The%20Power%20of%20Experimentation%20_01-11-09_.pdf.

Fetz, A., Filippini, M., 2010. Economies of vertical integration in the swiss electricity sector. Energy Econ. 32 (6), 1325–1330.

Filippini, M., 1995. Electric demand by time of use: an application of the household AIDS model. Energy Econ. 17, 197–204.

Fisher, F.M., Kaysen, G.S., 1962. The Demand for Electricity in the United States. North-Holland, Amsterdam.

Fraquaelli, G., Piancenza, M., Vannoni, D., 2004. Scope and scale economies in multi-utilities: evidence from gas, water and electricity combinations. Appl. Econ. 36 (18), 2045–2057.

Fraquaelli, G., Piancenza, M., Vannoni, D., 2005. Cost savings from generation and distribution with an application to italian electric utilities. J. Regul. Econ. 28 (3), 289–308.

Fuller, W.A., Battese, G.E., 1974. Estimation of linear models with crossed-error structure. J. Econ. 2, 67–78.

Gallop, F., Roberts, M., 1981. The source of economic growth in the U.S. electric power industry. In: Cowing, T., Stevenson, R. (Eds.), Productivity Measurement in Regulated Industries, Academic Press, New York.

Giles, D., Wyatt, N., 1989. Economies of Scale in the New Zealand Electricity Distribution Industry. Performance Measures and Economies of Scale In the New Zealand Electricity Distribution system. Ministry of Energy, Wellington, New Zealand.

Gilsdorf, K., 1994. Vertical integration efficiencies and electric utilities: a cost complementarity perspective. Q. Rev. Econ. Finance. 34 (3), 261–282.

Gilsdorf, K., 1995. Testing for subadditivity of vertically integrated electric utilities. Southern Econ. J. 62 (14), 126–138.

Goldfeld, S.M., Quandt, R.E., Trotter, H.F., 1966. Maximization by quadratic hill climbing. Econometrica 34 (3), 541–551.

Grace, R.C., Rickerson, W., Corfee, K., 2008. California Feed-in Tariff Design and Policy Options. Prepared for the California Energy Commission. Publication number: CEC-300-2008-009D. www.energy.ca.gov/2008publications/CEC-300-2008-009/CEC-300-2008-009-D.PDF.

Greene, W., 1993. Econometric Analysis, second ed. Macmillan Publishing Co., New York.

Greer, M., 2003. Can rural electric cooperatives survive in a restructured US electric market? An empirical analysis. Energy Econ. 25 (3), 487–508.

Greer, M., 2008. A test of vertical economies for non-vertically integrated firms: the case of rural electric cooperatives. Energy Econ. 30 (5), 679–687.

Greer, M., 2011. Electricity Cost Modeling Calculations. Elsevier, New York.

Griffin, J.M., 1979. Energy Consumption in the OECD: 1880-2000. Ballinger Publishing Co., Cambridge, MA. pp. 181–210.

Halvorsen, R., 1975. Residential demand for electric energy. Rev. Econ. Stat. 57, 12–18.

Hartman, R.S., 1979. Frontiers in energy demand modeling. Annu. Rev. Energy 4, 433–466.

Harunuzzaman, M., Koundinya, S., 2000. Cost Allocation and Rate Design for Unbundled Gas Services. The National Regulatory Research Institute, NRRI 00-08, Silver Spring, MD.

Hawdon, D., 1992. Is electricity consumption influenced by time of use tariffs? A survey of results and issues. Energy Demand: Evidence Expect, Surrey University Press, London.

Hayashi, P., Sevier, M., Trapani, J., 1985. Pricing efficiency under rate-of-return regulation: some empirical evidence for the electric utility industry. Southern Econ. J. 51 (3), 776–792.

Henderson, J., 1985. Cost estimation for vertically-integrated firms. In: Crew, M. (Ed.), Analyzing the Impact of Regulatory Change in Public Utilities. Lexington Books, Lexington, MA.

Herter, K., 2007. Residential implementation of critical-peak pricing of electricity. Energy Policy 35 (4), 2121–2130.

Herter, K., McAuliffe, P., Rosenfeld, A., 2007. An exploratory analysis of California residential customer response to critical peak pricing of electricity. Energy 32 (1), 25–34.

Hiebert, L.D., 2002. The determinants of the cost efficiency of electric generating plants: a stochastic frontier approach. Southern Econ. J. 68 (4), 935–946.

Hollas, D., Stansell, S., 1991. Regulation, ownership form, and the economic efficiency of rural electric distribution cooperatives. Rev. Reg. Stud. 21, 201–220.

Hollas, D.R., Stansell, S.R., Claggett, E.T., 1988. An examination of the effect of ownership form on price efficiency: proprietary, cooperative, and municipal electric utilities. Southern Econ. J. 55 (2), 336–350.

Hollas, D.R., Stansell, S.R., Claggett, E.T., 1994. Ownership form and rate structure: an examination of cooperative and municipal electric distribution utilities. Southern Econ. J. 61 (2), 519–529.

Houthakker, H.S., 1951. Some calculations of electricity consumption in Great Britain. J. R. Stat. Soc. 114, 351–371.

Houthakker, H.S., Taylor, L.B., 1970. Consumer Demand in the United States: Analysis and Projections, second ed. Harvard University Press, Cambridge, MA.

Huettner, D.A., Landon, J.H., 1978. Electric utilities: scale economies and diseconomies. Southern Econ. J. 44, 883–912.

Hyman, L.S., 1994. America's Electric Utilities: Past, Present and Future, fifth ed. Public Utilities Reports, Inc., Arlington, VA. p. 102.

Jacobsson, S., Lauber, V., 2005. Germany: from a modest feed-in law to a framework for transition. Switching to Renewable Power: A Framework for the 21st Century, Volkmar Lauber, ed. Earthscan, London. pp. 122–158.

Jara-Diaz, S., Ramos-Real, F.J., Martinez-Budria, E., 2004. Economies of integration in the Spanish electricity industry using a multistage cost function. Energy Econ. 26, 995–1013.

Joskow, P.L., 2001. California's Electricity Crisis. Updated September 28, 46–64.

Joskow, P.L., Schamalensee, R., 1983. Markets for Power: An Analysis of Electricity Utility Deregulation. MIT Press, Cambridge, MA.

Kahn, A.E., 1970. The Economics of Regulation: Principles and Institutions, vol. I: Economic Principles. John Wiley & Sons Inc., New York.

Kahn, A.E., 1971. The Economics of Regulation: Principles and Institutions, vol. II: Institutional Issues. John Wiley & Sons Inc., New York.

Kamerschen, D.R., Thompson, J.R., 1993. Nuclear and fossil-fuel steam generation of electricity: differences and similarities. Southern Econ. J. 60, 14–27.

Karlson, S.H., 1986. Multiple-output production and pricing in electric utilities. Southern Econ. J. 53 (1), 73–86.

Kaserman, D., Mayo, J., 1991. The measurement of vertical economies and the efficient structure of the electric utility industry. Southern Econ. J. 39 (5), 483–502.

Kelly, J., 2007. Renewed interest in demand response, but "wither the economic rational for efficient pricing?" USAEE Dialogue 15 (2), 13–16.

King, K., Shatrawka, P., 1994. Customer response to real-time pricing in great britain. In: Proceedings of the ACEEE 1994 Summer Study on Energy Efficiency in Buildings, Washington, DC.

Klein, A., Pfluger, B., Held, A., Ragwitz, M., Resch, G., Faber, T., 2008. Evaluation of different feed-in tariff design options. Best Practice Paper for the International Feed-in Cooperation, second ed. Funded by the Ministry for the Environment, Nature Conservation and Nuclear Safety (BMU). www.feed-in-cooperation.org/images/files/best_practice_paper_2nd_edition_final.pdf.

Kleit, A.N., Terrell, D., 2001. Measuring potential efficiency gains from deregulation of electricity generation: a Bayesian approach. Rev. Econ. Stat. 83 (3), 523–530.

Kolstad, C.D., Lee, J.-K., 1992. Dynamic specification error in cost function and factor demand estimation. College of Commerce and Business Administration, University of Illinois at Urbana-Champaign. Faculty Working Paper 92–0126.

Kwoka, J., 1996. Power Structure: Ownership, Integration, and Competition in the U.S. Electricity Industry. Kluwer Academic Publishers, Boston.

Laband, D., Lentz, B., 2005. Higher education costs and the production of extension. J. Agric. Appl. Econ. 37 (1), 229–236.

Lee, B., 1995. Separability test for the electricity supply industry. J. Appl. Econ. 10 (1), 49–60.

Levesque, C.J., 2001. Real-time metering: still as sweet with prices controlled? Publ. Utilities Fortn. 16.

Lucas, R.E., 1967. Optimal investment policy and the flexible accelerator. Int. Econ. Rev. 8 (1), 78–85.

Mak, J.C., Chapman, B.R., 1993. A survey of current real-time pricing programs. Electricity J. 6, 54–65.

Maloney, M., 2001. Economies and diseconomies: estimating electricity cost functions. Rev. Ind. Organ. 19 (2), 165–180.

Marshall, A. Principles of Economics. Macmillan, London, 1927.

Martins, R., Fortunato, A., Coelho, F., 2006. Cost Structure of the Portuguese Water Industry: A Cubic Cost Function Application. Universidade de Coimbra, Estudos Do GEMF, No. 9. http://gemf.fe.uc.pt/workingpapers/pdf/2006/gemf06_09.pdf.

Mayo, J.W., 1984. The multiproduct monopoly, regulation, and firm costs. Southern Econ. J. 51 (1), 208–218.

McCullough, R., 2003. Regional economic losses from enron's fat boy scheme. Memorandum, Portland, OR.

Meyer, R.A., 1975. Publicly owned versus privately owned utilities: a policy choice. Rev. Econ. Stat. 4, 391–399.

Moore, T., 1970. The effectiveness of regulation of electric utility prices. Southern Econ. J. 36, 365–375.

Morrison, C., 1988. Quasi-fixed inputs in U.S. and Japanese manufacturing: a generalized leontief restricted cost function approach. Rev. Econ. Stat. 70, 275–287.

Mundra, K., Russell, R., 2004. Dual Elasticities of Substitution. University of California at Riverside, Riverside, CA.

Nadiri, M.I., Rosen, S., 1973. A Disequilibrium Model of Demand for Factors of Production. New York. The American Economic Review, Volume 64, Issue 2, Papers and Proceedings of the Eighty Sixth Annual Meeting of the American Economic Association (May, 1974), 264–270.

National Rural Electric Cooperative Association. Electric Cooperatives: Meeting the Challenges of a Carbon-Constrained Future. http://www.nreca.org/Documents/PressRoom/REM_Meeting Challenge.pdf.

Neenan Associates, 2002. NYISO PRL Program Evaluation, pp. E-6.

Nemoto, J., Goto, M., 2004. Technological externalities and economies of vertical integration in the electric utility industry. Int. J. Ind. Organ. 22, 676–681.

Nerlove, M., 1963. Returns to scale in electricity supply. In: Christ, C. (Ed.), Measurement in Economics: Studies in Mathematical Economics and Econometrics in Memory of Yehuda Grunfeld. Stanford University Press, Stanford, CA, pp. 167–198.

Network for New Energy Choices, 2008. Freeing the Grid: Best and Worst Practices in State Net Metering Policies and Interconnection Standards. Network for New Energy Choices, New York.

Neuberg, L.G., 1977. Two issues in the municipal ownership of electric power distribution systems. Bell J. Econ. 8 (1), 303–323.

Newcomb, J., Byrne, W., 1995. Real-Time Pricing and Electric Utility Restructuring: Is the Future "Out of Control?". SIP, E Source, Boulder, CO.

Newey, W.K., McFadden, D., 1994. Large sample estimation and hypothesis testing. In: McFadden, D., Engle, R. (Eds.), Handbook of Econometrics, vol. 4. Elsevier, North-Holland, Amsterdam, The Netherlands, pp. 2113–2245 (Chapter 36). ISBN: 9780444887665.

Panzar, J., Willig, R., 1997. Economies of scale in multi-output production. Qual. J. Econ. 91 (3), 481–493.

Park, R.E., Action, J.P., 1984. Large business customer response to time-of-day electricity rates. J. Econ. 26, 229–252.

Patrick, R.H., Wolak, F.A., 1997. Estimating the Customer-Level Demand Under Real-Time Market Prices. Draft, Rutgers University, Newark, NJ.

Pescatrice, D., Trapani, J., 1980. The performance and objectives of public and private utilities operating in the united states. J. Public Econ. 13 (2), 259–276.

Piacenza, M., Vannoni, D., 2005. Vertical and horizontal economies in the electric utility industry: an integrated approach. HERMES Working Paper.

Pindyck, R.S., 1980. International comparisons of the residential demand for energy. Eur. Econ. Rev. 13 (1), 1–24.

Pindyck, R.S., Rotemberg, J.J., 1983. Dynamic factor demands and the effects of energy price shocks. Am. Econ. Rev. 73 (5), 1066–1079. (P&R83) http://www.jstor.org/view/00028282/di950035/95p0014g/0?currentResult=00028282.

Pindyck, R.S., Rubinfeld, D., 1991. Econometric Models and Economic Forecasts. McGraw-Hill, New York, NY.

Ramos-Real, F.J., 2004. Cost Functions and the Electric Utility Industry. A Contribution to the Debate on Deregulation www.grjm.net/documents/Divers-WP/ramos-real-2004.pdf.

Roberts, M., 1986. Economies of density and size in the production and delivery of electric power. Land Econ. 62 (4), 378–387.

Romano, J.P., Siegel, A.F., 1985. Counterexamples in probability and statistics. Chapman & Hall, Great Britain. ISBN: 0412989018.

Rothwell, G., Gomez, T., 2003. Electricity Economics: Regulation and Deregulation. IEEE Press, New York.

Salvanes, K., Tjotta, S., 1994. Productivity differences in multiple output industries: an empirical application to electricity distribution. J. Productivity Anal. 5, 23–43.

Schmalensee, R., 1978. A note on economies of scale and natural monopoly in the distribution of public utility services. Bell J. Econ. 9, 270–276.

Schwarz, P., 1984. The estimated effects on industry of time-of-day demand and energy electricity prices. J. Ind. Econ. 32, 529–539.

Sharkey, W., 1982. The Theory of Natural Monopoly. Cambridge University Press, Cambridge, UK.

Sing, M., 1987. Are combination gas and electric utilities multiproduct natural monopolies? Rev. Econ. Stat. 69 (3), 392–398.

Taylor, L.D., 1975. The demand for electricity: a survey. Bell J. Econ. 6 (1), 74–110.

Taylor, L., Blattenberger, G., Rannhacke, R., 1984. Residential energy demand in the united states: empirical results for electricity. In: Moroney, J.R. (Ed.), Advances in the Economics of Energy Resources, vol. 5. JAI Press, Greenwich, CT, pp. 103–127.

Thompson, H.G., 1995. Economies of scale and vertical integration in the investor-owned electric utility industry. NRRI Publication, Silver Spring, MD.

Thompson, H.G., 1997. Cost efficiency in power procurement and delivery service in the electric utility industry. Land Econ. 73 (3), 287–296.

Thompson, H.G., Wolf, L., 1993. Regional differences in nuclear and fossil-fuel generation of electricity. Land Econ. 69 (3), 234–248.

Tishler, A., 1991. Complementarity-substitution relationships in the demand for time differentiated inputs under time of use pricing. Energy J. 12, 137–148.

Tishler, A., 1998. The bias in price elasticity under separability between electricity and labor in studies of time-of-use electricity rates: an application to Israel. Energy J. 19, 217–235.

U.S. Bureau of the Census, 1975. Historical Statistics of the United States, Colonial Times to 1970, Bicentennial Edition, Part 2. U.S. Department of Commerce, Washington, DC.

U.S.D.A., Rural Utilities Service, Statistical Report Rural Electric Borrowers, 1998–2001. United States Government Printing Office, Washington.

U.S. Department of Energy, Energy Information Administration, 2009b. Electric Sales, Revenues and Price. www.eia.doe.gov/cneaf/electricity/esr/esr_sum.html.

U.S. Energy Information Administration, 2008, Emissions of Greenhouse Gases in the United States 2007. Office of Integrated Analysis and Forecasting, U.S. Department of Energy, Washington, DC.

U.S. Environmental Protection Agency, 2007. National Action Plan for Energy Efficiency.

Varian, H., 1992. Microeconomic Analysis. W.W. Norton & Company, New York.

Wade, S.H., 2003. Price and Responsiveness in the AEO2003 NEMS Residential and Commercial Buildings Sector Models. EIA/DOE, Washington, DC.

Walfridson, Bo., 1987. Dynamic Models of Factor Demand: An Application to Swedish Industry. Ph.D. Theses, University of Gothenburg, Gothenburg, Sweden.

Weisstein, E.W., Maximum Likelihood. From MathWorld–A Wolfram Web Resource http://mathworld .wolfram.com/MaximumLikelihood.html.

White, H., 1980. A heteroskedasticity-consistent covariance matrix estimator and a direct test for heteroskedasticity. Econometrics 48, 817–838.

Woo, C.K., 1985. Demand for electricity of small nonresidential customers under time-of-use (TOU) pricing. Energy J. 6, 115–127.

Yatchew, A., 2000. Scale economies in electricity distribution: a semiparametric analysis. J. Appl. Econ. 15 (2), 187–210.

Yoo, J., 1988. Economies of Scale and Scope for Gas-Electric Companies. http://warrington.ufl.edu/ purc/purcdocs/papers/8807_Yoo_Economies_of_Scale.pdf.

Zellner, A., 1962. An efficient method of estimating seemingly unrelated regressions and tests for aggregation bias. J. Am. Stat. Assoc. 57, 348–368.

Index

Page numbers in *italics* indicate figures, tables, examples and footnotes.

A

Accounting issues, 80
Advance Notice of Proposed Rulemaking
    (ANOPR), 83
Advanced metering infrastructure (AMI), 292,
    315
    technology, 194
AEP, *see* American Electric Power Company
Aggregation policies, 91
AIDS, *see* Almost Ideal Demand System
Allen–Uzawa elasticities of substitution, 226
Almost Ideal Demand System (AIDS), 277
American Clean Energy Leadership Act of
    2009, 86–87
American Electric Power Company (AEP), 53
American Energy and Security Act (ACES)
    of 2009, 85–86
American Recovery and Reinvestment Act
    (ARRA) of 2009, 85, 293–294
Ancillary services, 75
ANOPR, *see* Advance Notice of Proposed
    Rulemaking
A priori expectations, 146–147, 169–170
Asymptotic normality, 183
Autocorrelation, 108
Average cost (AC), 17–19, 152
    curves, *18*, *200*, 204–205
    and marginal curves, 192, *193*
    pricing, 2, 8, *8*, 244–246
Average incremental costs, 28–29, 125, 160
Average pricing structure, 97
Average system cost, 70
Averch–Johnson effect, 9–10, 32

B

Best linear unbiased estimators (BLUE), 104
Block rates, 251–254
BLUE, *see* Best linear unbiased estimators
Bonneville Power Administration (BPA), 45, 49
Bonneville Project Act of 1937, 49
Box Cox (BxCx) function, 267
BPA, *see* Bonneville Power Administration

C

California
    debacle, 291–292, 304, 314–316
    electric rates in, *304*
    electricity crisis, 304–306
    electricity demand in, 313–314
    market, 307–309
        structure and rules, 305–306
California Assembly Bill 1890, 304
California Public Utilities Commission
    (CPUC), 295, 303
Cap-and-trade system, 11–12
Carbon emissions, reducing, 7–8
    cost of, 10–12
    optimal rate/tariff design and tax credits,
        12–13
Carbon footprint, 56
Caveat, 206–207
CBL, *see* Customer baseline load
CES, *see* Constant elasticity of substitution
Classic linear regressions model, 206
Classical regression model, 163
Clean Air Act, *1*, 64
Clean Air Act Amendments (CAAA) of 1990,
    71
Cobb–Douglas
    cost function, 102–103
    cost model, 209–210
    elasticities of substitution for, 210–211
    functional form, 111–112
Competitive markets, 2, *2*
Composite cost function model, 153
Comprehensive design tool kit, 259
Concave and linear homogeneous input prices,
    199
Concave in input prices, 128
Consistent parameter estimation, 282–283
Consolidated Appropriations Act of 2008,
    84
Constant elasticity of substitution (CES),
    117, 133, 210–211, 296–299
    demand function, 297–299

Constant elasticity of substitution (CES) (*Cont.*)
  model, 272
    price and substitution elasticities using,
      300
Cooperatively owned firms, 6–7
Cost allocation, 238
  reforms, 88
Cost complementarity, 31, 103, 178–179
Cost concepts, 28
Cost function, 196
  Cobb–Douglas, 102–103
  subadditivity of, 31
Cost minimization, 101–102
Cost models, 101–104, 197–201
  basic cost model, *141–142*, 143–144
  Cobb–Douglas, 141–142, 209–210
  cost-share equations, 145–146
  cubic cost models, 168–171
  econometrics of
    examples, 105–107
    heteroscedasticity, 109
    impure heteroscedasticity, 109–111
    OLS estimation, 104
  estimation results of, 108–109,
    197–198
  homotheticity, 150
  multiple-output models, 171–173
  multiproduct cost
    concepts, 157–159
    functions, 153
    models, 155–157
  Nerlove cost model, 143–144
  Nerlove original data and, *144*
  price elasticities, 149
  priori expectations, 146–147
  quadratic cost functions, 161–164
  regression analysis and, 104–105
  single-output translog cost model, *147*
  translogarithmic cost function, 144–145
Cost shift variables, 124
Cost-of-service regulation, 236
Cost-share equations, 145–146
CPP, *see* Critical peak pricing
Critical peak pricing (CPP), 98, 295, 316
Cross-price effect, 102
Cross-price elasticities, 226, 273
Cross-subsidization, 236, 250–251
Cubic cost function, 17–19, 193,
  196, 200
Cubic cost models, 128–129, 168–171
  estimation results, *202*, 202–203
Customer baseline load (CBL), 256

D

Dead-weight loss, 240
Declining-block rate structure, 66
Degree of scale economies (SCE), 21, 28, 125,
  175–176
Demand equation, residential, 284
Demand response, 295
Demand-side bidding, 81
Demand-side management (DSM), 11, 98
Differential rates, 234
Discrete choice model, 276
Distributed generation, 293
Distributed lag
  models, 269
  polynomial, 269
DSM, *see* Demand-side management
Dual-cost function, 133
Dummy variable model, 207
Durbin–Watson test statistic, 108
Dynamic models, 274–275
Dynamic pricing, 98–100, 294, 311

E

Econometric issues, 285–287
  identification
    problem, *279*, 279–280
    and systems bias, 277–280, *278*
Econometric models, 265–266
Economic efficiency, 3
Economic Recovery Tax Act of 1981, 70
Economic theory, 12–13, 203, 323–324, *324*
Economics of Regulation (1970, 1971), *The*
  (Kahn), 3
Economies of density, 24
Economies of scale, 15, 24, 114, 140–141,
  153–154, 196–197, 199
  product-specific, 28–29, 160–161, 176–177
  single-output models, 22–24
  in transmission and distribution, 24
Economies of scope, 30, 127–128, 160–161,
  177–178, 203–204
  importance of, 30
  and subadditivity, 31–33
Economies of vertical integration, 33–34
Efficiency, 191
  measures, 204
Efficient pricing, 231–232
  theory of, 193–194
Efficient public utility pricing, 239–241
Elasticities, 116, 220–227
  cross-price, 226, 273

of demand, *266*
price, 263–265, *264–265*
generalized Leontief cost function, 220–224
Hicks-Allen, 272–273
long-run *vs.* short-run, 267–270
own-price, 225
price, 216, 225–226
and substitution, 272–274
substitution, 199, 226, 325–328
Elasticities of substitution, 148–149, *149*
Elasticity, 244
estimation results, 299–300
models for estimating price, 296–299
of substitution, 289–292
Electric Consumers Protection Act of 1986, 70
Electric cooperatives
and IOU, 136–138
institutional differences, 137
philosophical, 138
regulatory differences, 138
urban *vs.* rural, 136
Electric industry
average pricing structure, 97
CPP, 98
decoupling, 95
demand side management, 98
dynamic pricing, 99–100
electric-generating utilities, 97
EPA, 96
FTC, 59
natural monopolies, 58
rate-making mechanisms, 95
restructuring, 306–309, *309*
RGGI, 96–97
RTP, 99
smart grid, 98
time-of-use rates, 98
true-up mechanism, 95
utility's rates, 95
Electric market, competitive
electric power marketers, 56
ESCOs, 57–58
merchant generators, 53
transmission companies, 53
UDC, 56–57
Electric power marketers, 56
Electric Power Research Institute (EPRI), 257
Electric price elasticity, 310
estimation
difficulties in, 312–313
for time-of-use studies, *310*
Electric Rates (Watkins), 234

Electric utility, 40
customers, 257–258
research program, 258
Electric utility industry
ACES of 2009, 85–86
American Clean Energy Leadership Act of 2009, 86–87
ARRA of 2009, 85
Clean Air Act, 64
combination utilities, 69
Consolidated Appropriations Act of 2008, 84
cost allocation reforms, 88
economies of scale to, 21–22
EISA of 2007, 83
EPACT of 1992, 72–75
EPACT of 2005, 82–83
FERC, 65–66
Order 719, 84
Order 980, 84–85
Order 1000, 87
Orders 888 and 889, 75–78
grid regionalization and FERC Order 2000, 78–82
industry restructuring accelerates, 71–72
National Energy Conservation Policy Act, 66
Natural Gas Utilization Act of 1987, 71
NERC, 62–64
Public Utility Regulatory Policies Act, 66–69
PUHCA, 60–62
state regulations, 88–89
U.S. electric utility industry, 58
Electric-generating utilities, 97
Electricity
consumers of, 11
demand for, *284–285*
economies of scope and subadditivity, 31–33
log-linear demand model for, *287*
on-peak *vs.* off-peak, *298*
optimal pricing of
demand response, 235
differential rates, 234
EPACT of 2005, 235
PURPA, 234
time of use pricing, 232–236
two-rate system, 233
Electricity Cost Modeling Calculations (Greer, 2011), 135, 291
Electricity crisis
chronology of California, 304–306
factors precipitating, 304

Empirical estimation, 217–219, 326–327
Endogeneity problem, 281
End-user rates, 239
Energy
  charges, 239
  demand for, 265–266, *266*
  efficiency, 1, 12
Energy Economics in 2008, 6
Energy Independence and Security Act (EISA)
  of 2007, 83, 292–293
Energy Policy Act (EPACT) of 1992, 47,
  72–75, 82
Energy Policy Act (EPACT) of 2005, 82–83,
  235, 292–293
Energy price trends, *42*
Energy services companies (ESCOs), 57–58
EnronOnline (EOL), 309
Environmental Protection Agency (EPA), 1, 96
EPA, *see* Environmental Protection Agency
EPRI, *see* Electric Power Research Institute
Error components model, 207
ESCOs, *see* Energy services companies
Estimation methods, 283–285
EWGs, *see* Exempt wholesale generators
Exempt wholesale generators (EWGs), 72
Expenditure share models, 270–272
Expenditure system models, 276–277

F

Factor demand equations, 102, 121, 130
Factors of production, 231
Fair value accounting, *307*
Faruqui and Sergici study, 300–301
FDC, *see* Fully distributed cost
FDR, *see* Franklin Delano Roosevelt's
Federal electric utilities, 46
Federal Energy Regulatory Commission
  (FERC), 7, 39, 54, 65–66, 194, 309
  Order 719, 84
  Order 980, 84–85
  Order 1000, 87
  Order 2000, 78–82
  Orders 888 and 889, 75–78, 303
Federal laws, 292
  ARRA, 293–294
  policies and programs, 294–295
Federal power agencies, 45–47
Federal Trade Commission (FTC), 59
Feed-in tariffs, 94–95
FERC, *see* Federal Energy Regulatory
  Commission
Flexible functional forms, 144

Four-factor model, 212
Franklin Delano Roosevelt's (FDR), 47
FTC, *see* Federal Trade Commission
Fully distributed cost (FDC), 238–239
Functional forms, 111–112, 267
Functionalization, 237

G

Generalized least squares (GLS), 208
Generalized Leontief model, 272–273
  cost function, 216–217
    elasticities, 220–224
  substitution elasticities calculation, 325–326
Generalized method of moments (GMM),
  180–187
Generation and transmission (G&T), 6
Georgia Power, 311
GHG, *see* Greenhouse gas
GLS, *see* Generalized least squares
Green power, 91
Greenhouse gas (GHG), 1, 193–194
Grid regionalization, 78–82
G&T, *see* Generation and transmission

H

HCCME, *see* Heteroskedasticity-consistent
  covariance matrix estimation
Hessian matrix, 130, 199, *274*
Heteroscedasticity, 109, 318
  impure, 109–111
Heteroskedasticity-consistent covariance matrix
  estimation (HCCME), 203
Hicks–Allen elasticities, 273
  for generalized Leontief cost model, *221*
  of substitution, 120, 215–216, 272
Homogeneous of degree one in input prices,
  213
Homothetic functions, 150
Homotheticity, 150
Houthakker and Taylor model, 269

I

Implied elasticity, 318
Impure heteroscedasticity, 109–111
Incremental cost, 160
Independent power producers (IPP), 44
Independent system operators (ISO), 52, 54–55
Independent variables, 146
Instrumental variables (IV) approach, 282
Integration, 153–154
Interclass subsidization, 251
Interconnection standards, 91

Internal revenue code, 82
Interstate highway system, 86
Inverse elasticity rule, 244
Investor-owned utilities (IOUs), 4, 42–44,
        135, 194
    profit maximization, 5
    regulation of, 8–10
IOUs, *see* Investor-owned utilities
IPP, *see* Independent power producers
ISO, *see* Independent system operators
Isoquant, *118*
Iterated seemingly unrelated regression, 219
IV approach, *see* Instrumental variables
        approach

J

J test, 186–187
Joint cost allocation, 238
Joint municipal action agencies, 45

K

Koyck model, 274

L

Lagged endogenous (LE) model, 268
Lagrangian multiplier technique, 9
LDCs, *see* Local distribution companies
LE model, *see* Lagged endogenous model
Leontief cost function, 119–120, 273
    generalized, 117–118
Leontief production technology, 118–119
Linear homogeneity, 101, 161, 220
Linear model, *319*
Load factor, *238*
Local distribution companies (LDCs), 40
Local identification, 182
Log linear model, *319*
Logit cost share model, 272
Logit equation, 271
Logit model, 271
Log-linear demand model, for electricity usage,
        287
Long-run elasticities *vs.* short-run elasticities,
        267–270
Long-run price elasticity, 289

M

MACRS, *see* Modified Accelerated Cost
        Recovery System
Margin, *308*
Marginal cost (MC), 152
    curves, *18, 200, 205,* 253

pricing, 168, 231, *232*
    doctrine, 2–4
    for electric utilities, 13–14
Marginal rate of substitution, 210
Marginal rate of technical substitution (MRTS),
        9, 117, 297
Market participants, 42
Mark-to-market accounting, *307–308*
Mayo single-output quadratic cost model, *166*
Member coops, 6
Merchant generators, 53
Microeconomic model, 3
Minimum efficient scale (MES), 23, 114, *115*
Modified Accelerated Cost Recovery System
        (MACRS), 71
Monotonic in output, 110, 201
Morishima elasticities of substitution, 226–227
MRTS, *see* Marginal rate of technical substitution
Multipart tariffs, 251
Multiple product markets, 81
Multiple-output models, 171–173
    measures of efficiency for, 175
Multiple-output natural monopoly, 26
Multiproduct cost, 30
    concepts, 157–159
    functions, 153
    models, 155–157
    ray average costs, 157–159
Multiproduct cubic model, 201
Multiproduct natural monopoly, 27
Multisettlement markets, 81
Municipal utilities, 44–45

N

National Action Plan for Energy Efficiency, 10
National Association of Regulatory Utility
        Commissions conference, *191*
National Bureau of Economic Research, 210
National Energy Conservation Policy Act, 66
National Rural Electric Cooperative
        Association, 198
Natural Gas Utilization Act of 1987, 71
Natural monopoly, 58, 231
    conundrum, 15–16
    definition, 16–19
    multiple-output, 26
    single-output market, 17
Necessary condition, 19–21, 280
Negative semidefinite, 102, 130
NERC, *see* North America Reliability Council
Nerlove cost model, 143–144
    specification, *142*

Nerlove's Cobb–Douglas function, 112–114
    cost model, 141–142
    results, 114–115
Net metering, *92*, 92
Network economies, 16, 24–26
Nondecreasing in input prices, 102
Nonlinear estimation, 201
    least-squares, 162–164
Nonlinear quadratic cost model, 224–225
Nonutility power producers, 52
North America Reliability Council (NERC),
    62–64

O

OASIS, *see* Open Access Same-Time
    Information System
Off-peak electricity, *298*
OLS, *see* Ordinary least squares
Omitted-variable bias (OVB), *110*, 285–286
On-peak electricity, *298*
Open Access Same-Time Information System
    (OASIS), 76
Optimal pricing, electricity, 232–235
Optimal two-part tariff, *246*, 249–250
Optimization, 130
Order conditions, 182, 280
Ordinary least squares (OLS), 207, 270,
    281–282
    estimation, 104
OVB, *see* Omitted-variable bias
Own-price elasticities, 225, 266, 271, 273

P

Panel data, 206–207
Pareto optimal, 248
Pareto-efficient outcome, 231, *232*
PBFs, *see* Public benefit funds
Peak-load pricing, 321–324
Pearl Street-generating station, 58
Personal tax incentives, 94
Physical feasibility, 81
PMA, *see* Power marketing administrations
Polynomial distributed lag, 269
Power exchanges, 80
Power marketing administrations (PMA),
    49–50
Power pools, 52
Power supply cooperatives, 50
Price elasticities, 149, 216, 225–226,
    272–274, 310, 317–320
    of demand, 263–265, *264–265*, 284,
    287–289, 317–320

models for estimating, 296–299
    using CES model, 300
Price signals, 2–3
Private electric utilities, 134
Producers of electricity, 10
Product-specific returns to scale, *29*, 29,
    125–126, 160, 204
Profit-maximizing price, *242*
Promotional rate structure, 66
Property tax incentives, 94
Public benefit funds (PBFs), 92–93
Public electric utilities, 134
Public Utilities Fortnightly, 1
Public utility districts (PUDs), 45
Public Utility Holding Company Act (PUHCA)
    of 1935, 60–61
Public Utility Regulatory Policies Act (PURPA)
    of 1978, 40, 66–69, 234
Publicly owned firms, 5–6
Publicly owned utilities, 44–45
PUDs, *see* Public utility districts

Q

Quadratic cost functions, 161, 199
Quadratic cost model, 111, 123–124, 224
    multiple-output, 124
    nonlinear, 224–225
Quasiconcavity, in input prices, 227–228

R

RAC, *see* Ray average cost
Ramsey number, 244
Ramsey prices, 13, 241–244
Rate classes, determination of, 237–239
Rate design, 236–237
    costs, functionalization and classification of,
        237
    end-user rates, 239
    fixed costs, 238–239
    joint cost allocation, 238
    rate classes, determination of, 237–239
    total revenue requirements, 237
Rate structure options, 258–260
Rate-making issues, 13–14
Rate-making mechanisms, 95
Rate-of-return regulation, 9–10
Ray average cost (RAC), 27–28, *27*, 125, *126*,
    157–159
Ray cost output elasticity, 175
Real-time pricing (RTP), 99, 192, 256–257,
    311–312, 317–320
Real-time tariffs, 256

Rebate programs, 93
Regional Greenhouse Gas Initiative (RGGI),
    96–97
Regional planning, 55
Regional Transmission Organizations (RTOs),
    54–55
Regression analysis and cost modeling,
    104–105
Regulation, *240*, 240
Renewable energy, 294
    access laws, 93
    production incentives, 93
Renewable portfolio standard, 89–91, 195
Renewable technologies, *1*, *191*
Residential demand equation, 284
Residential pricing program, *322*
Residential responsive pricing (RRP) pilot,
    320–321
    group 2, 323, *324*
Restructuring
    of electric utility industry, 66, 69–71
    industry in 1980s and 1990s, 69–71
Retail rates, 56
Returns to scale, 112
    product-specific, 160
RGGI, *see* Regional Greenhouse Gas Initiative
Rotterdam model, 277
RRP pilot, *see* Residential responsive pricing
    pilot
RTOs, *see* Regional Transmission
    Organizations
RTP, *see* Real-time pricing
Rural electric cooperatives, 50–52, 105, 194,
    198
    cost studies on, 138–140
Rural Electric Magazine, 194
Rural Electrification Act, 136
Rural Utilities Service (RUS), 6, 49, 51, 116,
    198, *199*
    data provided by, 198
RUS, *see* Rural Utilities Service
RUS97_basic data set, 105, *105–106*

S

Seasonal two-part tariff, *254*
SEC, *see* Securities Exchange Commission
Second-order conditions, 101
Second-order Taylor series approximation, 211
Securities Exchange Commission (SEC), 60
Seemingly unrelated regression (SUR) method,
    283
Self-regulatory organization, 62

Separability, 35–36
    vertical integration and, 36–37
Serial correlation, 318
Shephard's duality theory, 117
Shephard's lemma, 102, 121, *121*, 129,
    224, 272
Short-run elasticities *vs.* long-run elasticities,
    267–270
Short-run price elasticity, 289
Simultaneous equations, 281–282, 284
    estimates of parameters in, 282
Single-output market, efficient industry
    structure in, 21–37
Single-output translog cost
    equation, 147–148
    model, *147*
2SLS method, *see* Two-stage least squares
    method
Smart grid
    assets, 293
    technology, 295–296
Specification bias, *110*
Standardized design process, 259
State regulations, 88–89
State regulatory commission, 11
State sales taxes, 94
Static stock models, 266
Structural models, 276, 282
Subadditivity, 28, 164
    economies of scope and, 31–33
Substitution elasticities, 226, 272–274,
    289–292, 325–328
    for Cobb–Douglas, 210–211
    cost models for, 296–297
    estimating price and, *298–299*
    for translog form, 215–216
    using CES model, 300
Sufficient condition, 19–21
Supply curves, for electricity utility,
    195, *195*
Supply equation, 284
Supply–demand system of equations, 280
Supply–demand time-series model, 279
SUR method, *see* Seemingly unrelated
    regression method
System benefit charges, 92–93

T

Tariff design and rate-making issues, 13–14
Tax incentives, 93–94
Tennessee valley authority (TVA), 46–49
Thermal efficiencies, 23

Time-of-use
  electric tariff, *195*
  pricing, 232, 254–256, 310
    effect on peak utility load, 313
    impact on peak demand, *290*
  rates, 98, 254
Time-series cross-sectional data, 206
Total distribution cost, 25
Total revenue requirements, 237
Traditional price regulation, *22*
Traditional regulatory system, 56
Translog form, substitution elasticities for,
  215–216
Translog share equations, *214*
Translogarithmic cost function, 103–104,
  120–123, 144–145, 211–215
Transmission access, 74
Transmission companies, 53
True-up mechanism, 95
TVA, *see* Tennessee valley authority
Two-part tariff, 246–247
  with customer classes, *248*
    optimal two-part tariff, 249–250
    utility rate, cross-subsidization, 250–251
Two-rate system, 233
Two-stage least squares (2SLS) method, 282

U

UDCs, *see* Utility distribution companies
United States electric industry
  regulation of, 7–8
  structure of, 4, *4*
U.S. Department of Energy (DOE), 293–294
U.S. electric utility industry, 58
  deregulation and, 41–42
  structure, 39–41

U.S. Energy Information Administration, 2009,
  191, *192*
Utility distribution companies (UDCs), 56–57
Utility green power consumer option, 92
Utility's rates, 95
Utility's revenue requirement, 236

V

Vertical integration
  definition, 35
  of electric utilities, 34–35
  and separability, 33–34, 36–37
Vertically integrated model
  federal power agencies, 45–47
  investor-owned utilities, 42–44
  IPP, 44
  municipal utilities and publicly owned
    utilities, 44–45
  nonutility power producers, 52
  power marketing administrations, 49–50
  power pools, 52
  rural electric cooperatives, 50–52
  TVA, 47–49

W

Wash trades, 308

Y

Young's theorem, 102

Z

ZEF, *see* Zellner efficient estimator
Zellner efficient estimator (ZEF), *215*,
  219–220
Zellner's method, 123
Zellner's seemingly unrelated regression, 219